# Authors

Kurt K. Allen

James T. Blodgett

Kelly S. Burns

Robert J. Cain

Sheryl L. Costello

Tom J. Eager

Jeri Lyn Harris

Brian E. Howell

Roy A. Mask

Willis C. Schaupp, Jr.

Jeffrey J. Witcosky

James J. Worrall

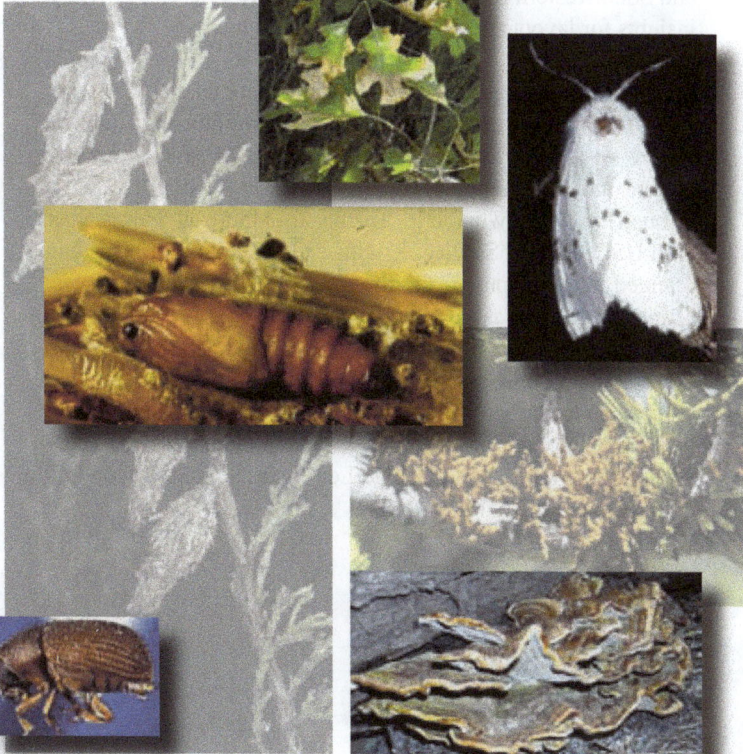

# Acknowledgments

Great appreciation is extended to the entomologists and pathologists who have studied and published on the often complicated life cycles, subtle taxonomic variations, and effective management strategies of the diverse organisms presented in this field guide. References are cited at the end of this guide.

The following individuals have all provided invaluable observations through personal communication and published materials: Whitney Cranshaw and William Jacobi with Colorado State University; David Leatherman and Ingrid Aguayo, formerly with Colorado State Forest Service; Mark Harrell with Nebraska Forest Service; John Ball with South Dakota State University; Robert Mathiasen from Northern Arizona University; Les Koch with the Wyoming State Forestry Division; Jose Negron with the USDA Forest Service, Rocky Mountain Research Station (RMRS); and John Schmid, formerly with RMRS. Field guides produced in the Pacific Northwest, Northern and Intermountain, and Southwestern Regions of the USDA Forest Service provided insights for layout and design.

Many of the photographs presented here were taken by Federal, State, and university pathologists and entomologists from the Rocky Mountain Region or adjacent USDA Forest Service regions and were made available by the University of Georgia through www.bugwood.org. Thanks to MaryLou Fairweather with USDA Forest Service, Southwestern Region, for providing scanned photographs from the *Field Guide to Insects and Diseases of Arizona and New Mexico*.

Special thanks also to Jose Negron and Bill Jacobi for their reviews and comments, to Justin Backsen and Diane Hildebrand for their help during manuscript preparation, and to Lindy Myers, Lane Eskew, and Connie Lemos for editing, layout, and publication of the field guide.

# Contents

## Diseases

**Insects**

# About This Field Guide

This field guide details the most commonly encountered diseases and insects of forest trees in the Rocky Mountain Region. Descriptions of diseases, insects, and physical injuries focus on the most diagnostic features of each. Color photographs, line drawings, and tables are used to illustrate and emphasize characteristics described in the text. Diseases and insects in plains hardwood trees are not covered in depth. Ornamental trees are sometimes affected by the diseases or insects included in this guide but may not be specifically mentioned as hosts.

This guide presents diseases and then insects. Entries are arranged according to the part of trees typically damaged by the agent described. The disease section describes dwarf and true mistletoes, decays (including root diseases and stem decays), cankers, wilts, rusts, foliage diseases, shoot blights, and abiotic injuries and miscellaneous diseases. The insect section describes bark beetles, defoliators, wood borers, sap-sucking insects, gall formers, mites, and bud and shoot insects.

At the end of the guide is a subject index, a host-pest index to damaging agents by tree species and part of the tree affected, and a glossary of terms. The host-pest index provides a rapid means of assessing the number and variety of agents described for each tree species.

This field guide applies to the USDA Forest Service's Rocky Mountain Region, which includes Colorado, Wyoming, South Dakota, Kansas, and Nebraska (fig. 1). Additional hosts, diseases, and insect pests may be encountered outside this Region that are not included here, and a few of the diseases and insects included in this guide may not be seen in other areas.

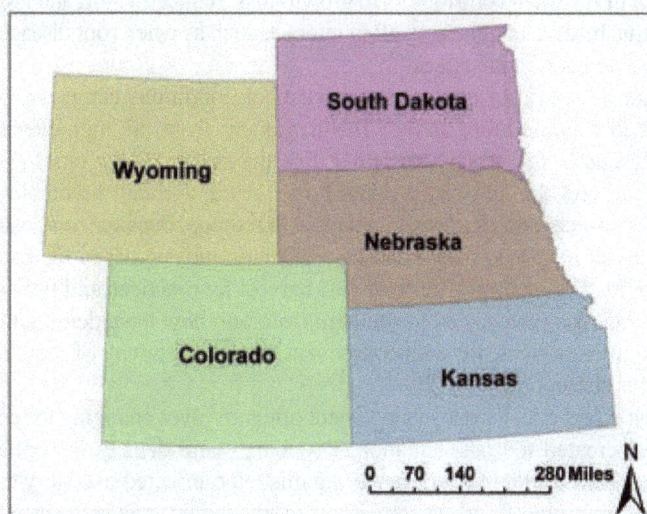

**Figure** 1. Rocky Mountain Region area covered by this Field Guide. *Image: Sheryl Costello, USDA Forest Service.*

Plant pathologists and entomologists are available to assist resource managers with identification of insects and pathogens encountered in the forests. A list of offices that provide this assistance is included at the back of this field guide along with instructions for specimen collection and shipping.

# Role of Diseases and Insects in Forest Ecosystems

Most forest insects and diseases are native to and important players in the forest ecosystems of the Rocky Mountain Region. Some insects and diseases reduce tree growth, cause mortality, reduce timber productivity, create hazardous situations in developed sites, and change wildlife habitat, fire hazard conditions, and overall watershed quality. In terms of mortality and growth loss, the impacts of diseases and insects on forests are far greater than those of fire or any other disturbance. Along with fire, insects and diseases are the major disturbance agents for changing forest age, density, composition, and structure on a stand or landscape level. They provide food and habitat for other wildlife, increase structural diversity of forests, and are important forest recyclers in the Rocky Mountains.

Typically, forest insects are at low or endemic levels. From time to time, significant outbreaks cause rapid changes across the forest. For example, bark beetles can build to high populations relatively fast and can cause widespread tree mortality, significantly altering forest conditions. Diseases are more likely to increase gradually or remain at similar levels over time, depending on forest conditions. Dwarf mistletoes and root diseases are ever-present in the forest and are some of the main contributors to growth loss, reduced vigor, and mortality in their hosts. Insects and diseases often interact, such as when root diseases predispose trees to bark beetle attack.

At times, natural or human-made events or conditions can cause insects or diseases to become more active. This can occur from the high-elevation sub-alpine forests to the urban forests that line the streets of our cities and towns. Fire, wind, and drought can weaken host trees and create favorable environments for insects and diseases to attack. Altered age, density, composition, and structure can also lead to greater susceptibility and increased disease and insect activity. For example, through past harvesting practices and fire exclusion, many ponderosa pine stands in southern Colorado now have dense understories of shade-tolerant white fir, with consequent increased activity of root diseases to which it is susceptible.

Forest insect and disease management often involves changing the conditions that have created the pest situation. Reducing stand density in conifer stands can lower bark beetle impact. Removing mistletoe-infected overstory trees from a regenerating pine stand or converting to even-aged management where that is consistent with the natural disturbance regime can protect regeneration and reduce mistletoe impacts in future stands. Increasing the diversity of cultivars that are planted along our city streets and windbreaks can also reduce pest impacts.

The main goal of most forest managers is to maintain a healthy forest that is capable of producing a variety of resources. Knowledge of the role that forest insects and diseases play is critical to achieving the desired conditions from the forest. This manual provides users with a tool to begin identifying and understanding forest insects and diseases that may be detrimental to management goals. Please refer to your State Department of Agriculture or your local pest control specialist for the rules, regulations, and applicable allowances concerning specific control measures.

# Diagnosing Tree Problems

Several steps must be taken in order to effectively diagnose tree problems. The following are general guidelines:

- Properly identify the tree. It is important to know exactly what species you are looking at. It is also important to know what the leaves, bark, trunk, and roots should look like under "normal" conditions.
- Carefully observe the symptoms expressed. Look for patterns on the tree and in the surrounding vegetation.
  - Check for host specificity. Biotic agents tend to affect one species, are clumpy in distribution, show progressive symptoms, and usually impact specific plant parts. Abiotic agents tend to affect many species relatively uniformly.
  - Carefully examine the types of symptoms and the part of the plant impacted.
  - Typical symptoms include: underdevelopment of tissues or organs (stunting and malformed leaves); overdevelopment of tissues or organs (galls, brooms, and stress cones); necrosis (death) of plant parts (wilting, dieback, and leaf spots); and alteration of normal appearance (chewing and chlorosis [pale coloration in leaves and flowers]).
  - Examine how the symptoms are distributed. If the entire tree is impacted, there is likely something wrong with the roots or stem, or there may be an environmental cause. Single or randomly scattered affected branches are often associated with insects or diseases.
- Determine the history of the tree and the site. Has the root system been disturbed? Have chemicals been used? Has there been any harvesting? Other site factors, such as changing water relations, extreme temperatures, or wind, may affect tree vigor.
- Look for signs of biotic agents such as fungal fruiting bodies, parasitic plants, larvae, or adult insects.
- Identify agents. Laboratory tests may be necessary in some circumstances.

**Sources of Additional Information**

Additional information on insects and diseases can be found online at:

- USDA Forest Service, Rocky Mountain Region: http://www.fs.usda.gov/goto/r2/fh;
- USDA Forest Service, Northern and Intermountain Region: http://www.fs.fed.us/r1-r4/spf/fhp/mgt_guide/;
- USDA Forest Service, Southwestern Region: http://www.fs.fed.us/r3/resources/health/field-guide/index.shtml;
- Colorado State University Extension: http://www.ext.colostate.edu/pubs/pubs.html#insects; and
- State Forestry Agencies and County Extension Offices.

Updated Forest Insect and Disease Leaflets can be found online at:

- USDA Forest Service, Forest Health Protection: http://www.fs.fed.us/r6/nr/fid/wo-fidls/index.shtml.

Images and publications can also be found online at:

- University of Georgia and USDA Forest Service online database: http://www.bugwood.org/.

# Diseases

# Introduction to Dwarf Mistletoes

## Parasitic vascular plants with conifer hosts

**Pathogens**—Dwarf mistletoes (*Arceuthobium* spp.) are parasitic plants of co-nifers that obtain almost all of their needs, including water, mineral, and carbon nutrients, from their hosts. Fulfilling these resource requirements stresses infect-ed trees, causing reductions in growth, cone, and seed production and, with high infection levels, mortality. Dwarf mistletoes are some of the most common and easily identified disease agents where they occur in the coniferous forests of the Rocky Mountain Region.

**Hosts**—Five species of dwarf mistletoes occur in this Region, each with a spe-cific set of susceptible hosts (table 1).

**Signs and Symptoms**—Dwarf mistletoes produce aerial shoots on branches or stems of infected trees (fig. 2). Shoots are nearly leafless and vary in color; yel-low, brown, purple, or green shoots are common. Plants size also varies within and among species. Douglas-fir dwarf mistletoe shoots are often shorter than the host's leaves, while southwestern dwarf mistletoe shoots are typically 3-6 inches (7-15 cm) long. When shoots are shed, characteristic basal cups remain (fig. 3). Infection with dwarf mistletoe also causes characteristic deformities in the host. Witches' brooms are areas of profuse, dense branching often induced by dwarf mistletoe infection (fig. 4). Branch swellings are often found in the immediate vicinity of local infections (fig. 2). Cankers (areas of dead cambium) are often associated with older infections. Dieback of the host's foliage from the top-down and eventual mortality is often observed in trees that have been infected for many

**Table 1.** Dwarf mistletoes and their hosts in the Rocky Mountain Region.

| Dwarf mistletoe (DM) | Primary host[a] | Other hosts[a] |
|---|---|---|
| Lodgepole pine DM<br>*Arceuthobium americanum* | Lodgepole pine | Secondary: ponderosa pine<br>Occasional: whitebark and limber pines<br>Rare: Engelmann and blue spruce, bristlecone pine |
| Limber pine DM<br>*A. cyanocarpum* | Limber pine,<br>Whitebark pine,<br>Bristlecone pine | Rare: ponderosa pine, lodgepole pine |
| Pinyon DM<br>*A. divaricatum* | Pinyon pine | None |
| Douglas-fir DM<br>*A. douglasii* | Douglas-fir | Rare: subalpine fir, Englemann and blue spruce |
| Southwestern DM<br>*A. vaginatum* subsp.<br>*cryptopodium* | Ponderosa pine | Occasional: bristlecone pine, lodgepole pine<br>Rare: limber and southwestern white pine, blue spruce |

[a] Hosts are in the following categories:
  Primary: more than 90% infection when close to heavily infected trees.
  Secondary: frequently infected (50-90% infection) when close to heavily infected principal hosts.
  Occasional: occasionally infected (5-50% infection) when close to heavily infected principal hosts.
  Rare: rarely infected (≤5% infection), even when close to heavily infected principal hosts.

　　　　USDA Forest Service RMRS-GTR-241. 2010.

**Figure 2.** Aerial shoots of American dwarf mistletoe plant on lodgepole pine. Note the swelling in the branch associated with the aerial shoots. *Photo: Brian Howell, USDA Forest Service.*

**Figure 3.** After aerial shoots are shed and basal cups remain as signs of dwarf mistletoe infection. *Photo: Kelly Burns, USDA Forest Service.*

**Figure 4.** Large witches' brooms formed on a ponderosa pine that is heavily infected with southwestern dwarf mistletoe. The top of the tree is dying back. *Photo: Bob Cain, USDA Forest Service.*

**Figure 5.** Lodgepole pine killed as a result of lodgepole pine mistletoe infection. Note the typical witches' brooms. *Photo: Jim Worrall, USDA Forest Service.*

years (fig. 5), depending on the species of dwarf mistletoe and host, the level of infection, and site factors. These symptoms on the hosts are often associated with mistletoe infection but may also be caused by other agents. Plants or basal cups should be present for positive identification of dwarf mistletoe infection.

**Disease Cycle**—Dwarf mistletoes have separate male and female plants (figs. 6-7). Seeds are produced annually on mature female plants. These are explosively released and typically fly less than 33 ft (10 m). Upon germination, the dwarf mistletoe plant produces a specialized root-like structure that contacts the phloem and xylem of the host, from which the parasite obtains nutrients and water. Aerial shoots appear 3-5 or more years after infection; the time period before shoots are visible is known as the latent period (fig. 8).

Dwarf mistletoes spread both within and between tree crowns. As a result of the explosive seed-dispersal mechanism, infections tend to build up initially in the lower portion of the crown and spread gradually upward. Lateral spread

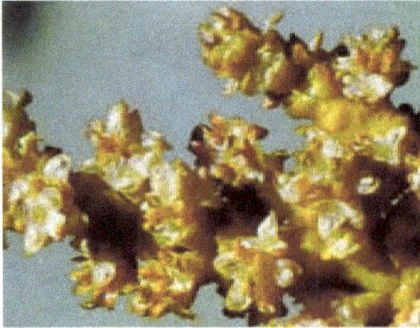

**Figure 6.** Male flowers on shoots of southwestern dwarf mistletoe. *Photo: Brian Howell, USDA Forest Service.*

**Figure 7.** Immature female berries on shoots of southwestern dwarf mistletoe. *Photo: Brian Howell, USDA Forest Service.*

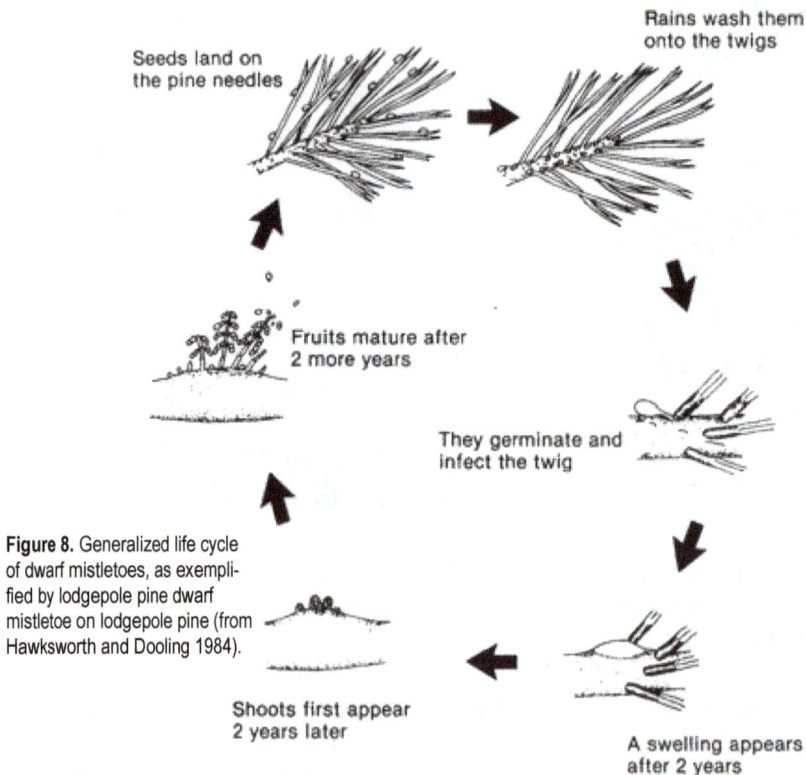

Seeds land on the pine needles

Rains wash them onto the twigs

Fruits mature after 2 more years

They germinate and infect the twig

**Figure 8.** Generalized life cycle of dwarf mistletoes, as exemplified by lodgepole pine dwarf mistletoe on lodgepole pine (from Hawksworth and Dooling 1984).

Shoots first appear 2 years later

A swelling appears after 2 years

of dwarf mistletoe through single-storied stands averages about 1.5 ft (0.5 m) per year. Spread is most rapid from infected overstory to adjacent regeneration. Long-distance seed dispersal by birds is not common but can introduce dwarf mistletoe to new areas.

**Impact**—As parasites, dwarf mistletoes cause significant changes in physiological processes and structural characteristics of infected trees, resulting in changes

**Figure 9.** Evidence of squirrel or porcupine feeding on sugar-rich phloem found near southwestern dwarf mistletoe infections. *Photo: Brian Howell, USDA Forest Service.*

in the structure and function of forest communities. Tree growth and vigor usually decline when more than half of the crown is parasitized. Most trees survive infection for decades, but small trees tend to decline and die more quickly than large ones. Tree mortality in areas with extensive infection is often three to four times higher than in uninfected areas. Bark beetles frequently attack heavily infected trees, especially during drought.

Extensive dwarf mistletoe infection greatly reduces forest productivity. However, infection has some benefits for wildlife. Large witches' brooms provide nesting and seclusion habitat for birds and small mammals. Snags created by dwarf mistletoe infection offer habitat for cavity-nesting birds. A few species are known to feed on shoots of dwarf mistletoes and the sugar-rich phloem found in and around infection sites (fig. 9).

**Management**—The first step when making management decisions in stands infected with dwarf mistletoe is to quantify the incidence and severity of infection. Although many systems have been used to rate levels of infection by dwarf mistletoe, one is now used almost universally: Hawksworth's 6-class dwarf mistletoe rating (DMR) system (fig. 10). Many disease parameters and management recommendations are provided in terms of DMR because this system has been used for many years. A tree's DMR ranges from 0 (uninfected) to 6 (over half the branches infected throughout the crown). Rate each third of the crown on a scale from 0 to 2, then sum the thirds for the tree rating (fig. 10). Binoculars should be used to enhance detection.

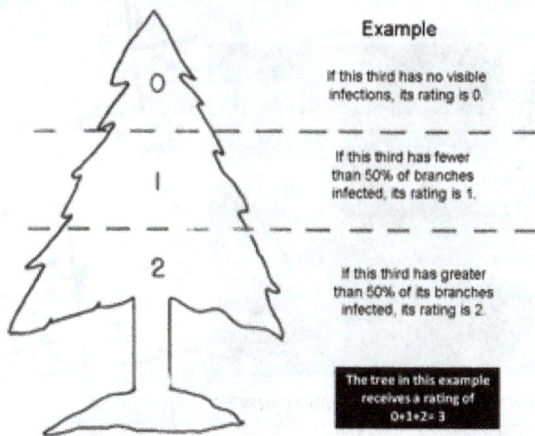

**Figure 10.** The 6-class Hawksworth dwarf mistletoe rating system (from Hawksworth and Wiens 1996).

Silvicultural control of dwarf mistletoes can be effective and should be considered for use in a variety of stand conditions and dwarf mistletoe infection levels. Because dwarf mistletoes require a living susceptible host, silvicultural options include pruning, harvesting, and favoring non-host species. Due to the explosive seed dispersal mechanism, implementing buffer strips around infection centers or around sanitized patches can also be effective. A thorough discussion of management options based on stand conditions and objectives is outside the scope of this manual but can be found on the Region 2 Forest Health Management website at: http://www.fs.fed.us/r2/fhm/bugcrud/DM_MgmtGuide_R2.pdf.

**References: 66, 70, 158**

# Douglas-Fir Dwarf Mistletoe

## Large brooms, small aerial shoots

**Pathogen**—Douglas-fir dwarf mistletoe (*Arceuthobium douglasii*) causes large brooms but has very small shoots. Aerial shoots are olive green, average $4/5$ inch (20 mm) long (maximum 3 inches or 8 cm) and $1/25$-$1/16$ inch (1-1.5 mm) in diameter, and have fan-shaped branching (fig. 11).

Douglas-fir dwarf mistletoe occurs throughout the range of Douglas-fir in the central and southern mountains of Colorado. It is absent from northern Colorado (except for the extreme northwest) and from the portion of Wyoming in the Rocky Mountain Region (fig. 12).

**Hosts**—Douglas-fir dwarf mistletoe primarily infects its namesake, although several true firs and spruces are occasional or rare hosts.

Figure 11. Douglas-fir dwarf mistletoe parasitizing Douglas-fir. Note that the mistletoe plants are growing with the branch (systemic infection). *Photo: Robert Mathiasen, Northern Arizona University.*

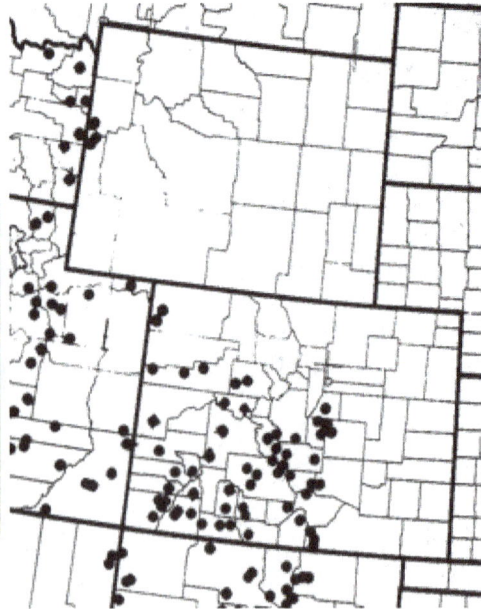

Figure 12. Distribution of Douglas-fir dwarf mistletoe in the Rocky Mountain Region (from Hawksworth and Wiens 1996).

**Figure 13.** Douglas-fir dwarf mistletoe plants, which are approximately the same length as the host's leaves. *Photo: USDA Forest Service.*

**Figure 14.** Large brooms are typical in Douglas-fir heavily infected with Douglas-fir dwarf mistletoe. *Photo: USDA Forest Service.*

**Figure 15.** Trees heavily infected with Douglas-fir dwarf mistletoe typically die from the top-down, causing dead (spike) tops. *Photo: USDA Forest Service.*

**Signs and Symptoms**—Douglas-fir dwarf mistletoe has the smallest shoots (fig. 13) of all mistletoes in the Region, but it can form the largest witches' brooms (fig. 14). Douglas-fir infections grow along with the infected host branches (systemic infection). Mistletoe shoots may be spread along young host branches or be aggregated near the annual bud scars. Because shoots are so small, they are normally detectable only in branches close to the ground. Witches' brooms, which are used for detection and rating, become noticeable about 10 years after infection and develop best in direct sunlight. Brooms occur mostly in the lower half of tree crowns. They can weigh hundreds of pounds, can break off of the tree, and are considered hazards in developed sites.

**Impact**—Dwarf mistletoe is the most detrimental disease of Douglas-fir. Damages typically associated with dwarf mistletoe infection are: growth reduction, spike tops (fig. 15), reduced cone and seed production, and mortality. Infections have reportedly increased in abundance since the late 1800s. In northern Idaho and western Montana, Douglas-fir stands have become more widespread due to fire suppression, a history of selective harvesting that removed pines and encouraged shade-tolerant species, and white pine blister rust, which largely eliminated western white pine.

Data on growth effects from western Montana indicate that light, medium, and severe infections caused decreases in basal area growth rate of 14, 41, and 69%, respectively. Effects on height growth were similar. Horizontal spread in single-storied stands is estimated at 1.5-2 ft (45-61 cm) per year. Upward spread in crowns is about 4-6 inches (10-15 cm) per year.

Please see the Introduction to Dwarf Mistletoes entry for disease cycle and management information.

**References: 59, 70**

# Limber Pine Dwarf Mistletoe

## Infects five-needle pines

**Pathogen**—Aerial shoots of limber pine dwarf mistletoe (*Arceuthobium cyanocarpum*) are yellow-green, 1 1/4-2 3/4 inches (3-7 cm) long, and up to 1/13 inch (2 mm) diameter. Branching is fan-shaped and shoots are densely clustered (fig. 16). Limber pine dwarf mistletoe generally occurs in five-needle pine (predominantly limber pine) stands along the Continental Divide in the Rocky Mountains but also occurs in other mountain ranges (fig. 17).

**Hosts**—Almost all of the five-needle pines in the Rocky Mountain Region, including limber, whitebark, and Rocky Mountain bristlecone pine, are primary hosts of limber pine dwarf mistletoe. The only endemic white pine that is not a host in nature is southwestern white pine, although it has been infected in greenhouse trials.

**Signs and Symptoms**—Signs of infection include aerial shoots and basal cups left

**Figure 16.** Limber pine dwarf mistletoe parasitizing limber pine. *Photo: Brian Howell, USDA Forest Service.*

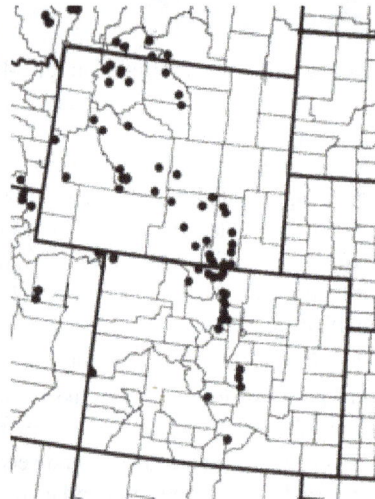

**Figure 17.** Distribution of limber pine dwarf mistletoe in the Rocky Mountain Region (from Hawksworth and Wiens 1996).

after shoots have fallen off branches. Limber pine dwarf mistletoe causes small, tightly clustered witches' brooms (fig. 18). Other symptoms of infection include swelling of branches at infection sites, dieback, and eventual mortality of heavily infected trees.

**Impact**—Limber pine dwarf mistletoe causes extensive mortality of limber pine in many parts of the Rocky Mountains and can also cause mortality in other hosts when infection levels are high. It is the most important native disease of high-elevation white pines in the West; only white pine blister rust is more damaging. Lateral spread in single-storied stands is estimated to be 1.5-2 ft (45-61 cm) per year.

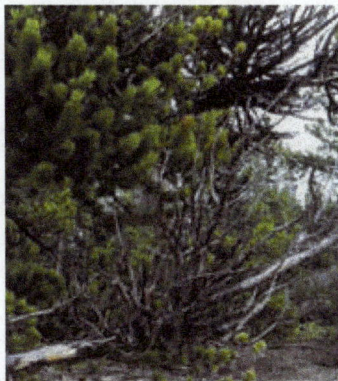
**Figure 18.** Tightly clustered brooms are a symptom of limber pine dwarf mistletoe. *Photo: Brian Howell, USDA Forest Service.*

Please see the Introduction to Dwarf Mistletoes entry for disease cycle and management information.

**References: 70, 171**

# Lodgepole Pine Dwarf Mistletoe
## Common cause of brooming in lodgepole pine

**Pathogen**—Lodgepole pine dwarf mistletoe (*Arceuthobium americanum*) is the most widely distributed, one of the most damaging, and one of the best studied dwarf mistletoes in North America. Aerial shoots are yellowish to olive green, 2-3 1/2 inches (5-9 cm) long (maximum 12 inches [30 cm]) and up to 1/25-1/8 inch (1-3 mm) diameter (figs. 19-20). The distribution generally follows that of its principal host, lodgepole pine, in the Rocky Mountain Region (fig. 21).

**Hosts**—Lodgepole pine dwarf mistletoe infects primarily its namesake, but, as noted in table 1, ponderosa pine is considered a secondary host of this species. However, lodgepole pine dwarf mistletoe can sustain itself and even be aggressive in pure stands of Rocky Mountain ponderosa pine in northern Colorado and southern Wyoming sometimes a mile or more away from infected

**Figure 19.** Flowering male lodgepole pine dwarf mistletoe plant parasitizing lodgepole pine. *Photo: Brian Howell, USDA Forest Service.*

**Figure 20.** Female lodgepole pine dwarf mistletoe plant with immature fruit parasitizing lodgepole pine. Note the basal cups left behind where old shoots have fallen off. *Photo: Brian Howell, USDA Forest Service.*

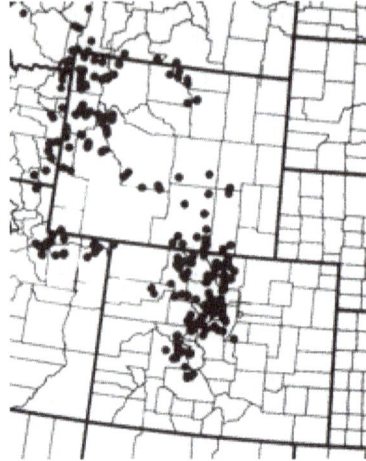

**Figure 21.** Distribution of lodgepole pine dwarf mistletoe in the Rocky Mountain Region (from Hawksworth and Wiens 1996).

lodgepole pine. This infection generally occurs in areas outside the range of ponderosa pine's usual parasite, southwestern dwarf mistletoe.

**Signs and Symptoms**—Signs of infection are shoots and basal cups (fig. 20) found at infection sites. Symptoms include witches' brooms, swelling of infected branches, and dieback. Lodgepole pine dwarf mistletoe infections grow systemically with the branches they infect, sometimes causing large witches' brooms with elongated, loosely hanging branches.

**Impact**—Heavily infected trees experience reduced diameter and growth, reduced cone production, and eventual mortality (fig. 22).

Spread rate in even-aged stands can be about 1.7 ft (50 cm) per year in open stands and 1.2 ft (36 cm) per year in dense stands. Intensification (increase in number of infections over time) occurs most quickly in stands 15-60 years old in Colorado. During that time, DMR increased one class in 14 years (see "Management" in the Introduction to Dwarf Mistletoes entry). A feature of

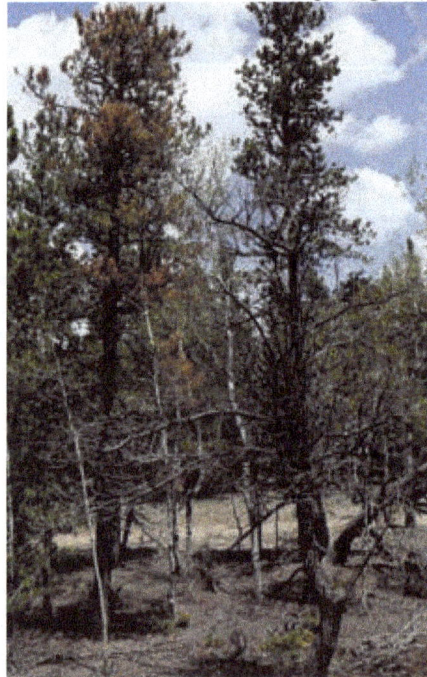

**Figure 22.** Dead and dying lodgepole pine heavily infected by lodgepole pine dwarf mistletoe. *Photo: Brian Howell, USDA Forest Service.*

this species that is potentially useful in management is that the upper elevational limit is usually about 600-650 ft (185-200 m) below the upper elevational limit of lodgepole pine for a given latitude. Experiments have shown that the mistletoe can survive at higher elevations but it cannot reproduce because the fruit is killed by early autumn frosts before it can mature.

Please see the Introduction to Dwarf Mistletoes entry for disease cycle and management information.

**References: 68, 70**

# Pinyon Dwarf Mistletoe

## Disease of pinyon on Colorado's Western Slope

**Pathogen**—Aerial shoots of pinyon dwarf mistletoe (*Arceuthobium divaricatum*) are olive green to brown, about 3-5 inches (8-13 cm) long, and up to $1/6$ inch (4 mm) diameter. Shoots often have a long, thin, and spreading appearance (fig. 23). Branching is fan-shaped.

Pinyon dwarf mistletoe is found throughout the pinyon range in the western quarter of Colorado but is absent in pinyon stands east of the Continental Divide (fig. 24).

**Hosts**—Pinyon dwarf mistletoe infects only pinyons.

**Signs and Symptoms**—Signs of infection include aerial shoots and basal cups. Symptoms include witches' brooms, swelling of infected branches, and dieback. This dwarf mistletoe may not result in well-developed witches' brooms, but those that do develop are usually small.

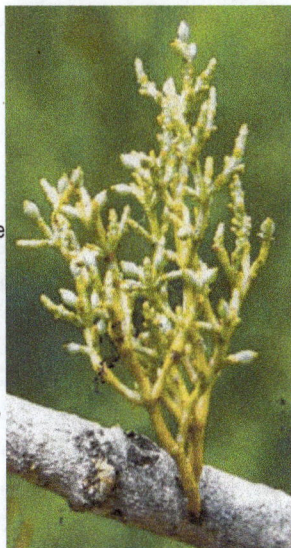

Figure 23. Female pinyon dwarf mistletoe plant parasitizing piñon pine. Note the olive-green color of shoots. *Photo: Robert Mathiasen, Northern Arizona University.*

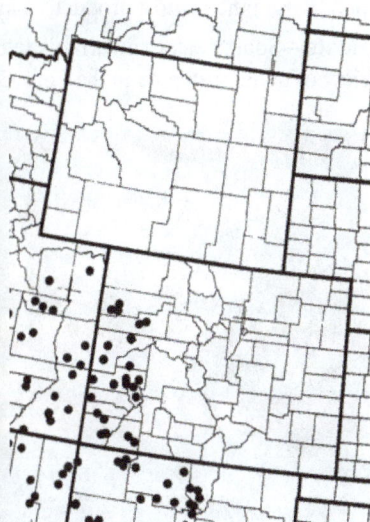

Figure 24. Distribution of pinyon dwarf mistletoe in the Rocky Mountain Region (from Hawksworth and Wiens 1996).

**Impact**—Pinyon dwarf mistletoe is considered less lethal than other dwarf mistletoes of the region. However, growth loss and mortality can be significant when infection is severe (DMR 5 or 6).

Other dwarf mistletoes greatly reduce seed production of their hosts, but such effects on pinyon are unknown. Reduced seed production could be particularly important in pinyon because the nuts are collected for food by humans and wildlife, and they are necessary for species reproduction.

Please see the Introduction to Dwarf Mistletoes entry for disease cycle and management information.

**References: 70, 118**

# Southwestern Dwarf Mistletoe

## Infects ponderosa pine in Colorado

**Pathogen**—Aerial shoots of southwestern dwarf mistletoe (*Arceuthobium vaginatum* ssp. *cryptopodum*) vary in color from orange to reddish brown to almost black. Shoots are the largest of dwarf mistletoes in this Region and are approximately 4 inches (10 cm) long (maximum 11 inches or 27 cm) with a basal diameter of $1/13$-$3/8$ inch (2-10 mm) (fig. 25). This species is unusual among dwarf mistletoes in temperate regions in that seed germination occurs immediately after dispersal in the fall rather than in the following year. Within the Rocky Mountain Region, southwestern dwarf mistletoe is found in southern Colorado on the Western Slope extending into northern Colorado on the Front Range (fig. 26). No dwarf mistletoe occurs in the Black Hills National Forest, where ponderosa pine is most productive in the Region.

**Hosts**—Southwestern dwarf mistletoe primarily infects the Rocky Mountain variety of ponderosa pine in the Four Corners states (Colorado, Utah, Arizona, and

**Figure 25.** Male southwestern dwarf mistletoe plant parasitizing Rocky Mountain ponderosa pine. *Photo: Brian Howell, USDA Forest Service.*

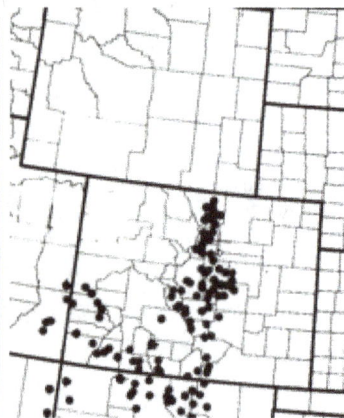

**Figure 26.** Distribution of southwestern dwarf mistletoe in the Rocky Mountain Region (from Hawksworth and Wiens 1996).

**Table 2.** Expected half-life (time, in years, for half the trees to die) of ponderosa pine infected with southwestern dwarf mistletoe at Grand Canyon National Park (ref. 67).

| Initial DMR | 4-9 inches DBH | >9 inches DBH |
|---|---|---|
| 0-1 | ND[a] | ND[a] |
| 2-3 | 30 | 57 |
| 4-5 | 17 | 25 |
| 6 | 7 | 10 |
| Total | | 14 |

[a] No decrease in longevity detected; half-life too long to estimate.

New Mexico) with a small distribution in west Texas. Occasionally, southwestern dwarf mistletoe will infect bristlecone pine and lodgepole pine. It rarely infects limber pine, southwestern white pine, and blue spruce.

**Signs and Symptoms**—Signs of infection include aerial shoots and basal cups, and symptoms include witches' brooms, swelling of infected branches, and dieback (fig. 27).

**Figure 27.** Heavily infected southwestern ponderosa pine with characteristic broomed branches and top dieback. *Photo: Brian Howell, USDA Forest Service.*

**Impact**—Damage is usually greater along the Front Range than in southwestern Colorado. Witches' broom development can be weak, but large and robust brooms with thick, distorted branches are common in older infectons. Mortality from southwestern dwarf mistletoe was quantified in a 32-year study at Grand Canyon National Park. Ninety percent of uninfected or lightly infected (DMR 0-1 at the start) trees survived the entire study period. Of heavily infected trees (DMR 6), only 5% over 9 inches (23 cm) diameter at breast height (DBH) survived, and none survived in the 4-9 inches (10-23 cm) size class. Intermediate infection levels were associated with intermediate mortality levels. Infection intensified during the study, so much so that most trees that died were in DMR class 6 by the time of death. Based on the data, the authors estimated the half-life of trees (time in which half the trees are expected to die) by DMR class, as described in table 2.

Estimates of spread rate in single-storied stands vary. Recent estimates of 2-3 ft (61-91 cm) per year indicate that southwestern dwarf mistletoe has one of the faster spread rates. Earlier estimates were about 1.3 ft (0.4 m) per year in open stands and 0.9 ft (0.3 m) per year in dense stands. Spread from overstory to understory is faster in ponderosa than in lodgepole pine.

Please see the Introduction to Dwarf Mistletoes entry for disease cycle and management information.

**References: 12, 65, 67, 70, 115**

# Juniper Mistletoe

## Minor effects on junipers

**Pathogen**—Juniper mistletoe (*Phoradendron juniperinum*) is the only member of the true mistletoes that occurs within the Rocky Mountain Region (fig. 28).

**Hosts**—Within the Rocky Mountain Region, juniper mistletoe is found in the pinyon-juniper woodlands of southwestern Colorado (fig. 29) and can infect all of the juniper species that occur there.

**Signs and Symptoms**—Juniper mistletoe plants are generally densely branched in a spherical pattern and are green to yellow-green (fig. 30). Unlike most true mistletoes that have obvious leaves, juniper mistletoe leaves are greatly reduced, making the plants look similar to, but somewhat larger than, dwarf mistletoes. However, no dwarf mistletoes infect junipers in the Rocky Mountain Region.

**Figure 28.** Juniper mistletoe plants on one-seed juniper in Mesa Verde National Park. *Photo: USDA Forest Service.*

**Disease Cycle**—Juniper mistletoe plants are either male or female. The female's berries are spread by birds that feed on them. As a result, this mistletoe is often found where birds prefer to perch—on the tops of taller trees (fig. 28), near water sources, etc. When the seeds germinate, they penetrate the branch of the host tree. In the branch, the mistletoe forms a root-like structure that is used to gather water and minerals. The plant then produces aerial shoots that produce food through photosynthesis.

**Impacts**—Impacts associated with juniper mistletoe are generally

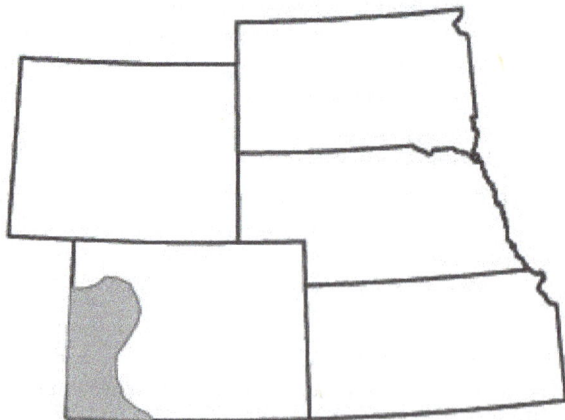

**Figure 29.** Distribution of juniper mistletoe in the Rocky Mountain Region (from Hawksworth and Scharpf 1981).

USDA Forest Service RMRS-GTR-241. 2010.

minor. While the true mistletoe plants do receive some small proportion of their carbon nutrition from their hosts, they are considered only "water and mineral" parasites. Unlike dwarf mistletoes, true mistletoes produce most of their own food through photosynthesis. However, during periods of drought stress, when

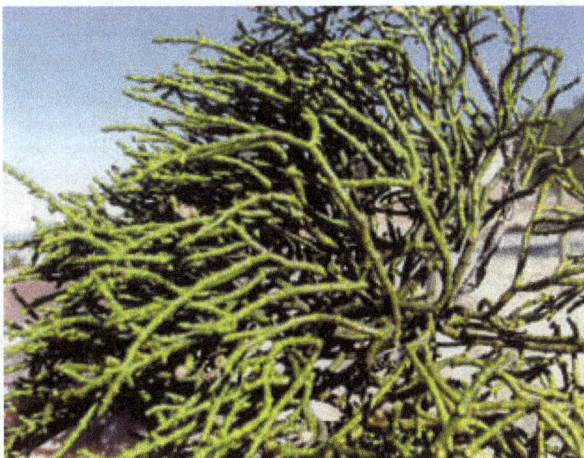

**Figure 30.** Closeup of juniper mistletoe on juniper branch. *Photo: Robert Mathiasen, Northern Arizona University.*

the host trees have shut down their transpiration to conserve water, juniper mistletoe plants continue to transpire, causing further drought stress on the host.

**Management**—Juniper mistletoe is generally not considered worth managing as the impacts are minor. If management is desired, infected branches can be effectively pruned because juniper mistletoe requires a living host to survive. If mistletoe shoots are removed, they will resprout.

**References: 53, 69**

# Introduction to Decay Diseases

## Wood decays of roots and stems

Many fungi decay wood of live trees in the Rocky Mountain Region. The diseases are often called heart rots, but frequently the decay is not restricted to heartwood. They are generally divided into stem decays (trunk rots) and root and butt rots.

**Decay Types**—Each fungus causes one of two types of decay: white rot or brown rot. These are easily distinguished (table 3, fig. 31) and are features used to diagnose the disease.

**Stem Decay Disease Cycle**—Stem-decay fungi release wind-disseminated spores. Some infect through wounds, but other more specialized pathogens do not require obvious wounds. They can infect small twig or branch stubs, and then either grow into the stem and inner wood or become dormant and wait until the tree grows around them and they become embedded in heartwood. When sufficient food is available through decay, a fruiting body (conk) is produced. Spores from the conk repeat the cycle.

**Table 3.** Features of white and brown rot.

|  | Chemistry | Color | Texture | Other |
|---|---|---|---|---|
| White rot | All wood components removed, either simultaneously or lignin preferentially, in early stages | Generally white, but can be yellowish to reddish brown | Varies: spongy, stringy; some types are called laminated, pitted, or pocket rot based on texture/appearance | Texture types vary among white rot fungi; some produce zone lines in wood; there may be mats or rhizomorphs; pocket rots may have black flecks in pockets |
| Brown rot | Cellulose and hemicellulose chains broken early, then removed; lignin remains | Brown, often with a sheen on split surfaces early on | When advanced, wood shrinks with cubical checks and can be crumbled to a powder | Decay is fairly uniform; some fungi produce white mats or felts or wispy fine cords in checks, causes rapid strength loss |

**Root Disease Cycle**—Root and butt rots often have a more complex disease cycle. As with stem decays, wind-blown spores can be the initial inoculum. Spores may infect wounds in the butt or root crown, or they may per-colate through the soil to infect roots. In general, these pathogens kill and decay roots, decay inner wood in the butt, and often kill sapwood and cambium in the root crown.

An important difference from stem decays is that most, but not all, of these diseases can also spread locally. The pathogen grows from infected roots to roots of neighboring healthy trees at contacts and grafts or, in one case, through soil. Root disease centers result in the stands characterized by older mortality in the middle and more recent mortality, symptomatic live trees, and apparently healthy trees toward the outside. The pathogen may survive for many years in dead root systems, infecting future tree generations. In many cases, this local spread is much more common than spores initiating new infections.

**Impact**—Impacts of decay diseases vary greatly, depend-ing on the disease type and the specific disease. They may include: loss of fiber to decay (cull), growth loss, direct mortality, predisposition to bark beetles, uprooting or snapping of live trees, and provision of wildlife habitat (figs. 32-36).

These diseases may affect various values, including timber, wildlife, aesthetics, and recreation. In developed sites, the potential for failure of live trees can significant-ly threaten safety and property.

**Figure 31.** Dead, decaying stem with white rot in the outer wood and brown rot in the inner wood caused by two different fungi. The white rot is stringy-fibrous; the brown rot breaks easily across the grain, has no fibrous strength, and is crumbly when dry. *Photo: Jim Worrall, USDA Forest Service.*

**Figure 32.** Irregular snap of live spruce due to butt rot. *Photo: Jim Worrall, USDA Forest Service.*

**Figure 33.** Uprooting of blue spruce brought up large root plate. This is typical of windthrow, but some roots are rotted by *Armillaria ostoyae*. *Photo: Roy Mask, USDA Forest Service.*

**Figure 34.** Aspen stem with aspen trunk rot (note the conks) has begun to fail. *Photo: Jim Worrall, USDA Forest Service.*

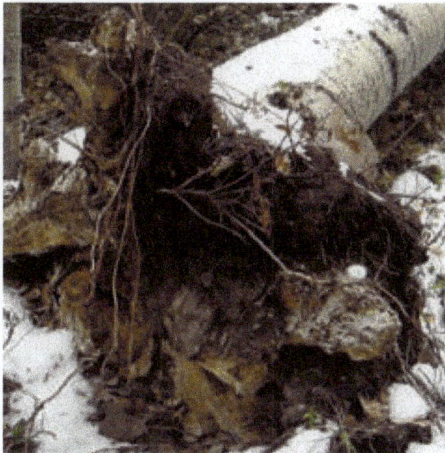

**Figure 35.** Typical failure associated with mottled root rot of aspen: roots are stubbed near the root collar and are almost all rotted. Also note the conk above the root crown. *Photo: Jim Worrall, USDA Forest Service.*

**Figure 36.** Snapping of spruce with white pocket rot; only a few centimeters of sound wood remain. *Photo: Jim Worrall, USDA Forest Service.*

Three important factors that contribute to the abundance and severity of these diseases are stand age, stand composition, and a history of wounding. Age, in particular, is closely correlated with amount of stem decay. Species composition is a major influence because, though some pathogens are fairly host-specific, in general there is a range of susceptibility of tree species to opportunistic, wound-infecting decay fungi. Ponderosa pine may represent the most resistant extreme, while true firs and aspen are among the Region's most susceptible species. A history of wounding is also associated with these opportunistic decay fungi in that there is generally a correlation between the size of wounds and their likelihood of infection.

Wood-decay diseases in this Region are listed in tables 4 and 5. Note that one root disease, black stain root disease, is not listed; it is the only major root disease in the Region that does not involve wood decay. These lists are not exhaustive, but include most decays that have been documented in the Rocky Mountain Region.

**References: 170, 192, 194**

**Table 4.** Root and butt rots of Region 2. Host abbreviations: DF = Douglas-fir, ES = Engelmann spruce, F = true firs, P = pines, QA = aspen, S = spruces, SAF = subalpine fir, WF = white fir. Uncommon hosts are in parentheses.

| Pathogen | Disease | Common hosts | Decay | Indicators |
|---|---|---|---|---|
| Armillaria solidipes and other species (honey mushroom) | Armillaria root disease | All | White rot; stringy-spongy, wet, zone lines | Basal resinosis, crown fading, mushrooms, fans |
| Coniophora puteana | | ES, SAF | Brown rot; thin, pale, brown cords in checks | Butt may show checks or collapse, indicating partial failure; fruiting inconspicuous after tree dies |
| Flammulina populicola | | QA | White rot; yellow, stringy | |
| Fomitopsis pinicola | Brown crumbly rot | S | Brown rot; thick fungal mats | Conk |
| Ganoderma applanatum (artist's conk) | White mottled rot | QA | White rot with mottled white/light tan areas; infrequent zone lines | Conks usually present |
| Heterobasidion annosum | Annosus root rot | P | White rot; may appear laminated, stringy, or with pits/pockets | Disease center, conks |
| Heterobasidion parviporum | Annosus root rot | WF (SAF, S) | White rot; may appear laminated, stringy, or with pits/pockets | Disease center, conks |
| Lentinellus montanus | | SAF, (ES) | White rot | None; fruits during snowmelt on downed logs |
| Onnia tomentosa/leporina | Red root (tomentosus and circinatus root rots) | S (other conifers) | Reddish stain becoming white pocket rot | Basal resinosis, conks |
| Phaeolus schweinitzii (cow-pie fungus) | Schweinitzii butt rot | DF (other conifers) | Brown rot | Possible ephemeral fruiting on or around tree |
| Phellopilus nigrolimitatus | Big white pocket rot | S (other conifers) | White pocket rot with large pockets | Usually none |
| Pholiota alnicola | | SAF, ES | White rot; stringy | Mushrooms |
| Pholiota squarrosa (scaly Pholiota) | | SAF, QA, (ES) | White rot; in aspen a gray-brown stain becoming soft and light tan, then stringy | Mushrooms |
| Pleurotus populinus | | QA | White rot | Mushrooms |
| Sistotrema raduloides | | QA | White rot | None |
| Vesiculomyces citrinus | | SAF, ES | White rot, yellowish, stringy | None; usually fruits on downed trees |

Table 5. Stem decays of Region 2. Host abbreviations: DF = Douglas-fir, ES = Engelmann spruce, F = true firs, LPP = lodgepole pine, PP = ponderosa pine, QA = aspen, S = spruces, SAF = subalpine fir, WF = white fir. Uncommon hosts are in parentheses.

| Pathogen | Disease | Common hosts | Decay | Indicators |
|---|---|---|---|---|
| Amylostereum chailletii | | SAF, S | Whiter rot; stringy | None (fruiting inconspicuous, ephemeral, and uncommon) |
| Antrodia serialis | | SAF | Brown rot | None (usually fruits after tree dies) |
| Cryptosphaeria lignyota | Cryptosphaeria canker | QA | White mottled rot | Canker |
| Dichomitus squalens | red ray rot | PP | White pocket rot, but sometimes difficult to recognize as such; decay often in radial, star-like pattern | Dead, often fallen branches with conks |
| Echinodontium tinctorium (Indian paint fungus) | rust-red stringy rot | WF (other conifers) | | Conks |
| Fomitiporia hartigii | | SAF | White rot | Conks may appear on undersides and base of branches |
| Laurilia sulcata | | ES | White pocket rot; yellowish, may be wet and spongy | |
| Peniophora polygonia | | QA | White rot; yellow-brown, stringy | |
| Phellinidium ferrugineofuscum | | ES | White rot; laminated, may have small pits with black flecks and white transverse streaks | Conks are rare |
| Phellinus tremulae | aspen trunk rot (white trunk rot) | QA | White rot; firm to spongy, yellowish tan in some areas | Conks, bird cavities |
| Porodaedalea pini | red ring rot | ES, LPP, DF, SAF | White pocket rot; sometimes with abundant zone lines; decay may progress into roots | Conks, punk knots |
| Stereum sanguinolentum (bleeding Stereum) | red heart rot | SAF, ES | White rot; initial red stain, becomes light brown, dry, friable, with white fungal sheets when advanced | Usually none, fruits on slash and logs |
| Veluticeps abietina/fimbriata | | S, F | Brown pocket rot | Conks |

# Annosus Root Disease

## Infects fresh stumps and spreads root-to-root

**Pathogen**—Recent studies show that annosus root disease is caused by two closely related fungi, both formerly known as *Heterobasidion annosum*. The two pathogens are *H. annosum* (in the strict sense) and *H. parviporum*. These are North American variants of species in Europe and Asia. The North American variants may be recognized and named as separate species in the future.

**Hosts**—*Heterobasidion annosum* is a pine specialist. In this Region, it has been found to cause disease only on pines and eastern redcedar and only in the Bessey District of the Nebraska National Forest. However, it occurs in Arizona, New Mexico, Idaho, and the Midwest, so it may occur undetected elsewhere in the Region.

*Heterobasidion parviporum* favors spruce and fir species but has been found only in mixed conifer forests within the range of white fir in southern Colorado. It is common on white fir and occurs occasionally on associated subalpine fir, Douglas-fir, and blue and Engelmann spruce. It has not been found in our spruce-fir forests outside the range of white fir in this Region.

**Signs and Symptoms**—In some cases, resin flow may be evident near the root collar as the tree defends itself against attack. Diseased pines may eventually show crown thinning and yellowing. In pines, the disease is most active in the sapwood, killing tissues as it progress. In other hosts, the fungus grows first in inner wood once it reaches the root collar, so butt rot is a more prominent feature of the disease.

Decay may be preceded by a pink to dull violet stain of the wood. Later, small, poorly defined pockets or pits are often evident. Small black flecks can often be found in well-developed pockets, and wood may separate along the annual rings (laminated rot). Finally, the pockets are lost as the entire mass of wood becomes spongy or stringy.

Conks are frequently found in white fir disease centers but usually in protected, moist microsites such as under litter, inside hollow stumps, and even down in hollow root channels. Perennial and tough, they can be up to a foot wide. Depending on where they form, they may have an irregular brown cap or bracket or be completely flat on the substrate. The pore surface is whitish with small pores; the flesh is creamy tan (figs. 37-38). In some cases, especially with *H. annosum*, only tiny "popcorn" conks may be found. Fresh conks have a strong mushroom aroma.

**Figure 37.** The pore surface of *Heterobasidion parviporum* from a white fir stump. *Photo: Jim Worrall, USDA Forest Service.*

**Disease Cycle**—The disease cycle in pines begins with freshly cut stumps. Wind-blown spores infect stumps of live trees within a few weeks of cutting. The fungus grows down into the stump roots. Where there are root contacts, the fungus may grow across and infect neighboring trees, eventually creating a root disease center. Centers typically have a stump in the middle, old dead and downed trees

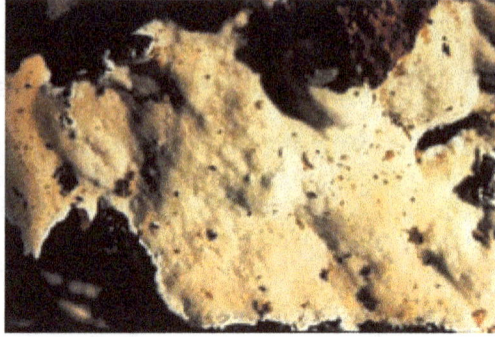

Figure 38. Closer view of the pore surface of *Heterobasidion parviporum* from a white fir stump. *Photo: Jim Worrall, USDA Forest Service.*

nearby, recent mortality farther out, and live trees that may have crown symptoms on the outside. The fungus fruits on stumps and infected trees, produces spores, and completes the cycle. The fungus may survive many years in dead root systems and can infect successive tree generations.

The disease cycle may work similarly in true firs, but evidence suggests that stump infection may not be the only way for new disease centers to be initiated. New infections may occur through basal scars and even through direct infection of roots by spores in the soil.

**Impact**—In pines, *H. annosum* impact in this Region is geographically restricted and is not a significant concern, although the pathogen could conceivably move into and cause substantial damage to important pine forests elsewhere in the Region. In white fir, disease centers and mortality are common and the impact is substantial. Ecologically, the abundance of white fir and this disease in formerly pine-dominated forests is due to fire exclusion and early harvest of seral species. Thus, restoration to a more open, pine-dominated forest maintained by fire would greatly reduce the disease's impact.

In pines, the sapwood and cambium are often killed before extensive decay occurs, and trees tend to die standing. In firs and spruces, especially in larger trees, extensive root and butt rot often occur to the point that live trees may fail mechanically before dying (figs. 39-41). However, the fir engraver, *Scolytus ventralis*, is attracted to diseased firs and may kill them before direct mortality or failures occur.

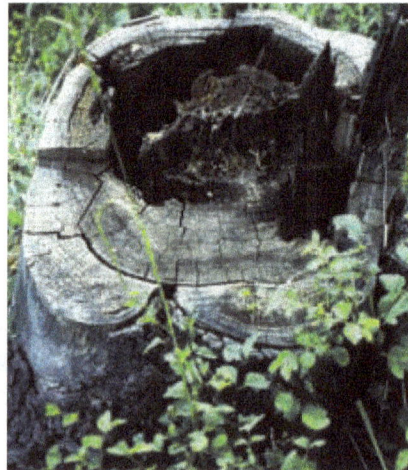

Figure 39. Typical pattern of decay from *Heterobasidion parviporum* in a white fir stump in Colorado. *Photo: Jim Worrall, USDA Forest Service.*

Figure 40. Butt-rotted, live white fir after failing due to *Heterobasidion parviporum* in Colorado. *Photo: Jim Worrall, USDA Forest Service.*

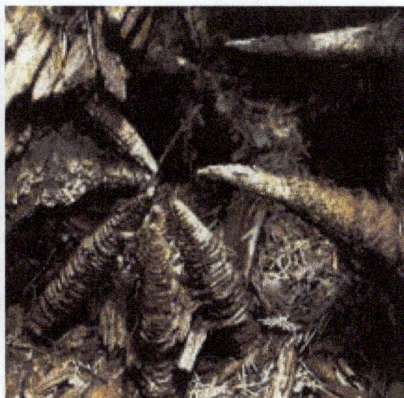

Figure 41. Decayed hollow in white fir stump with residual branch traces, caused by *Heterobasidion parviporum* in Colorado. *Photo: Jim Worrall, USDA Forest Service.*

**Management**—Management of annosus root disease is based on two approaches: using resistant species and preventing primary infection. (See comments on white fir ecology under "Impact.")

- Manipulating species composition. Recent advances in understanding the pathogen species and their host specialization provide greater opportunity for management through species composition. Where pines or eastern redcedar are infected, other species may be planted or favored, and should generally be resistant. Where white fir is infected, species other than true firs and spruces will likely be successful.

- Chemical protection of stump tops. When applied shortly after cutting, borax powder (available commercially as Sporax or Tim-bor) effectively prevents establishment of *H. annosum* and *H. parviporum* in stump tops (figs. 42-43). This prevents establishment of new disease centers but will neither eradicate existing infections nor prevent wound infection on residual trees.

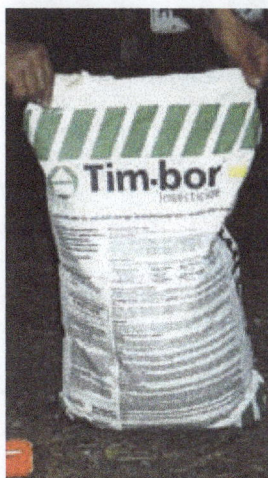

Figure 42. A commercial borax product. *Photo: Pete Angwin, USDA Forest Service.*

Figure 43. Commercial borax product application on a stump. *Photo: Pete Angwin, USDA Forest Service.*

- Biological control of stump tops. Several benign fungi are aggressive stump colonizers and can colonize the wood before the pathogen, preventing the pathogen from effective establishment. They are applied as stump top treatments, like borax. The most widely tested fungus is *Phlebiopsis gigantea*.

**References: 90, 158, 168, 193**

# Armillaria Root Disease

## Most important root disease in the Rocky Mountain Region

**Pathogen**—Armillaria root disease is caused by many species of *Armillaria*, a genus of mushroom-forming fungi. *Armillaria solidipes* (= *A. ostoyae*) is the most common species in the Rocky Mountain Region, but there are indications that additional species are present in Wyoming.

**Hosts**—Armillaria root disease has a very wide host range and has been found on almost all common tree species and in all major forest types in the Rocky Mountain Region. It is also very widespread, probably occurring in all forests of the Region. It seems to be most abundant in spruce-fir and mixed conifer forests.

**Signs and Symptoms**—Signs are physical evidence of the pathogen and are diagnostic (unique to the disease) when reliably identified:

- Mushrooms. Mushrooms are the most conspicuous sign of the pathogen (fig. 44). However, they are only abundant in certain years and only last for a few weeks, usually in late summer to fall. The other signs are almost always present with the disease.

**Figure 44.** Mushrooms of *Armillaria ostoyae* can be abundant for short periods in late summer to fall during wet years. Mushrooms are usually in clusters around the base of a tree. Caps are honey-brown with small scales; gills are white and attached to the stem, which is white to scurfy brown at the base with a poorly-formed ring. *Photo: William Jacobi, Colorado State University.*

- Rhizomorphs. These brown to black, root-like fungal organs are several millimeters in diameter, branched, and composed of fungal hyphae aggregated within a protective sheath. They grow in soil or along roots and infect when they encounter susceptible hosts.
- Mycelial fans. These whitish, fan-like layers of fungal tissue grow between bark and wood as the fungus invades roots and lower stems (figs. 45-46).

Symptoms may not appear until very late in disease development and are not diagnostic when they do appear. Symptoms may include:

- Resinosis. Resin exudation near the soil line (above or below) is a good symptom (figs. 46-47). It occurs mostly in the resinous conifers but not in all cases. It is evidence of the host attempting to defend itself.

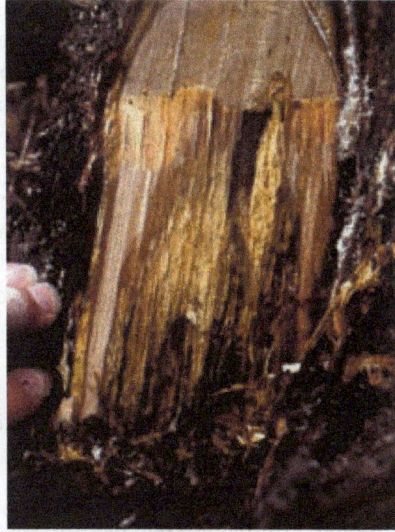

**Figure 46.** Symptoms and signs of Armillaria root disease: mycelial fan barely visible in the bark on the left (near fingers); resinosis apparent in upper right; decay in a root near the soil line. *Photo: Jim Worrall, USDA Forest Service.*

**Figure 45.** Mycelial fans under the bark are the characteristic sign of Armillaria root disease. *Photo: USDA Forest Service training slide set.*

- Wood decay. Openings or uprooting of the tree may reveal decay in the wood of roots or lower stem. Otherwise, decay is hidden and difficult to detect. The decay is a white rot and often contains zone lines (fig. 48).
- Crown symptoms. These include slower terminal growth, thin or yellow foliage, branch dieback, and stress crop of cones (particularly in Douglas-fir).

**Disease Cycle**—The pathogen survives many years in dead roots, perpetuating the disease on a site. It has two main ways of spreading. One is local spread: the fungus grows across root contacts and grafts between infected and healthy roots or, in some cases, via rhizomorphs that grow through soil.

**Figure 47.** Resinosis is a symptom often caused by Armillaria root disease near the soil line. *Photo: Jim Worrall, USDA Forest Service.*

**Figure 48.** Armillaria decay with zone lines. *Photo: Jim Worrall, USDA Forest Service.*

This can result in gradually expanding disease centers (fig. 49). However, many of the Region's subalpine forests are completely colonized, so infection is spatially random rather than occurring in discrete centers. The fungus also spreads is by producing mushrooms with wind-dispersed spores. Spores are rarely successful, and it is thought that this mode of spread is uncommon, especially in the arid West. Small, stressed, or highly susceptible trees may succumb quickly. Older, vigorous, or somewhat resistant trees may survive for many years with the disease progressing slowly or remaining isolated in initially infected roots.

**Impact**—Armillaria root disease causes mortality, growth reduction, cull, and predisposition to other lethal agents. Root diseases lead to stand heterogeneity, creating openings and patches of coarse woody debris, uneven age classes, and altered composition. Mortality is often through mechanical failure of live trees, which can be dangerous in developed sites (figs. 50-51). Impacts are not quantified in the Rocky Mountain Region (though annual volume losses in subalpine fir alone in Colorado were estimated at 10,500 m³ [370,800 ft³]), but because the disease is widespread in most forest types, impacts are undoubtedly substantial. In lodgepole pine, mortality is common up to about age 20, when the mortality rate greatly decreases (fig. 52). In most other species, impacts seem to be more severe in mature trees. Subalpine fir is often killed quickly, but Engelmann spruce often survives until

**Figure 49.** Armillaria root disease center with standing mortality. *Photo: Jim Worrall, USDA Forest Service.*

**Figure 50.** Vehicle crushed by a tree infected with Armillaria root disease that failed while green. *Photo: Jim Worrall, USDA Forest Service.*

**Figure 51.** Aftermath of an Armillaria root disease center in a campground. *Photo: Pete Angwin, USDA Forest Service.*

mechanical failure. In the Front Range and Black Hills, 62% and 75% of ponderosa pine attacked by mountain pine beetle were also infected by Armillaria root disease. Similar interactions probably occur in spruce-fir forests.

**Management**—Management of Armillaria root disease is difficult. The following approaches used in other areas are impractical in this Region's forests but can offer general guidance:

**Figure 52.** Armillaria in young lodgepole pine.
*Photo: Pete Angwin, USDA Forest Service.*

- Manage for resistant species. Almost all tree species in the Rocky Mountain Region are susceptible, and few are suitable for a given site. However, lodgepole pine becomes resistant at about 20 years and aspen is usually little affected, so some heavily impacted stands can be converted by favoring resistant species in thinning or during regeneration.
- Tailor management in disease centers. In subalpine forests in this Region, the disease is typically random rather than aggregated in discrete disease centers. However, where centers are recognized, management can be tailored to them.
- Stump extraction. In the Pacific Northwest, long-term experiments have demonstrated that removing stumps during regeneration can significantly reduce root disease in the future stand. This is sometimes used operationally there, but has not been attempted in the Rocky Mountain Region.
- Biological. Research suggests that competitive fungi might be manipulated to reduce the food available to the pathogen and perhaps even replace it. However, these approaches are not operational at this time.

**References: 94, 154, 199, 200**

# Black Stain Root Disease

## Mortality centers in pinyon

**Pathogen**—Black stain root disease is caused by the fungus *Leptographium wageneri* var. *wageneri*. This variety infects only pinyons, including two-needle as well as singleleaf pinyon in other Regions.

**Hosts**—The disease causes expanding patches of mortality in many pinyon stands of the western slope of the southern and middle Rocky Mountains as far north as Idaho (figs. 53-54). It has not been found east of the Continental Divide. Two other varieties do not occur in this Region: *L. wageneri* var. *ponderosum* infects lodgepole, Jeffrey, and ponderosa pine; and *L. wageneri* var. *pseudotsugae*

**Figure 53.** Dead and dying pinyon near the edge of a mortality center caused by black stain root disease. The smaller trees in the foreground may not be affected yet because their smaller root systems are not yet contacting infected roots. *Photo: Jim Worrall, USDA Forest Service.*

infects Douglas-fir. Both those varieties occur primarily in the northern Rocky Mountains, British Columbia, and Pacific coast states.

**Signs and Symptoms**—In advanced disease, foliage is sparse and sometimes chlorotic (yellow). A mortality center is often evident, with old snags near the

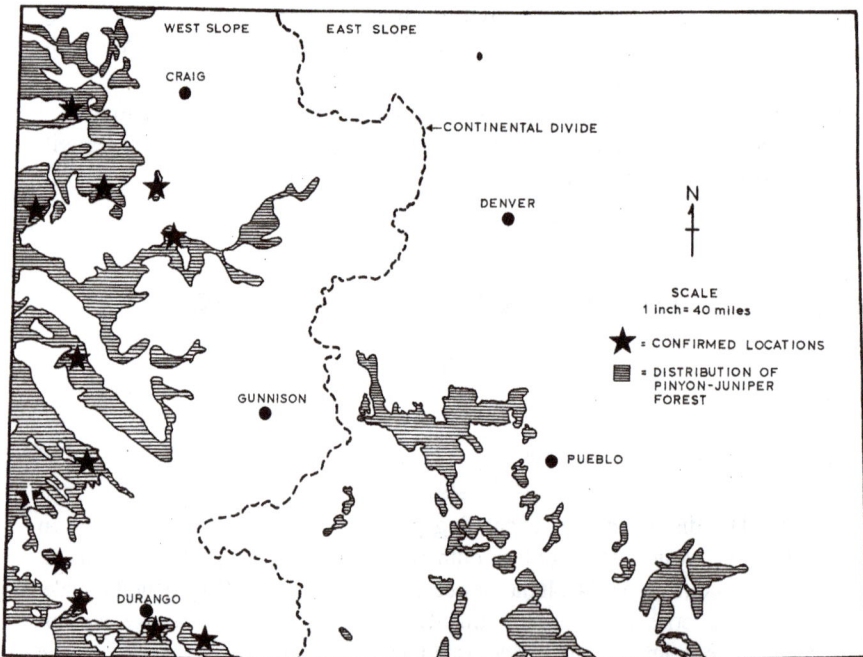

**Figure 54.** Distribution of black stain root disease in Colorado (from Landis and Helburg 1976).

USDA Forest Service RMRS-GTR-241. 2010.

center, recent mortality farther out, and symptomatic, live trees at the edge (fig. 53).

An intense black stain can be found in wood of the roots, root collar, and often the lower stem. Unlike the pattern of blue stain, which progresses radially along the rays to the inner sapwood, black stain progresses longitudinally and somewhat tangentially. Longitudinally, it forms long streaks following the wood grain (fig. 55). In cross section, it appears as arcs following short segments of annual rings (fig. 56).

**Disease Cycle**—Unlike blue-stain fungi, which colonize rays, this pathogen colonizes the tracheids (water-conducting and structural elements) of the sapwood and causes a wilt disease. It aggressively invades living wood of the roots, root collar, and lower stem until trees are killed. It does not decay wood.

Where roots grow closely together, the fungus can grow from tree to tree. In this way, mortality centers expand. Rate of radial expansion of disease centers in pinyon is about 3.3-6.6 ft (1-2 m) per year, but they do not expand indefinitely. It is not clear why expansion eventually ceases.

Other varieties of the pathogen are transmitted long distances by insect vectors, primarily root weevils and bark beetles. The vectors are attracted to dying roots, so the disease is associated with stand disturbance, especially thinning and road construction. As the insects burrow into the soil looking for those roots, they tend to graze on intermingled roots of other trees, thereby inoculating the pathogen. A vector has not been identified in pinyon, but it is strongly suspected that there is one and that the relationship to stand disturbance is similar to that of the disease caused by the other varieties.

Fruiting of the fungus is almost microscopic and difficult to observe in the field. Minute black stalks are formed in

**Figure 55.** Black stain in the lower stem of diseased pinyon. Stem with the wood exposed in tangential view. *Photo: Jim Worrall, USDA Forest Service.*

**Figure 56.** Black stain in the lower stem of diseased pinyon. Stem with the wood exposed in transverse view. *Photo: Jim Worrall, USDA Forest Service.*

cavities under the bark, such as insect galleries. Sticky masses of spores are produced on the stalks. Insects that contact the fruiting structures are readily coated with spores.

**Impact**—In Colorado, disease frequency is clearly associated with soil depth, precipitation, site quality, and tree size and density. Stands with large or dense trees may facilitate root-to-root spread of the pathogen because of more frequent root contacts. However, rate of disease expansion was not related to density in one study. Because the pathogen is restricted to pinyon, mixed stands inhibit tree-to-tree spread of the disease.

In the Rocky Mountains, the disease is most active in southwestern Colorado, especially in the Four Corners area, but it occurs all along the Western Slope (fig. 54). Over the long term, black stain is an important disturbance agent in regulating structure and composition of pinyon-juniper stands. It creates structural heterogeneity in the forest, with openings that provide habitat diversity for other plants and animals. In this sense, the disease promotes old-growth characteristics of pinyon-juniper stands.

The pathogen dies quickly after host tissue dies and after a stand replacing fire. Disease incidence then increases with time since regeneration. Just as fire makes the disease less likely, the disease makes fire more likely. One of the major impacts of the disease is an increase in dead fuels. Thus, a fire-disease feedback loop can be envisioned that contributes to disease reduction and forest regeneration. The pinyon engraver commonly attacks trees with black stain. Under endemic beetle populations, black-stain centers are probably a major source of beetles.

**Management**—No control approach has been effective in pinyon. Cutting and burning killed or symptomatic trees does nothing to stop the disease, but because pinyon engraver often invades diseased trees before death, sanitation may be effective in preventing beetle outbreaks. Any replanting should be with species other than pinyon. Trenching around a disease center to sever all media through which the pathogen may grow has been attempted but was unsuccessful, probably because the pathogen is often in at least two trees beyond the last symptomatic one. Another approach is removing all pinyon from in and around disease centers, waiting four years, and replanting pinyon. This could be effective because the pathogen does not survive long in roots once they are dead. If implemented, at least three healthy-looking trees beyond the center should be removed.

Any discretionary pinyon cutting would best be done in the fall or winter. Vectors of the other forms of black stain root disease, which are attracted to fresh cuts, are active in spring and early summer; presumably, the same is true of the pinyon form.

**References: 73, 92, 101, 110, 188**

# Coniophora Root and Butt Rot

## Brown rot with gray-brown cords and mycelium

**Pathogen**—Several species of *Coniophora* may cause this disease in the Rocky Mountains, but they are difficult to distinguish, and there is no evidence that they differ in pathology or ecology. The most common is *C. puteana*, and in a broad sense this name can be used for all of them.

**Hosts**—Engelmann spruce and subalpine fir are probably the most common hosts, but the disease has been found in lodgepole pine, and likely all of the Region's conifers can be infected.

**Signs and Symptoms**—The fungus causes a brown cubical rot in the roots and base of the stem. Frequently, there are no external indicators, but there may be cavities near the root collar that reveal the decay. When decay is advanced and little sound wood remains, the root collar may crack open due to wind. Mechanical failure of live trees may eventually occur at the roots or lower stem (fig. 57).

When the decay becomes exposed, characteristic mycelium and mycelial cords are the most frequent diagnostic signs (figs. 58-60). Some pieces of decayed wood may have a thin, velvety, gray-brown mycelium on the surface. More commonly, gray-brown to grayish white mycelial cords are present in the decayed wood and under the bark of dead roots. These are similar to but less well-defined and less organized than the black rhizomorphs of *Armillaria* species.

**Figure 57.** Engelmann spruce tree that failed due to Coniophora root and butt rot. *Photo: Jim Worrall, USDA Forest Service.*

The fruiting body is rarely seen and normally appears after the tree is down. It is a thin layer with an irregular surface (no

**Figure 58.** Gray-brown mycelium associated with brown cubical rot. *Photo: Jim Worrall, USDA Forest Service.*

**Figure 59.** Gray-brown mycelium and mycelial cords. Cords are often finer and more delicate than these. *Photo: Jim Worrall, USDA Forest Service.*

**Figure 60.** Mycelial cords. Cords are often finer and more delicate than these. *Photo: Jim Worrall, USDA Forest Service.*

pores), brown in the center and whitish at the margin, and a few inches in diameter (figs. 61-62).

Other diseases causing brown root and butt rot can be distinguished from this one by carefully inspecting the decayed wood. *Phaeolus schweinitzii*, the cowpie fungus, has no cords associated with the decay. *Fomitopsis pinicola*, the red-belt fungus, may occasionally infect the butts of conifers. It has no cords but has distinct, thick, creamy white mycelial mats that form in splits of the decayed wood.

**Disease Cycle**—The disease cycle has not been well studied. Basal wounds and fire scars are thought to be infection courts, at least in some cases. Infection centers have not been observed, and there is apparently no evidence for root-to-root spread from an infected tree to a neighbor.

**Impact**—The incidence and impact of Coniophora root and butt rot have not been quantified in the Rocky Mountain Region. In pristine Norway spruce forests of northern Finland and northwestern Russia, Coniophora root and butt rot is the most important mortality agent, causing death by mechanical failure.

**Figure 61.** Young fruiting body. *Photo: Jim Worrall, USDA Forest Service.*

**Figure 62.** Young fruiting body. Note the fine cords associated with the fruiting body. *Photo: Jim Worrall, USDA Forest Service.*

In northwestern Ontario, it is one of the three most important of 21 decay fungi that contribute to black spruce mortality. In a large study of white spruce, black spruce, and balsam fir in the same area, Coniophora root and butt rot was the fourth most frequent of 30 identified root diseases.

From inoculations of white spruce, *C. puteana* infected more trees and grew faster than *Onnia tomentosa*. During the 3 years after inoculation, *C. puteana* infected more trees and grew faster in those that died than in those that remained alive. This suggests that host vigor plays a role in suppressing infections by this fungus.

**Management**—Because it is difficult to detect in standing trees, management of this disease is not often a consideration. Where it is detected in developed sites or in timber stand improvement projects, infected trees should be removed. Otherwise, it is not considered a threatening disease but a native part of ecosystem function.

**References: 111, 182, 183, 184**

# Root Diseases with White Pocket Rots

## Big white pocket rot and red root rot

**Pathogen**—Several pathogens cause a white pocket rot in the roots and butts of conifers in the Rocky Mountain Region. However, discussed in this entry are *Phellopilus* (*Phellinus*) *nigrolimitatus*, which causes a root disease called big white pocket rot and two species, *Onnia* (*Inonotus*) *tomentosa* and *leporina* (formerly misnamed *O. circinata*), which cause a disease known as red root rot, or tomentosus and circinatus root rot, respectively. The following two pathogens (described separately in this guide) also cause white pocket rots and may be confused with the fungi described in this section: *Porodaedalea* (*Phellinus*) *pini* often grows down into roots and causes red ring rot in conifer stems; annosus root rot may also appear as a white pocket rot.

**Hosts**—Big white pocket rot and red root rot can infect most conifers, but they infect mostly spruce species in this Region.

**Signs and Symptoms**—Big white pocket rot (caused by *Phellopilus nigrolimitatus*) usually has no external indications. The fruiting body (conk) is uncommon and usually forms after the tree is dead. Conks are perennial, are flat on the bark, and often have a small shelf or cap at the upper end that is somewhat soft and spongy. The pore surface is cinnamon-colored and smooth with very small pores. The internal flesh is brown and typically has one or more black lines.

Red root rot (caused by *Onnia* species) usually has no external symptoms, but there may be some basal resinosis. Fruiting is uncommon but may be abundant in certain years. Fruiting bodies arise from buried roots with a short, more or less central stem or appear directly on the tree, usually at the root collar (figs. 63-64). Conks are annual and soft to leathery. The cap is circular with a sunken center, up to 4 inches (11 cm) in diameter, yellowish brown, and velvety. The lower, pore surface is pale brown but darkens with age. The two species of *Onnia* are virtually indistinguishable in the field, but conks of *O. leporina* are more likely to be on the base of the tree and in fewer numbers than those of *O. tomentosa*.

These diseases all cause a white pocket rot. Big white pocket rot has ellipsoidal pockets up to about $3/8$ by 1 inch (10 by 25 mm), rectangular and white, separated by pale brown, firm wood (figs. 65-66). Pockets in the other decays

**Figure 63.** *Onnia tomentosa*, showing caps from above. *Photo: Jim Worrall, USDA Forest Service.*

Figure 64. *Onnia tomentosa*, showing caps from the sides and small patches of fungal mycelium on roots. *Photo: Jim Worrall, USDA Forest Service.*

Figure 65. Big white pocket rot. Pockets are up to 3/8 inch (1 cm) wide and many have white cellulose remains. *Photo: Jim Worrall, USDA Forest Service.*

Figure 66. Big white pocket rot. Decayed wood has weathered, leaving relatively sound wood between the pockets. *Photo: Jim Worrall, USDA Forest Service.*

Figure 68. Red root rot caused by *Onnia* sp. Stain preceding decay in the butt of a sapling in British Columbia. *Photo: Jim Worrall, USDA Forest Service.*

are about 1/13 by 3/8 inch (2 by 10 mm) (fig. 67). Early decay caused by the *Onnia* species is indicated by a reddish stain of the wood (fig. 68). Roots in the soil may have small patches of golden brown fungal tissue on the surface (fig. 64). Zone lines could be present in decays by any of these fungi.

**Disease Cycle**—Initial infection is by spores from conks. The infection site is unknown but could be roots in the soil or basal stem wounds. Decay progresses for

Figure 67. Red root rot caused by *Onnia* sp. Advanced decay. *Photo: Borys M. Tkacz, USDA Forest Service, Bugwood.org.*

many years before fruiting occurs. In the case of big white pocket rot, fruiting does not occur until after the tree is dead and down. With red root rot, there may be direct growth of the fungus tree-to-tree between contacting roots, leading to mortality centers, but that has not been shown with big white pocket rot.

**Impact**—Big white pocket rot is associated with old-growth stands and large-diameter trees. It causes extensive wood decay of the roots and butt and can grow far up the stem, leading to mechanical failure of live trees. In Scandinavia, *Phellopilus nigrolimitatus* is a red-listed (protected) species. Studies have linked it to old-growth conditions and suggest that forest management reduces its population. This may be a consideration if management is considered in old-growth stands.

Red root rot is a damaging disease of spruces and other hosts elsewhere, causing large mortality centers. Although infected trees can be found in the Rocky Mountain Region and fruiting is sometimes abundant, major damage and mortality centers have not been documented thus far.

**Management**—Big white pocket rot can lead to tree failure in developed sites with large trees. Because of the difficulty of detection and damage that can be done by large trees, sounding and increment coring should be done carefully to detect the disease. For timber management in old-growth stands, consider the general abundance of the fungus, its presence in the area, and its potential sensitivity to management. However, there are no data indicating how common or rare the fungus is in this Region. Red root rot is normally not damaging enough to consider during timber management, but as with all root rots, it can be important and contribute to hazard when it occurs in developed sites.

**References: 57, 158, 169**

# Schweinitzii Root and Butt Rot

## Red-brown cubical root and butt rot of conifers

**Pathogen**—Schweinitzii root and butt rot is caused by *Phaeolus schweinitzii*, also known as the velvet-top or cowpie fungus.

**Hosts**—All conifers are probably susceptible to the disease, but the most common host in the Rocky Mountain Region is Douglas-fir. Infrequent hosts include lodgepole pine and Engelmann spruce.

**Signs and Symptoms**—Trees infected with *P. schweinitzii* rarely display outward symptoms unless they are in the advanced stages of the disease, so diagnosis often occurs after the tree loses structural support and topples or is windthrown. Possible symptoms may include thinning crowns, poor shoot growth, and/or branch dieback. Wood decay may be visible in openings on the stem or nearby stumps (figs. 69-70). Incipient decay is yellow to red and dry. Advanced decay is red-brown and cubical and sometimes has thin, resinous felts present in the cracks.

**Figure 69.** *Phaeolus schweinitzii* decay in a Douglas-fir stem. *Photo: Jim Worrall, USDA Forest Service.*

Occasionally, fruiting bodies can be seen on the ground emerging from diseased roots of stumps or living trees (figs. 71-73). Infrequently, they emerge directly from the tree's base or stump. Fruiting bodies are annual, spongy, and mushroom-like with large, irregular pores on the undersurface. Caps are red-brown and velvety, margins are yellowish brown, and undersides are green when fresh, becoming brown with age. As they dry, they become entirely brown and brittle and resemble cow pies. Caps are usually 5-10 inches (13-25 cm) in diameter with short stems.

**Figure 70.** Brown cubical rot typical of *Phaeolus schweinitzii*. *Photo: Joseph O'Brien, USDA Forest Service, Bugwood.org.*

**Figure 71.** Pore surface of young *Phaeolus schweinitzii* conk. *Photo: Jim Worrall, USDA Forest Service.*

**Figure 73.** Old *Phaeolus schweinitzii* conk that is entirely brown, dry, and brittle. *Photo: USDA Forest Service, Bugwood.org.*

**Figure 72.** Typical *Phaeolus schweinitzii* conk with brown, velvety surface; yellowish margin; and greenish undersurface. *Photo: Jim Worrall, USDA Forest Service.*

**Disease Cycle**—Spread of *P. schweinitzii* occurs primarily by means of wind-dispersed spores produced in conks. Root-to-root infection may occur, but it appears to be very uncommon. Therefore, diseased trees are dispersed in stands rather than in discrete disease centers. The fungus gains entry through basal wounds, particularly fire scars or damaged roots. Conks are produced annually from decaying wood. The fungus can persist for many years in stumps and dead trees.

**Impact**—Schweinitzii root and butt rot is a major disease of mature Douglas-fir. Decay is generally confined to the heartwood and is found in the roots and lower 10 ft (3 m) of the stem. Wood loses its structural integrity rapidly as decay progresses, and susceptibility to breakage and windthrow increases. Infected trees may become more susceptible to Douglas-fir beetle or Armillaria root disease.

**Management**—Butt rot can be detected by sounding the lower stem with an ax and coring wood. Management strategies include avoiding wounding, removing infected trees, and harvesting on shorter rotations. Remove trees showing evidence of schweinitzii root and butt rot in recreation areas and other developed sites.

**References: 10, 159**

# White Mottled Rot

## Root rot that topples live aspen

**Pathogen**—White mottled rot is caused by the fungus *Ganoderma applanatum*. The fruiting body of the fungus is known commonly as "artist's conk" for reasons explained below.

**Hosts**—Although *G. applanatum* occurs on many tree species (often on dead trees) across the northern hemisphere, it occurs primarily on living aspen in the Rocky Mountains. It can persist for some time after the tree dies and may also occur on cottonwoods.

**Signs and Symptoms**—The fruiting body is a conk (shelf fungus) that occurs at the base of infected aspens, usually within about 12 inches (30 cm) of the soil line (figs. 74-75). It is initially a white bulge, and the margin remains white when actively growing. It becomes more or less flat and may grow up to about 12 inches (30 cm) wide and project out the same distance. The upper surface is irregular-shaped and brown to gray; the lower surface is pure white when fresh with fine pores. Although not every infected tree has conks, most advanced infections produce conks before the tree falls over or dies. Conks are the only useful indicator of infection in live, standing trees.

Where the fresh white underside of a conk is bruised, as with a finger or stick, it immediately and permanently becomes dark brown. Artists may use this reaction to draw pictures, leading to the common name of the fungus, "artist's conk"

**Figure 74.** Conk of the pathogen *Ganoderma applanatum*. The margin is white and blunt and the underside is white when actively growing. *Photo: Jim Worrall, USDA Forest Service.*

**Figure 75.** Another conk of the pathogen *Ganoderma applanatum*. The margin is white and blunt and the underside is white when actively growing. *Photo: Mary Lou Fairweather, USDA Forest Service.*

(fig. 76). If the conk is then left to dry, the contrast remains and the surface is no longer sensitive.

The fungus causes extensive decay of the roots and butt of the tree (fig. 77). Decayed wood is mottled with alternating, small, white and light tan areas. The wood eventually becomes spongy, and black zone lines may develop in it.

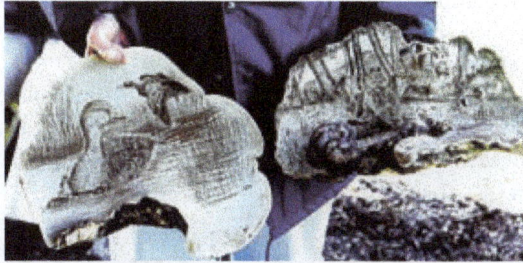

**Figure 76.** Artwork drawn using only a stylus and the fresh, sensitive pore surface that turns dark brown where bruised. *Photo: Jim Worrall, USDA Forest Service.*

**Disease Cycle**—Infection may occur from airborne spores or, in some cases, from contacts with infected roots of neighboring trees. Little is known about how or where spores cause infection, but it may occur at small wounds in the roots or root collar. Root-to-root spread, leading to disease centers or groups of infected trees, seems to occur primarily on the best aspen sites with deep soil.

After the tree is infected for several years, the pathogen has gathered sufficient resources to reproduce. Conks are produced near the base of the tree. Microscopic, airborne spores are released from the pores on the underside of conks.

Conks can produce huge numbers of spores. In only 4 by 4 inches (10 by 10 cm)

**Figure 77.** Typical failure of live aspen due to white mottled root rot. Roots usually break near the root collar and no significant root plate or ball is lifted from the soil. *Photo: Jim Worrall, USDA Forest Service.*

of conk surface, it has been estimated that 4.65 billion spores can be produced and released into the air in a 24-hour period.

**Impact**—White mottled rot is by far the most important root disease of aspen in the West. Infected trees usually fall over while alive, often with healthy-looking crowns. The disease is typically associated with about 90% of windthrow of live aspen trees. It can be particularly important in developed sites as it contributes to hazard trees. Small trees or trees on poor sites with dry, shallow soils may be killed before decay leads to windthrow.

The disease seems to be most common and damaging on moist sites with good aspen growth. It is not known if this is because of favorable physiology of vigorous trees or favorable soil characteristics of good aspen sites. In particular, root-to-root spread, leading to disease centers, occurs primarily on good aspen sites.

**Management**—Practical methods for preventing or reducing the incidence of white mottled root rot are not available. In developed sites with aspen, hazard tree inspectors should be trained to recognize conks. In the long run, Regional policy strongly encourages shifting vegetation away from aspen in developed sites.

Because the disease kills and decays roots, there is concern that it may affect the success of suckering after cutting infested stands. However, studies are needed to determine whether this occurs and how large the effect is.

**References: 79, 143**

# Aspen Trunk Rot

## Common conk on aspen

**Pathogen**—Aspen trunk rot (also called white trunk rot) is a stem decay (heart rot) of living aspen. It is caused by the fungus *Phellinus tremulae*.

**Hosts**—This pathogen occurs only on living aspen.

**Signs and Symptoms**—The fruiting body is a conk (shelf fungus). The top of the conk generally slopes down and the bottom slopes up, so it is roughly triangular in profile (fig. 78). It is hard and woody, black and cracked on top, and purplish brown with tiny pores on the bottom. Inside is a granular core at the point of attachment with a hard flesh layer above and a tube layer below. The conks are perennial and may live up to about 20 years. A new tube layer is produced each year at the bottom, but the layers are indistinct and difficult to count.

Decayed wood is firm to spongy, fibrous, yellowish tan, and often has a sweet, wintergreen odor. Decay columns rarely become hollow. Diffuse zone lines are scattered through the decayed wood (figs. 79-80), particularly in earlier stages near the edge of the decay column. The decay is a white rot. Because cellulose is only degraded as it is used by the fungus (unlike brown rot), decayed wood has some intact cellulose until very late stages and remains fibrous. Therefore, it can be used in limited proportions in products such as waferboard.

**Disease Cycle**—The pathogen infects branch stubs or small dead branches, eventually growing into the inner wood. It may infect wounds but does not require substantial wounds. It does not colonize dead trees and dies soon after the host dies. After decaying for perhaps 5 years or more, it begins to grow out to the surface along branch traces to produce conks. Microscopic spores produced in conks are airborne and travel long distances. Spores can cause infection if they land on a suitable point.

**Impact**—Aspen trunk rot is the most common stem decay of aspen in North America (but in a study in Colorado, incidence of *Peniophora polygonia* was slightly higher). More importantly, it decays the greatest volume of wood. Infected trees lose an average of 70% of wood volume in cull. Stand age and site quality can be important. On good sites with deep soils and adequate moisture, incidence of decaying trees increased linearly with age from near 0% at 40 years to 91%

**Figure 78.** Fruiting body (conk) of *Phellinus tremulae*, the cause of white trunk rot of aspen. There are often multiple conks on one tree. Also, note the cavity excavated by a woodpecker. Such cavities are almost invariably excavated in aspens with white trunk rot, sometimes below the conk so that the cavity has an awning. *Photo: Jim Worrall, USDA Forest Service.*

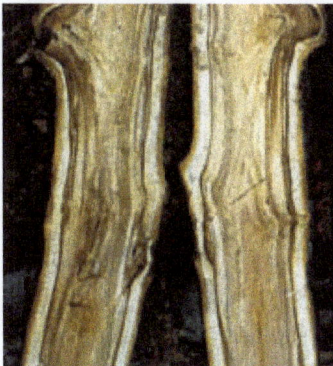

at 160 years. Cull increased to roughly 50% at 160 years. On poor sites with dry, shallow soils, decay was higher, and less dependent on age. The pathogen may kill slow-growing trees directly by growing into and killing older sapwood. Tree breakage is another impact of particular concern in developed sites.

**Figure 79.** Longitudinal section showing the extensive column of decay and zone lines. *Photo: Jim Worrall, USDA Forest Service.*

**Figure 80.** Cross section from a different tree that shows the bulge in the column where a conk was produced. *Photo: Jim Worrall, USDA Forest Service.*

*Phellinus tremulae* provides important habitat for cavity-nesting birds (fig. 78), although in the long term it can also contribute to deterioration of aspen stands and loss of aspen cover type. Red-naped sapsucker, Williamson's sapsucker, downy woodpecker, and hairy woodpecker nest primarily in aspen in many areas. Nests are almost always excavated in trees with decay. Aspen may be important for nesting because it has stem decay much more frequently than other common tree species of the Rocky Mountains.

**Management**—Compared to other decays, conks are reliably produced and are useful indicators for detecting and estimating decay: 75-85% of trees with cull due to aspen trunk rot have conks. Perhaps one additional tree may be infected, then, for every three or four trees with conks. Average cull for trees with conks is 82%, but only 40% for infected trees without conks. On a linear basis, decay generally extends 8-12 ft (2.4-3.7 m) in each direction from conks, and cull increases with number of conks.

Harvesting stands before decay becomes severe is probably the most effective management approach in stands managed for timber. Clonal variation in susceptibility to decay has been demonstrated. Clones differed not only in percent decay but also in position of rot columns and type of rot. Focusing management on clones with low levels of decay should result in future stands with the same resistance.

Partial cutting in aspen stands is strongly discouraged. Stands often deteriorate rapidly within 5 years after partial cutting. Wounding and subsequent canker infections, sunscald, and boring insects weaken and kill residual trees. Clearcutting is most likely to result in prolific sprouting of aspen. It interrupts succession to conifers and consequent loss of aspen cover type.

**References: 4, 36, 75, 79, 82, 179, 181**

# Brown Crumbly Rot

## Common saprot of conifers

**Pathogen**—Brown crumbly rot is caused by *Fomitopsis pinicola* (= *Fomes pinicola*). The fungus is a very common decayer of conifers and is known as the red-belt fungus.

**Hosts**—*Fomitopsis pinicola* is found on dead conifers and aspen. Occasionally, it may be found on dead parts of living trees.

**Signs and Symptoms**—The red-belt fungus forms perennial conks that are corky and shelf-like. Tops of conks are leathery to woody and range in color from cream, gray-brown, reddish brown, and dark brown to almost black. They often have a reddish orange growing margin (red belt) and a whitish to cream-colored pore surface (figs. 81-82). Conks vary in size and can be slightly over a foot (30 cm) in diameter.

Young conks are amorphous masses of white or cream-colored tissue (fig. 83). Later, the fruiting body develops into the typical perennial shelf conk

**Figure 81.** Red belt conks often have a distinctive red band along the perimeter when mature. *Photo: Susan K. Hagle, USDA Forest Service, Bugwood.org.*

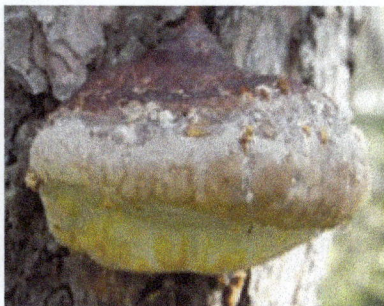

**Figure 82.** The distinctive red band might also be absent. *Photo: James T. Blodgett, USDA Forest Service.*

from 3-18 inches (7.6-45.7 cm) wide with a light lower surface of minute pores filled with reproductive spores. A reddish orange band typically forms just inside the white to cream-colored edge of the conk's upper surface—the point where a new pore layer is produced annually.

Initially, the wood becomes yellowish to pale brown and later develops into the crumbly brown cubical rot (fig. 84). In later stages, the wood develops cubical cracks, in which white sheets of mycelium can develop.

**Disease Cycle**—Spores are disseminated by wind. The spores typically germinate on and colonize dead trees, but the airborne spores can infect and colonize living trees through wounds and broken tops.

**Impact**—Brown crumbly rot is one of the most common and important wood decays in North America. It performs an important ecological role in the degradation of woody material necessary for nutrient recycling in a forest. It is almost exclusively a saprobe, rotting dead trees and stumps. Both sapwood and heartwood are readily decayed. Rarely, this fungus causes a heartrot of living conifers where a large wound has allowed this very weak pathogen to enter the heartwood. Ponderosa pines killed by fire or beetles are often invaded by *F. pinicola* and decay rapidly, reducing salvage potential.

**Figure 83.** The distinctive band is usually absent on immature conks. *Photo: John W. Schwandt, USDA Forest Service, Bugwood.org.*

**Figure 84.** Brown cubical rot with white mycelial mats, caused by *Fomitopsis pinicola*. *Photo: James T. Blodgett, USDA Forest Service.*

**Management**—Injuries caused by falling trees or mechanized equipment during thinning can offer the fungus an opportunity to become established in standing trees. Therefore, injury prevention can reduce volume loss.

*Fomitopsis pinicola* is an important component of the coniferous forest ecosystem because it decays dead trees and logging slash, leaving a lignin-rich residue that is a major organic component of the upper soil layers. This residue enhances water-holding and cation-exchange capacities of soil. The residue is also very resistant to further decomposition and, therefore, enhances carbon sequestration by forests. Decayed snags and fallen trees are habitat for many amphibians, birds, and mammals.

**References: 1, 57, 72, 125, 145**

# Gray-Brown Saprot
## The pouch fungus

**Pathogen**—Gray-brown saprot is caused by *Cryptoporus volvatus* (= *Polyporus volvatus*), the pouch fungus.

**Hosts**—Gray-brown saprot is common on most recently killed conifers except cedar and five-needle pines. Conks are frequent on dead ponderosa pine.

**Signs and Symptoms**—Fruiting bodies of the pouch fungus are often scattered on the bark surface of trees killed by bark beetles and fire (fig. 85). Rounded, white, or tan conks that are about 1 inch (2.5 cm) in diameter form on the outer bark (figs. 85-86). This fungus causes a grayish to gray-brown-colored white rot of the sapwood (fig. 87).

**Disease Cycle**—Fruiting occurs within 1-3 years after a tree dies. Conks often emerge through holes in the bark produced by bark beetles or woodborers. Conks are leathery and scaled at first with an air space and a pink pore layer inside. At maturity, the conks break open on the underside, releasing spores. Spores are disseminated by wind, and there is evidence that bark beetles may acquire spores as they fly due to electrostatic forces. New conks may be produced on a dead tree each year for up to 3 years. Conks live only one summer and deteriorate on the tree within 2 years.

**Impact**—The pouch fungus is a common saprotting organism of recently killed trees. It can cause extensive sapwood decay, and it rapidly colonizes the outer sapwood (fig. 87).

**Management**—Prompt salvage is the only means of damage control. Timber volume loss caused by pouch fungus decay can be substantial. The rapid decay of sapwood necessitates timely harvest of bark beetle and/or fire-killed trees. Occasionally, pouch conks appear on trees with green foliage following a light-intensity fire or bark beetle strip attack. Usually, these trees die by the following year and should be marked for salvage removal if consistent with the project guidelines.

**References: 1, 57, 145, 175**

Figure 86. Close-up view of the pouch fungus. *Photos: Susan K. Hagle, USDA Forest Service, Bugwood.org.*

Figure 87. Saprot caused by pouch fungus in a tree attacked by Douglas-fir beetle. *Photo: Russ Hogan, Bugwood.org.*

Figure 85. Numerous conks of the pouch fungus on boles. *Photo: Sandy Kegley, USDA Forest Service, Bugwood.org.*

# Red Ray Rot

## Difficult to detect in living trees

**Pathogen**—The fungus *Dichomitus squalens* (= *Polyporus anceps*) causes red ray rot, sometimes known as red rot.

**Hosts**—The hosts include ponderosa and pinyon pine. This decay is common on ponderosa pine in the Black Hills.

**Signs and Symptoms**—*Dichomitus squalens* rarely produces a flat, annual fruiting body on the underside of dead branches or stems with intact bark (fig. 88). The pore surface is white when fresh and ages to yellow. The indicators of decay are difficult to identify on standing trees and include decayed branch stubs and fruiting on downed branches. The fungus causes a white pocket rot. Like other wood decays, it has two distinct stages: incipient and advanced.

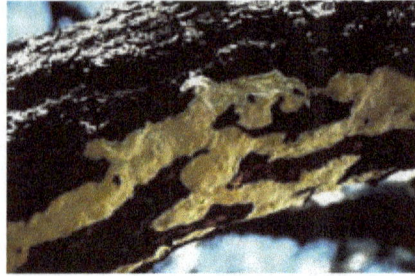

**Figure 88.** Conks of *Dichomitus squalens* underneath a dead ponderosa pine branch. *Photo: Southwestern Region, USDA Forest Service.*

The incipient stage is characterized by a reddish brown discoloration of the affected wood, with no obvious changes in structure or strength. The advanced stage includes a radial pattern of stain and decay (fig. 89). This is characterized by small, often poorly defined, white pockets in the discolored wood, accompanied by progressive changes in structure and reduction in strength. As decay progresses, the pockets become more and more numerous until they merge and give the affected wood the appearance of a fibrous white mass (fig. 90). Eventually, the white, lint-like

**Figure 89.** Decay caused by *Dichomitus squalens* in old growth ponderosa pine. *Photo: Jim Worrall, USDA Forest Service.*

material disappears, leaving the bleached, grayish brown, decayed wood in either a stringy or a somewhat amorphous condition.

Both stages of red ray rot are usually visible in a board sawed from a decayed log. At the point where rot started in the trunk heartwood, advanced decay often forms a cavity. Extending in both directions from this point are more or less continuous columns of advanced decay, bordered by incipient decay.

**Disease Cycle**—The spores are dispersed by wind, land in cracked bark crevices of dead

**Figure 90.** Decay caused by *Dichomitus squalens* in a branch. *Photo: Jim Worrall, USDA Forest Service.*

branches, and germinate to colonize the area between the bark and wood and eventually the heartwood. The fungus fruits abundantly on the lower side of decaying dead material in close contact with the ground. The flat, white fruiting bodies appear about 4 years after infection and then develop annually during the rainy season for about 6 years.

**Impact**—*Dichomitus squalens* is the most common decayer of ponderosa pine in the Black Hills. It has been found in Colorado, but seems to be relatively rare there. It is a decayer of slash (a saprobe) as well as a decayer of live trees. It causes a significant amount of cull in live trees greater than 150 years old. Red ray rot may provide habitat for cavity nesting birds and other wildlife.

**Management**—Timber volume loss caused by red ray rot can be considerable. Reducing tree wounds might reduce the probability of infection, but this fungus can enter through dead branches. Trees infected by this fungus should be marked for removal to improve the residual stand and to release healthy trees, if consistent with the project guidelines.

**References: 56, 57, 159**

# Red Ring Rot

## White pocket rot of conifers

**Pathogen**—Red ring rot is a wood-decay disease of the inner wood of stems of living conifers. It is caused by the fungus *Porodaedalea* (*Phellinus*) *pini*.

**Hosts**—Most conifers in the Rocky Mountain Region can be infected. Engelmann spruce, lodgepole pine, subalpine fir, and Douglas fir are commonly infected in some areas of the Region.

**Signs and Symptoms**—The fruiting body is a conk (shelf fungus), but it occurs less commonly on infected trees than *Phellinus tremulae* does on aspen. It usually occurs at branch stubs or knots (fig. 91). The upper surface is reddish brown to blackish, concentrically furrowed, and somewhat hairy near the margin. The lower surface is usually sloped, yellowish brown, and covered with circular to irregular-shaped pores. The overall texture is tough and corky.

**Figure 91.** Conk of *Porodaedalea* (*Phellinus*) *pini* on Engelmann spruce beneath a branch stub. Many infected trees do not have conks. *Photo: Jim Worrall, USDA Forest Service.*

In some hosts, punk knots, which are also definite indicators of the disease, occur more commonly than conks. Punk knots are swollen or sometimes sunken knots that are resinous and do not callus over normally (figs. 92-93). They are filled with fungal tissue. To investigate a suspicious knot, shave it with a hatchet or stout knife to see if it has the characteristic reddish brown fungal tissue in it, perhaps soaked partly with resin.

**Figure 92.** Punk knots and decay caused by *Porodaedalea (Phellinus) pini.* A slightly swollen, resinous knot in oblique view. *Photo: Jim Worrall, USDA Forest Service.*

Decay caused by *Porodaedalea pini* is also fairly unique and diagnostic. In early stages, infected rings may become reddish to purple in tangential arcs, giving the disease its name. Later, a white pocket rot develops (figs. 94-95). The defect is sometimes called white speck. Pockets are mostly hollow but are delignified and contain white residual cellulose. There may also be black specks in the pockets; these are chemical by-products of delignification. Borders between the pockets may be relatively undecayed. In some cases, the wood may also have distinct black zone lines throughout (fig. 95); this decay may be caused by a closely related species that has

**Figure 93.** Front view of punk knot of *Porodaedalea (Phellinus) pini* after shaving, revealing resin and reddish, golden brown fungal tissue that has replaced the branch trace. *Photo: Jim Worrall, USDA Forest Service.*

**Figure 94.** Typical white pocket rot of *Porodaedalea (Phellinus) pini* with white pockets bordered by reddish, relatively undecayed wood. *Photo: Jim Worrall, USDA Forest Service.*

yet to be described. *Phellopilus nigrolimitatus* causes a similar decay that may be confused with red ring rot, but the pockets tend to be larger.

**Disease Cycle**—Spores produced in conks are disseminated by wind. Large wounds are apparently not the usual site of infection as is true of some fungi. Spores that land on a suitable small wound or twig stub may infect and grow into the inner wood. When decay is sufficient to provide enough resources, a new conk may be produced. Time from infection to conk production may be 10-20 years or more.

**Figure 95.** Very advanced white pocket rot of *Porodaedalea (Phellinus) pini* with black zone lines. *Photo: Jim Worrall, USDA Forest Service.*

**Impact**—Red ring rot is the most common decay in Engelmann spruce and causes the largest decay columns. It is also important in lodgepole pine and is the second most important decay in subalpine fir. In a study of those three species, *Porodaedalea pini* caused 64% of all defect, including non-decay defects. It causes extensive cull, especially in old stands. Decay may extend 4 ft (1.2 m) above and 5 ft (1.5 m) below conks or punk knots. Decay tends to occur in the lower stem and may even develop into the large roots. It is not restricted to the heartwood and may develop outward into the sapwood. It sometimes leads to mechanical failure of live trees, causing hazard in recreation sites.

Decayed trees may provide nesting, denning, or hiding habitat for animals. Although most cavity nesting by birds in this Region is apparently in aspen, various animals may take advantage of advanced decay to excavate the inner wood. Old dead and downed trees may still be useful for such purposes. Dead trees usually decay from the outside-in and often do not provide such habitat.

**Management**—Where emphasis is on timber management, trees with indicators should be removed during any entry. Indicators and the amount of decay and cull associated with them were studied in refs. 80 and 86. Besides conks and punk knots, indicators include forks and dead rust brooms. If decay is frequent, consider reducing rotation age to minimize losses. Prevent injuries to trees during logging to prevent new infections of decay fungi.

**References: 80, 86, 194**

# Rust-Red Stringy Rot and Red Heart Rot

## Indian paint fungus and bleeding Stereum in firs

**Pathogen**—Rust-red stringy rot is caused by *Echinodontium tinctorium*. It is one of the few pathogens that has a common name—Indian paint fungus. Red

**Figure 96.** Indian paint fungus; upper view of conk. Note that the conk was removed with the branch stub at which it fruited. *Photo: Jim Worrall, USDA Forest Service.*

**Figure 97.** Indian paint fungus; lower view of conk. *Photo: Jim Worrall, USDA Forest Service.*

heart rot is caused by *Stereum* (*Haematostereum*) *sanguinolentum*, also known as the bleeding Stereum.

**Hosts**—Indian paint fungus attacks firs, hemlocks, and less commonly other species in western North America, but it is primarily found on white fir in the Rocky Mountain Region. Bleeding Stereum occurs primarily on subalpine fir and Engelmann spruce in this Region, though it can also attack other conifers.

**Signs and Symptoms**—Indian paint fungus produces conks frequently at branch stubs, at wounds, and even inside hollow stems. The conk is hard and woody, hoof-shaped, and perennial (figs. 96-98). The upper surface is blackish and rough with crevices. The lower surface is grey-brown with hard, blunt, thick teeth. The inside is brilliant brick- or rust-red, but it fades over time after exposure. This tissue was ground into a powder and used as paint by some Native Americans, giving the fungus its name. The only other indicator of this disease is punk knots known as "rusty knots." Although there is no swelling as with punk knots caused by *Porodaedalea pini*, the interior of the knot shows the rust-red color characteristic of the decay.

Bleeding Stereum fruits frequently on logs and slash, but infrequently on live trees. Conks are small, thin, inconspicuous, and leathery (fig. 99). Portions are appressed to the bark, often with a projecting cap. The lower surface has neither pores

**Figure 98.** Indian paint fungus; inner view of conk. *Photo: Jim Worrall, USDA Forest Service.*

**Figure 99.** Small, leathery fruiting bodies of the bleeding Stereum. *Photo: Jim Worrall, USDA Forest Service.*

nor teeth; it is fairly smooth. The upper surface is slightly hairy with concentric zones of different shades of brown, orange, and grey; the lower surface is also shades of light brown with a white margin. If the conk is wounded when fresh and moist, it bleeds red liquid, giving the fungus its name. There are no other external indicators, except wounds that may be points of infection.

Rust-red stringy rot begins as golden tan patches in the wood. Fine radial tunnels may develop, presumably following the rays. Rust-red streaks and patches begin to appear; the wood becomes noticeably soft and eventually has a tendency to separate along the annual rings (laminated rot). Finally, it becomes stringy and darker brown with rusty patches (fig. 100).

Wood in the early stage of red heart rot is water-soaked and reddish brown (figs. 101-102). Thin, white fungal tissue may appear, and as decay advances, the wood becomes drier, soft, and finely stringy.

**Figure 100.** Longitudinal section through stem, branch, and conk of Indian paint fungus, showing discolored, decayed wood. *Photo: Kelly Burns, USDA Forest Service.*

**Disease Cycle**—Spores of Indian paint fungus do not require wounds to infect a tree. Small wounds or tiny, shade-killed twigs may be infected. The fungus becomes dormant after infection. Eventually, after it is incorporated into heartwood, the fungus may resume growth and initiate decay. Conks form years later and spores produced on the teeth are wind-blown to start the cycle over.

Red heart rot is frequently associated with conspicuous trunk wounds

**Figure 101.** Log with decay and stain of red heart rot. *Photo: Rocky Mountain Region, USDA Forest Service.*

**Figure 102.** Balsam fir stem with advanced red heart rot in cross section. *Photo: Jim Worrall, USDA Forest Service.*

and broken tops, so presumably those are common sites of infection. Fruiting is uncommon on infected trees but common on logs and slash. Apparently, the spores come from those substrates, and parasitism of live trees may be a reproductive "dead end" for the fungus.

**Impact**—Rust-red stringy rot can form huge, long decay columns in old white fir. There may be less than 1 inch (2.5 cm) of sound shell left on a 30-inch (76-cm) DBH tree. Decay extends 10-16 ft (3.0 to 4.9 m) beyond a conk, so a single conk indicates extensive decay. In addition to substantial cull, the disease poses significant hazards in developed sites. Hazard tree inspectors dealing with white fir should carry binoculars to look for conks, carefully sound the trees, and core if needed.

Red heart rot is the most important stem decayer in subalpine fir and the second most important in Engelmann spruce.

**Management**—Rust-red stringy rot can be detected by conks, rusty knots, sounding, and coring. In timber management, remove infected trees. Cull factors are available based on indicators for cruising. In developed sites, hazard should be mitigated immediately, either by closing affected sites or by removing the tree. Infection sites are not conspicuous wounds, so there is no practical way to prevent infections.

Red heart rot is more difficult to detect, as the only indicators are usually potential sites of infection. Decay columns are usually not hollow so sounding may be less definitive. Infected trees should be preferentially removed where appropriate. Avoid wounding trees because red heart rot is likely to infect such wounds in spruce-fir stands.

**References: 43, 81, 86**

# Stem Decays of Hardwoods in the Plains

## Numerous decay fungi, numerous hosts

**Pathogen**—Many fungi decay wood in the roots, butts, and stems of hardwoods in the Great Plains of South Dakota, Nebraska, Kansas, and Colorado. Three stem-decay fungi are presented here: *Phellinus igniarius*, *Fomitiporia* (*Phellinus*) *punctata*, and *Perenniporia fraxinophila* (table 6).

**Hosts**—*Phellinus igniarius* has a wide host range, infecting species in over 20 genera of hardwoods. It is common in birch, but has also been found in ash, black walnut, poplars, buckthorn, and willows.

*Fomitiporia punctata* also has a wide host range. In a survey of North Dakota windbreaks, plantings, and natural stands, it was found on live willow, ash, *Prunus*, *Rhamnus*, *Caragana*, and *Syringa*.

*Perenniporia fraxinophila* infects primarily ash species, with a few records on other hardwood genera and even junipers.

**Table 6.** Descriptions of conks of three stem-decay fungi.

| Decay fungus | Shape | Surface features | Other features |
|---|---|---|---|
| *Phellinus igniarius* | Projects up to 4 3/4 inches (12 cm); upper surface curved down; rarely almost flat. Lower surface is generally flat or angled slightly upward. | Upper surface gray to black, hairless, becoming deeply cracked and crusty. Lower surface pale to dark; cinnamon to purplish brown. Pores on lower surface are fine. | Conk is hard and woody. Interior is dark reddish brown; tubes with bits of white tissue. |
| *Fomitiporia punctata* | Spreads flat on the surface, but often with a cap on the upper side. | Margin is at first yellowish brown, becoming black and cracked. Pore surface is yellowish to grayish brown. Pores are very fine. | Interior is dark reddish brown. |
| *Perenniporia fraxinophila* | Varies from projecting up to 2 3/4 inches (7 cm) with a distinct cap to completely flat on the surface. | Upper surface may be reddish brown initially, becoming cracked, crusty, and grayish black. Pore surface is ivory to buff; larger pores than the other species. | Interior is buff to pale yellowish brown, corky. |

**Signs and Symptoms**—Conks, the spore-producing fruiting bodies, are evidence of infection, but they are not always present on infected trees. Conks may form anywhere on the stem or branches (fig. 103), but are most common at branch stubs, in cankers, and near cavity openings.

In addition to conks, symptoms may be evident. Openings leading to internal hollows (cavities) may form, especially at branch stubs or openings created by cavity-nesting birds (fig. 104). *Fomitiporia punctata* also causes cankers, or patches of killed bark—such a disease caused by wood-decay fungi is called a canker rot. The fungus decays wood inside the stem but grows out along branch traces and kills the cambium around branch stubs, which results in death of the overlying bark. These fungi all cause a uniform white rot of the wood.

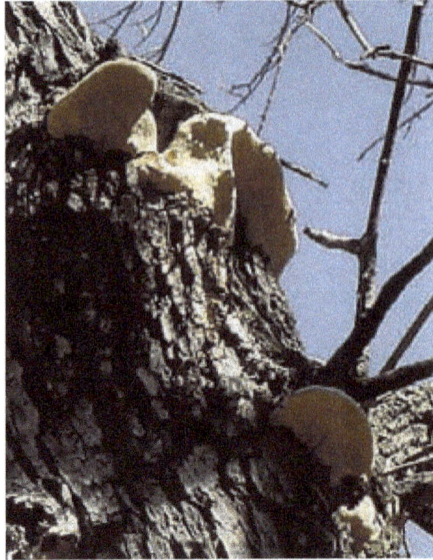

Figure 103. *Perenniporia fraxinophila* on green ash. *Photo: James T. Blodgett, USDA Forest Service.*

**Disease Cycle**—For most stem-decay diseases, the precise point of infection is uncertain. Branch stubs and even small dead twigs can be infected in some cases, but wounds, especially large ones, greatly increase the likelihood of stem decay. Spores germinate at the point of infection, and the fungus may grow down a branch stub or into a wound surface to reach the inner wood. The fungus decays the wood for some years and then returns to the surface, often along a branch trace, to produce a conk. Spores are released from the conk and are blown by wind to reach a new infection point.

**Impact**—Decay columns slowly expand, in some cases shrinking sapwood area and causing physiological stress and slower growth. Decay weakens wood, which can lead to structural failure of stems and branches. In developed areas, this creates hazard to people or structures. The softened

Figure 104. Two cavity nest openings on ash. Stem-decay fungi create conditions necessary for excavation of cavity nests. *Photo: Jim Worrall, USDA Forest Service.*

**Figure 105.** Conks of *Fomitiporia punctata*. *Photo: Mike Schomaker, Colorado State Forest Service, Bugwood.org.*

**Figure 106.** Small conk of *Phellinus igniarius*. *Photo: Jim Worrall, USDA Forest Service.*

inner wood also creates habitat suitable for excavation of cavity nests by birds (fig. 104).

*Fomitiporia punctata* is the most important stem-decay fungus on green ash in North Dakota, and it also occurs in Nebraska (fig. 105). *Phellinus igniarius* occurs throughout the Great Plains on a variety of hosts (fig. 106). In Nebraska, *Perenniporia fraxinophila* is the most important stem-decay fungus in green ash, fruiting on about 5.5% of trees in windbreaks (as of 1978) and about 10% of the green ash in woodlands (as of 1981). Incidence of the latter fungus increased consistently with tree diameter, and infected trees had an average of nine conks. The incidence of infection and amount of decay in infected trees increased with tree age.

**Management**—Wound prevention is the primary preventive approach to stem decays. Wounds from vehicles, machinery, livestock, and fire increase the likelihood of stem decay. Prune early when branches are small. Pruning should be done properly to minimize damage and maximize the ability of callus to cover the wound quickly. Studies have not shown any benefit from wound dressing treatments and, in some cases, have shown negative effects, so they are not recommended. Where hazard to people and property is a consideration, removal of infected trees is recommended.

**References: 14, 138, 139, 141, 142, 180**

# Introduction to Cankers and Overview of Aspen Cankers

## Diseases of the cambium and bark

Cankers are diseases of the bark. They differ from wounds in that the cause, usually a fungus, remains active for some time, causing progressive expansion of the lesion over time. Where the cambium and inner bark die, the sapwood beneath also dies. If a canker grows all the way around a stem or branch, it is said to be girdled, and everything distal to the girdle dies. Branches fade and die (branch flagging) after girdling. Diffuse cankers continue to grow, and the tree does not have time to produce callus. Target cankers grow short distances periodically, and the host produces callus each year in an expanding concentric pattern.

Canker impacts include: deformation of stem growth, leading to loss of wood value and a weak point that may snap; creation of infection court for wood-decay fungi; and mortality.

Some diseases include canker development as one of their symptoms but are classified primarily as other disease types. This group includes some stem-decay fungi that may come to the surface and kill cambium and bark (canker-rots) and some rusts and mistletoes.

**Aspen Cankers**—Aspen has more important canker diseases than any other tree species in the Rocky Mountain Region (figs. 107-111). Descriptions of the most important ones are available, but here we provide an overview and guide to identification of the aspen cankers common in the Rocky Mountains.

Sooty-bark canker is the most widespread, damaging canker of aspen in Colorado and probably other Rocky Mountain states. Typically, sooty-bark canker kills more mature aspen trees than any other agent (this is not the case during unusual episodes of aspen mortality such as from 2004 to 2009). In surveys, sooty-bark canker was responsible for 55% of mortality. Next in importance is Cryptosphaeria canker, causing 26% of mortality. Black canker is probably next most common but usually does not directly kill trees. Cytospora canker is quite common, but attributing mortality to it is more problematic because it is frequently associated with stress caused by other diseases and factors.

**Figure 107.** Typical sooty-bark canker on aspen (from Johnson and others 1995).

USDA Forest Service RMRS-GTR-241. 2010.

**Identification**—Generally, well-developed cankers have sufficiently distinct features that they can be readily identified in the field. However, small, young cankers often do not have all the typical features and can be difficult to identify. Following are brief descriptions that should enable identification using distinguishing features that are usually available in well-developed cankers. A more detailed comparison is presented in table 7.

- *Sooty-bark canker*: cankers expand rapidly so that callus formation is rare; alternating, concentric zones of light and dark where the periderm remains present versus where it falls off, revealing the sooty black inner bark; small ($1/25$-$1/13$ inch [1-2 mm] in diameter) light gray, usually shriveled, cup-shaped fruiting bodies (fig. 107).

- *Cryptosphaeria canker*: cankers become very long and narrow with orange-brown margins; inner bark becomes black but usually has small ($1/50$-$1/13$ inch [0.5-2.0 mm]) whitish lenticular spots; black, pimple-like fruiting bodies may be peppered on the dead bark in patches (fig. 108).

**Figure 108.** Typical Cryptosphaeria canker on aspen. *Photo: Jim Worrall, USDA Forest Service.*

- *Black canker*: cankers are diamond-shaped or oval, often with flaring margins; bark usually sloughs off to expose wood at canker center; wood and especially bark at margins are often dark to black; concentric annual callus rings, often quite narrow, are visible on the wood surface (fig. 109).

- *Cytospora canker*: canker margins are discolored orange; bleeding of black liquid may be present, especially in spring; older dead bark is not much altered but may become brownish and shriveled or sunken; pimple-like fruiting, often with white heads, eventually appears in some spots (fig. 110).

- *Hypoxylon canker*: cankers often have irregular shape; salt-and-pepper checkered appearance; fruiting, especially the small, gray-black, bumpy stromata, is almost always present; this is the least common of the five cankers described here (fig. 111).

**Figure 109.** Typical black canker on aspen. *Photo: Jim Worrall, USDA Forest Service.*

**Table 7.** Comparison of aspen cankers.

| | Canker type | Surface appearance | Fruiting | Mortality |
|---|---|---|---|---|
| Sooty-bark | Diffuse canker; grows rapidly | Alternating zones with periderm present versus exposed; black inner bark | Gray, cup-shaped; shriveled, 1/64-1/16 inch (0.5-2 mm) diameter common | Primary killer |
| Cryptosphaeria | Perennial target or diffuse; grows much faster longitudinally than tangentially | Long, narrow canker, dead bark adheres tightly; margin is orange-brown | Pimple-like, black heads with submerged pseudostroma; common | Primary killer |
| Black | Perennial target canker; grows slowly | Bark sloughs off and/or flares out; concentric annual callus ridges in wood are exposed | Minute, black perithecia in spring; rarely seen | Usually does not directly kill trees |
| Cytospora | Diffuse canker; grows rapidly on severely stressed trees | Orange discoloration at margin; bleeding, especially in spring | Pimple-like, often with white heads; common | Secondary killer; attacks stressed trees |
| Hypoxylon | Diffuse canker | Irregular shape; dead bark sloughs off in patches leading to a checkered pattern, yellowish-orange margin | Gray conidial pillars beneath blistered periderm; perithecia in small gray-black stromata | Primary killer but uncommon in Rocky Mountains |

Figure 110. Typical Cytospora canker on aspen. *Photo: Jim Worrall, USDA Forest Service.*

Figure 111. Example of Hypoxylon canker on aspen. *Photo: Rocky Mountain Region Archive, USDA Forest Service, Bugwood.org.*

**Management**—Because wounds are important infection courts for most of the canker pathogens, avoiding wounding is important, especially during the growing season. This is a major reason why partial cutting in aspen is strongly discouraged. It is also a major reason why developing campgrounds in aspen stands is strongly discouraged in Regional policy.

For timber management, information on the incidence and types of cankers in managed stands can be used in prioritizing stands for treatment. Trees can survive with black canker for many years, but most of the other cankers may be indicators of stands that are in the process of overstory mortality. Cytospora canker is usually evidence of some other stress factor at work.

**References: 76, 79, 96, 100**

# Black Canker

## A common, slowly developing target canker of aspen

**Pathogen**—Black canker is caused by the fungus *Ceratocystis populicola* (part of the *C. fimbriata* complex). The disease is sometimes called Ceratocystis canker or target canker of aspen.

**Hosts**—The *C. fimbriata* complex occurs all over the world on diverse hosts, causing diverse types of diseases. The form that causes black canker on aspen apparently occurs occasionally on other *Populus* species and has been introduced to eastern Europe where it is more lethal than in America.

**Signs and Symptoms**—Black canker may occur anywhere along a stem or branch. Cankers are typically diamond-shaped or oval, and the margins are often flared out (figs. 112-114). They are generally blackish and the bark tends to break off as the cankers develop, leaving an exposed canker face. Narrow, concentric, annual callus ridges are visible in the wood of the canker face. Fruiting is microscopic and usually not seen.

**Disease Cycle**—The fungus may infect at wounds, and when the host is dormant, the fungus kills a patch of cambium and inner bark. When the host is active, it produces callus that partially grows over the wood under the killed bark. When the host is again dormant, the fungus resumes growth, killing the callus and an additional bit of bark and cambium. This sequence continues for many years, resulting in successive, concentric rings of callus that are visible in the wood inside the canker. Because of the concentric rings, this type of canker is called a target canker. As the callus is produced, it tends to alter the form of the stem, leading to flaring of the margins.

The pathogen produces minute, black perithecia in cankers, but they are not always produced and are difficult to observe. They are produced in spring at the border of the canker on tissues dead for at least a year. *Ceratocystis* species generally have adaptations for dispersal by insects, and this appears to be

**Figure 112.** Large, black canker that shows diamond shape, flaring margin, lack of bark over canker, annual callus rings in wood, and that cankers are not always black. *Photo: Jim Worrall, USDA Forest Service.*

**Figure 113.** Another large black canker. *Photo: Jim Worrall, USDA Forest Service.*

USDA Forest Service RMRS-GTR-241. 2010.

**Figure 114.** Closeup of black canker, showing the narrow, annual callus rings. *Photo: Jim Worrall, USDA Forest Service.*

functional for black canker. Nitidulid beetles are attracted to the aroma of the fungus and may acquire the sticky spores from the fruiting bodies, which can be deposited again when the beetles visit fresh aspen wounds.

**Impact**—The incidence of infection varies greatly among stands, but it is not known if this is due to variation in resistance or due to stand/site factors that favor infection. Black canker develops slowly and normally does not kill trees directly (figs. 115-116). However, it has a number of impacts:

• Cankers deform the stem, resulting in cull if the trees are harvested for wood products.

• Cankers can predispose stems to snapping, resulting in premature mortality.

**Figure 115.** Black canker in heavily-affected aspen stand. *Photo: Jim Worrall, USDA Forest Service.*

CANKERS

- Cankers frequently serve as points of infection for wood-decay fungi, further affecting the tree, causing additional cull, and increasing the likelihood of snapping and mortality.
- Cankers and associated decay create hazards to people and property in developed sites.

**Management**—Practical means of preventing infection are not known. Certainly, wounding should be avoided, but natural infection courts apparently lead to heavy infection in some stands. In timber stands, partial cutting of aspen is strongly discouraged, because the residual stand often deteriorates in 5-10 years. Therefore, where the disease threatens management objectives, early harvest/regeneration should be considered.

**References: 7, 79, 96, 97, 100**

**Figure 116.** Black cankers on aspen. The disease develops slowly and usually does not kill trees directly. *Photo: Jim Worrall, USDA Forest Service.*

# Black Knot of Cherry and Plum

## Enlarged black growths on stems

**Pathogen**—Black knot is caused by the fungus *Apiosporina morbosa*.

**Hosts**—Hosts of black knot are cherry and plum species.

**Signs and Symptoms**—Elongated swellings or knots (about 1-8 inches (2.5-20 cm) long and 1/3-1 inch (8-25 mm) thick) on twigs, branches, and small stems are commonly seen on susceptible hosts. The knots start out greenish in color and soft but become hard and black over time (fig. 117).

**Disease Cycle**—Spores are discharged from the fruiting bodies (in knots) in the

**Figure 117.** Swellings on branches caused by black knot. *Photo: Joseph O'Brien, USDA Forest Service, Bugwood.org.*

spring as new growth starts to elongate. Spore discharge ends at about the time terminal growth stops. Rain is required for spore discharge, and spores are carried by both rain and wind to new infection sites. The fungus enters unwounded twigs. Infections are severe under moist conditions in the spring at the time of spore discharge.

Knots start to appear a few months after infection. Some appear in late summer, others not until the following spring. One to 2 years are required for the knots to produce fruiting bodies.

**Impact**—Fruit production can be significantly reduced in heavily infected cherries and plums. In heavy infections, small branches and sprouts can die over time (fig. 118).

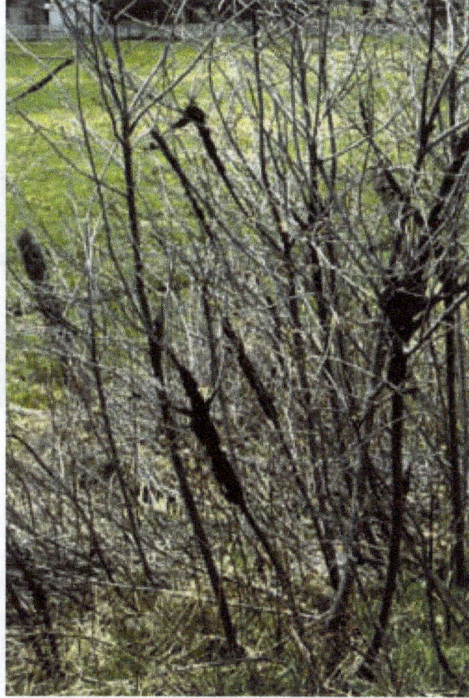

Figure 118. Black knot on plum. *Photo: William M. Brown, Jr.; Bugwood.org.*

**Management**—Management is generally not needed for black knot. When necessary, prevention of infection is the most efficient means of control. Removing infected wild hosts surrounding plantings will reduce infections in plantings. Pruning the knots can also reduce infections. Cuts should be made well below the knot (3-4 inches [7.6-10 cm]) as the fungus may extend beyond the swelling. Fungicides can also be applied in addition to the sanitation efforts. There is considerable variation in cultivar susceptibility to black knot.

**Reference: 140**

# Botryodiplodia Canker

## Difficult to detect on most hardwood species

**Pathogen**—Botryodiplodia canker is caused by fungal species in the genus *Botryodiplodia*, including *B. hypodermia* and *B. theobromae*. Botryodiplodia is the asexual form of the fungus. The sexual states for some of the *Botryodiplodia* species are *Botryosphaeria* spp.

**Hosts**—Hosts of *Botryodiplodia* species in the Rocky Mountain Region include elm, oak, and sycamore. Botryodiplodia canker is one of several hardwood canker diseases.

**Signs and Symptoms**—The signs are small, black, pimple-like fruiting bodies (pycnidia) produced on dying and dead bark at the canker margin (fig. 119). However, it is difficult to identify the pathogen by signs. Several fungi produce small, black, pimple-like fruiting bodies and occur on hardwood hosts. Microscopic examination of the spores is necessary for correct identification.

**Figure 119.** Signs and symptoms of Botryodiplodia canker on an elm. *Photo: William Jacobi, Colorado State University, Bugwood.org.*

Symptoms include discolored and cracked bark (fig. 119). Outer bark may become loose. Inner bark, cambium, and sapwood turn reddish brown to dark brown. The discoloration ends at the canker margin. Cankers on trees with rough bark can be detected only after removing bark to expose dead inner bark, cambium, and sapwood. Cankers may appear sunken and may be surrounded by callus tissue.

Girdled stems and branches die. The leaves above the cankers wilt, turn yellow to brown, and die, and adventitious shoots may form below the canker.

**Disease Cycle**—Spores (conidia) are released during rain and are dispersed by wind, insects, pruning tools, and rain droplets. Wounded bark can be infected throughout the growing season during moist conditions but most abundantly in spring. Cankers develop at wound sites and can girdle and kill the trees. Fruiting bodies are produced on dying and dead bark at the canker margin. The fungus overwinters in the bark as mycelium or as fruiting bodies. Cankers develop quickly during dry periods and during hot summer months. Fruiting bodies can be produced yearlong but are mainly produced in the fall on dying and recently dead bark.

**Impact**—These diseases cause branch dieback and tree mortality. Botryodiplodia canker diseases are important diseases of hardwoods in windbreaks.

**Management**—Preventing wounds is the best way to minimize cankers. Cankered branches should be pruned in the winter to reduce inoculum (fungus available to initiate new infections). Severely diseased trees and branches should be removed. Pruning cuts should be made well below canker margins and pruning tools should be disinfected. Because this disease responds to water stress, management of competing vegetation, stand thinning, planting techniques that reduce drought, and selection of a tree species compatible with a site will reduce losses.

**References: 114, 137**

# Cryptosphaeria Canker

## Long, narrow canker kills aspen

**Pathogen**—Cryptosphaeria canker is caused by the fungus *Cryptosphaeria lignyota* (formerly known as *C. populina*). The anamorph (asexual stage) is *Libertella* sp. The disease is also sometimes called "snake canker."

**Hosts**—The disease occurs on quaking aspen and other *Populus* species in the Rocky Mountains. The pathogen has been found throughout much of North America and in Europe.

**Signs and Symptoms**—Cryptosphaeria cankers are usually much longer and narrower than other cankers of aspen (fig. 120). Cankers may only be 2-4 inches (5-10 cm) wide and 10 ft (3 m) long. They may ultimately extend for much of the stem length or occur primarily in the portion of the stem with live branches.

Bark recently killed near the margin of the canker becomes discolored light brown to orange. Within a year or two, dead inner bark becomes stringy, black, and sooty, much like sooty-bark canker, but usually contains small (1/50-1/13 inch [0.5-2.0 mm]), lenticular, light-colored areas (fig. 121). Annual callus formation may be visible. Dead bark adheres tightly to cankers.

**Figure 120.** Cryptosphaeria canker killing an aspen. The canker winds around this tree all the way up to the live crown. *Photo: Jim Worrall, USDA Forest Service.*

**Figure 121.** Light-colored lenticular spots that frequently appear in older dead bark of canker caused by *Cryptosphaeria lignyota*. Live bark is at the left-most margin of the cut. *Photo: Jim Worrall, USDA Forest Service.*

# CANKERS

The disease is actually a canker-rot. The pathogen colonizes the heartwood and sapwood, causing discoloration and decay. Discoloration may have hues of gray, brown, yellow, orange, and even pink and extends up to 3.3 ft (1 m) or more beyond the canker. The decay is brownish and mottled and is probably a type of soft rot similar to white rot, as is typical of wood-decaying ascomycetes. The fungus grows out from the wood to the cambium and bark, causing the canker.

Mortality usually occurs in small trees after only a few years and before the canker girdles the stem. Killing of sapwood apparently causes more damage than the canker itself. Large trees may be killed more slowly as their branches are engulfed in the canker. Especially on large trees, branches may be infected first, from which the pathogen can spread to the stem.

*Cytospora chrysosperma* frequently infects and fruits at the margin of Cryptosphaeria cankers and quickly colonizes trees after they die, so the cause of death may be misattributed to Cytospora canker. For this reason, and because the pathogen may not fruit reliably, especially on small trees, diagnosis should be made by looking for the lenticular, light-colored areas in bark and the staining in the sapwood.

**Disease Cycle**—The fungus frequently infects trees through wounds. As described above, it appears to progress primarily through wood, then move out to cambium and bark. Black perithecia of the pathogen may appear scattered on bark that has been dead for a year or two (fig. 122), although fruiting may not occur on small trees. Beneath the surface, the perithecia are embedded in a widely spread pseudostroma.

Ascospores are forcibly ejected from perithecia during wet weather and presumably dispersed by wind. Conidia can also be produced but may function only in fertilization.

**Impact**—Cryptosphaeria canker is an important killer of aspen. In one Colorado survey, 83% of 30 surveyed sites had Cryptosphaeria canker, and 26% of aspen mortality was attributed to the disease. Unlike sooty-bark canker, it seemed to kill mostly small to mid-size trees. It is estimated that about 8% of aspen in Colorado have decay caused by *C. lignyota*. Cryptosphaeria canker is one of a number of wound-infecting cankers that leads to mortality following partial cutting of aspen stands and also in developed recreation sites established in aspen.

**Figure 122.** Perithecia (fruiting bodies) of *Cryptosphaeria lignyota*. *Photo: Jim Worrall, USDA Forest Service.*

**Management**—Wounding should be prevented. However, many infections are due to natural infection courts, and practical means of preventing such infection are not known. In timber stands, partial cutting of aspen is strongly discouraged because the residual stand often deteriorates in 5-10 years. Therefore, where the disease threatens management objectives, early harvest or regeneration should be considered. Similarly, development of recreation sites in aspen stands is strongly discouraged by Regional policy because cankers such as this one cause rapidly increased mortality following development and use.

**References: 77, 78, 100, 181**

# Cytospora Canker of Aspen

## Orange weeping bark

**Pathogen**—Cytospora canker of aspen is most commonly caused by *Valsa sordida*, a name that applies to the sexual fruiting stage as well as the whole fungus. However, the fungus is frequently called *Cytospora chrysosperma*, because this name applies to the asexual, or conidial, fruiting stage, which is most commonly seen. Other, related species of Cytospora-like fungi, such as *Leucostoma niveum* (asexual stage *Leucocytospora nivea*) can occur on dying and dead bark and may cause cankers in some cases.

**Hosts**—*Valsa sordida* primarily causes disease on *Populus* species. Related fungi occur on many other hosts.

**Signs and Symptoms**—The small, pimple-like fruitbodies are embedded and break through the bark surface. They may be so numerous and small that they roughen the bark surface like coarse sandpaper. The exposed surface of the asexual fruiting body (the disk) is gray to black and prominent or reduced in *Valsa sordida* and is usually white but sometimes gray-brown in *Leucostoma niveum*. During wet weather when temperatures are above freezing, conidia may be produced from the fruitbodies in a sticky matrix. If the bark is moist but not wet, fine, curly, orangish tendrils of spores project from the fruiting bodies, extruded like toothpaste from a tube. Otherwise, amber- to orange-colored spore masses may adhere to the bark surface (fig. 123). Sexual fruitbodies are superficially similar to those that produce conidia.

The most common symptom is rapidly spreading necrosis (death) of bark. On

**Figure 123.** Pycnidia of *Cytospora* sp. with orange spore masses. *Photo: Jim Worrall, USDA Forest Service.*

smooth bark, this appears first as an orange discoloration that may be accompanied by exudation of a brown liquid (fig. 124). The inner bark turns from green to brown to black, and bark begins to slough off after 2-3 years. Cankers may be diffuse, continuing their spread until the tree is dead without stopping or becoming sunken. Once any point on the stem or branch is girdled (killed all the way around), everything above that point dies. On trees of intermediate susceptibility, cankers may be annual, stopping growth after one season and becoming sunken and callused over as the tree grows around them. In rare cases, the canker can resume growth in successive dormant seasons, approximating the appearance of a perennial target canker.

**Disease Cycle**—*Cytospora* species are usually opportunistic pathogens. They are quick to attack plants that are stressed by heat, drought, winter injury, and other diseases and insects. They often attack and kill stressed trees that may otherwise have survived the stress and recovered. They also colonize dying or recently dead tissues as saprobes. *Valsa sordida*, the primary cause of Cytospora canker of aspen, is one of the more aggressive species in the group.

**Figure 124.** Cytospora canker of aspen. Bleeding is usually not this copious and may be absent. *Photo: Rocky Mountain Region Archive, USDA Forest Service.*

Most canker pathogens, including *Cytospora* species, have traditionally been thought to infect primarily through wounds. More recent information suggests that they can sometimes infect and inhabit apparently healthy bark and buds, thus being in a position to rapidly colonize and kill weakened tissues. Such infection by a pathogen without symptom production is termed a latent infection. In addition to wounds, infection may occur through buds, nodes, and lenticels.

Once the pathogen has infected and is causing disease, it kills the cambium and living tissues in the bark. As a result, the underlying wood dies. Cankers typically develop when the host is dormant, but can develop during the growing season when the host is severely weakened. Cankers develop in a temperature range of 36-86 °F and may grow as fast as 1.6 inches (4 cm) per day.

Microscopic conidia are produced by the millions in the asexual fruitbodies. Conidia are dispersed by rainsplash and incidentally by insects or other animals. Later, ascospores may be produced. Ascospores may be extruded and dispersed like conidia, but under some conditions, they are forcibly ejected into the air when mature and are dispersed by wind.

**Impact**—Mortality is the major impact of Cytospora canker. Although stress usually precedes severe infection, disease impact can be substantial because trees may recover from stress in the absence of the canker.

The canker also impacts regenerating stands. A study of mortality greater than 90% in 4- to 5-year-old stands in 1988 found rapidly expanding cankers with consistent *C. chrysosperma* fruiting. Symptoms were reproduced in greenhouse inoculations, but it was not known what conditions led to such severe disease in these cases.

**Management**—Avoiding wounds and stress will reduce the likelihood of Cytospora canker in individual trees. Clearcutting, prescribed fire, or wildfire will stimulate regeneration and will give the best chances for maintaining aspen on the site.

**References: 23, 89, 158, 162**

# Cytospora Canker of Conifers

## Branch flagging in stressed conifers

**Pathogen**—Cytospora canker of conifers is caused by *Valsa kunzei* (= *Leucostoma kunzei*) (asexual stage is *Cytospora kunzei*, = *Leucocytospora kunzei*). The disease is commonly referred to as Cytospora, Valsa, or Leucostoma canker.

**Hosts**—Many coniferous species are hosts, primarily spruce species. Common hosts in the Rocky Mountain Region include Colorado blue spruce, white spruce, Engelmann spruce, and Douglas-fir. The disease is particularly prevalent in windbreaks and ornamental plantings.

**Signs and Symptoms**—From a distance, the most obvious symptom is dead or dying branches, particularly older branches (fig. 125). Cankers are diamond-shaped and are usually very resinous—clear amber resin exudes from canker margins and eventually hardens to a conspicuous white crust (this symptom is less prominent on Douglas-fir). Cankers on trunks become sunken in the middle with expanding, flared edges (fig. 126). Branch cankers are common on all conifer hosts, and Engelmann spruce and Douglas-fir are also susceptible to trunk cankers. When a branch is girdled, foliage becomes discolored, dies in spring and summer, and brown needles are shed the following winter (fig. 127). The disease often progresses upward in tree crowns. Small (1/25-1/13 inch [1-2 mm]), black fruiting bodies are sometimes visible around the edges of cankers. During moist weather, yellow tendrils of spores ooze from pycnidia (fig. 128).

**Figure 125.** Typical symptomatic branch flagging caused by Cytospora. *Photo: Joseph O'Brien, USDA Forest Service, Bugwood.org.*

**Figure 126.** Sunken stem canker with flared edges on Engelmann spruce. *Photo: Kelly Burns, USDA Forest Service.*

**Figure 127.** Branch canker, showing white, resinous pitch. *Photo: Michael Kangas, North Dakota State Forest Service, Bugwood.org.*

**Figure 128.** Cytospora spore tendrils exuding from pycnidia. *Photo: Michael Kangas, North Dakota State Forest Service, Bugwood.org.*

**Disease Cycle**—Cytospora canker is a disease of stressed conifers. The pathogen overwinters in cankered bark. Ascospores and conidia are dispersed during wet weather in the spring, summer, and fall by splashing rain, wind, and insects. Infection occurs through wounds; tiny openings in the bark created by environmental stresses such as ice or snow may also serve as infection courts. Symptoms develop shortly after infection but latent infections are not uncommon. The fungus can remain dormant on the outer bark until an infection court becomes available. Pycnidial and later perithecial stromata form around the edges of old cankers, and the fungus colonizes adjacent healthy tissue.

**Impact**—Cytospora canker causes branch and stem cankers, which can deform stems. It is one of the most common and damaging diseases of planted spruce. Colorado blue spruce sustains the most damage east of its natural range. The pathogen is opportunistic and invades trees weakened by other factors such as drought, hail, or insects. Older branches are more susceptible, so infections generally start in the lower crown and progress upward. Mortality is rare.

**Management**—The incidence and severity of Cytospora canker may be reduced by maintaining tree vigor (fertilize, water, control insects), reducing injuries to the stem and bark, and pruning and destroying infected branches.
**Reference: 158**

# Hypoxylon Canker

## Uncommon but locally important canker of aspen

**Pathogen**—Hypoxylon canker is caused by the fungus *Entoleuca* (*Hypoxylon*) *mammatum*.

**Hosts**—Hypoxylon canker occurs on trembling aspen and a few other aspens. It occurs over most of the range of trembling aspen in North America and also in Europe.

**Signs and Symptoms**—Stem cankers are often centered on dead branches, from which they commonly enter the stem. Young cankers, and edges of older cankers, are yellowish orange to orange-brown with an irregular margin. In recently killed portions of a canker, black and cream mottling may be found in the inner bark, cambium, and outer sapwood. In a year or two, the bark begins to blister and the periderm (the thin outer layer of bark) falls in patches, giving the canker a mottled or salt-and-pepper appearance from a distance (fig. 129). Eventually, most of the outer layer falls off, leaving the blackened inner bark on the surface.

Patches of tiny, gray pillars, like short, coarse hairs, are revealed where the periderm falls off. They can often be found earlier by peeling back loose, blistering periderm. The sexual stage will eventually be present in slightly older parts of the canker. It is composed of raised stromata that are initially gray and become black when mature, are about $1/8$-$1/2$ inch (3-13 mm) in diameter, and each has several to 30, pimple-like, embedded perithecia (figs. 130-131). Small, white mycelial fans develop under the bark just behind the advancing edge of the canker.

**Disease Cycle**—The infection court is uncertain, but it appears that dying year-old twigs and small wounds can be infected. Various insects, including the poplar borer and the aspen tree hopper, make wounds that can also serve as infection courts.

Airborne spores that land on the infection court germinate during moist weather and grow into living bark. Two years may elapse before symptoms develop. In the year after a visible canker begins to form, the pillars are formed. The pillars

**Figure 129.** Hypoxylon canker killing an aspen. Note the mottled, salt-and-pepper appearance of the upper, younger portion of the canker where outer bark is missing in patches. *Photo: Jim Worrall, USDA Forest Service.*

**Figure 130.** A large stroma of Hypoxylon canker with multiple perithecia. *Photo: Jim Worrall, USDA Forest Service.*

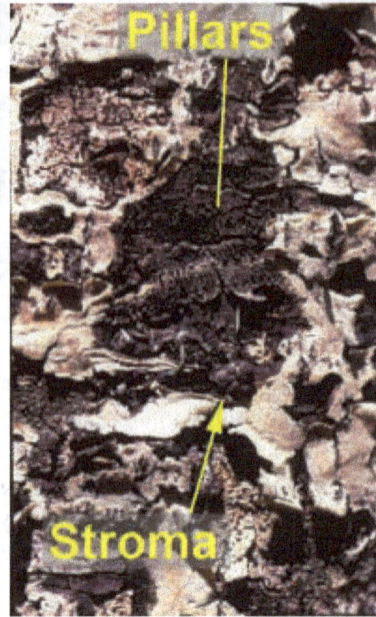

**Figure 131.** Conidial pillars and stromata of Hypoxylon canker. Generally stromata are produced each year where pillars were produced the year before. *Photo: Jim Worrall, USDA Forest Service.*

grow up to blister the periderm, which flakes off. This exposes the spores that are produced all over the surfaces of the pillars. These spores are not infectious; they function only in fertilizing other cankers.

Stromata, containing the perithecia, are produced in the following year. Spores are forcibly ejected from tiny holes at the tips of the perithecia and are wind-dispersed. Additional cycles of pillar and stromata occur each year until the tree is killed and the fungus dies soon after.

**Impact**—Hypoxylon canker is widespread geographically, but is the least common of the five major cankers of aspen in Colorado and, presumably, in the rest of the southern Rocky Mountains. East of the Great Plains, however, it is often the most destructive disease of aspen. Even in the southern Rocky Mountains, localized areas may suffer serious mortality from this disease. Trees of all sizes may be attacked. Damage is most severe in open, widely spaced stands and near stand edges or openings. There is some indication that good aspen sites with deep soils and adequate moisture have higher levels of Hypoxylon canker than poor sites with dry, shallow soils, but this has not been verified. Mortality may result from girdling or from snapping due to wood decay beneath the canker.

**Management**—The canker is favored by stand openings and stands of low density, so maintaining continuous aspen cover should reduce infection. Susceptibility to the disease varies clonally, so consider avoiding further management of heavily infected stands.

**References: 76, 100, 146, 158**

# Sooty-Bark Canker

## Normally the No. 1 killer of Rocky Mountain aspen

**Pathogen**—Sooty-bark canker is caused by the fungus *Encoelia pruinosa* (formerly known as *Cenangium singulare*). It is sometimes called Cenangium canker.

**Hosts**—Sooty-bark canker is found primarily on aspen. Other *Populus* species may be attacked occasionally in other areas.

**Signs and Symptoms**—Sooty-bark canker begins as a small, sunken, oval patch of dead bark. It may grow as much as 3.28 ft (1 m) long and 12 inches (30 cm) wide each year. Each year's canker growth typically has an alternation of light and dark zones, caused by the appearance of the thin, light layer of outer bark contrasting with the black inner bark where the outer bark is gone (figs. 132-133). The black inner bark becomes dry and powdery, leading to the term "sooty." Eventually, the canker girdles the tree and it dies.

The fruiting bodies are apothecia and often appear in large numbers on killed bark. They are cup-shaped, up to $1/8$ inch (3 mm) in diameter, and have very short stalks (fig. 134). They are gray with a fine granular surface (pruinose). When dry, they are shriveled and inrolled.

When the bark is completely weathered away, a final sign of sooty-bark canker is small patches (several millimeters in diameter) of black fungal material on the wood, separated by plain wood. The pattern is often likened to that on a leopard's fur (fig. 135).

**Disease Cycle**—Cankers can be initiated at points with no apparent wounds, but the fungus does infect stem wounds. However, the type, season, and size of wounds that are suitable infection courts are not known. The inner bark and cambium are rapidly invaded and killed. As described under "Signs and Symptoms" above, vertical growth is somewhat faster than horizontal growth. Callus is not produced unless the disease is stopped, which is unusual.

The pathogen fruits within a year or two of bark death. Ascospores are forcibly shot from apothecia into the air under moist conditions. Spores germinate and infect through suitable infection courts.

**Impact**—Sooty-bark canker is the most lethal canker on aspen in the West, and it is the most widespread canker in Colorado and probably other areas of the West. Under normal circumstances, it is the primary killer of aspen in Colorado and probably other parts of the West. This is not the case during unusual

**Figure 132.** The barber-pole pattern of sooty-bark canker. Each light and dark pair represents one year of growth. *Photo: Thomas Hinds, USDA Forest Service.*

# CANKERS

**Figure 133.** Another example of the barber-pole pattern of sooty-bark canker. The pattern may be more obscure than these examples. *Photo: Jim Worrall, USDA Forest Service.*

**Figure 134.** Apothecia (fruiting bodies) of *Encoelia pruinosa*, the sooty-bark canker pathogen. Apothecia are light gray, cup-shaped, and inrolled and shriveled during dry weather. *Photo: Jim Worrall, USDA Forest Service.*

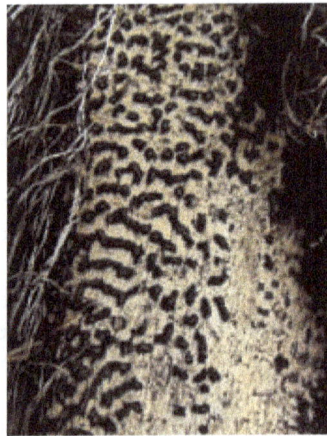

**Figure 135.** The "leopard spotting" that often remains on wood after the bark is weathered away from the sooty-bark canker. *Photo: Jim Worrall, USDA Forest Service.*

events such as sudden aspen decline, a mortality event of 2004-2009 that was associated with drought stress.

The disease usually affects the larger trees in a stand. It kills trees in only 3-10 years unless its development is arrested, which is uncommon. Damage from this canker is most common in stands where the incidence of wounding is high.

**Management**—As with the other aspen cankers, few practical approaches to preventing infection and canker development are known. Vigorous trees can be attacked and killed, so maintaining host vigor will not affect the disease. Avoiding wounds should reduce the likelihood of infection.

Silvicultural approaches can reduce the long-term impact. Because sooty-bark canker tends to attack older, larger trees, managing aspen in rotations of 80-100 years should be effective. Clearcutting, prescribed fire, and wildfire can all improve aspen condition by stimulating dense suckering before diseases like sooty-bark canker lead to gradual stand deterioration and, in some cases, conversion to other vegetation types.

**References: 79, 96, 100**

# Introduction to Wilt Diseases

## Disruption of water-conducting tissues

In a healthy tree, water in the soil enters roots and is transported through vessels or tracheids in the xylem to leaves. Wilt diseases disrupt this flow of water in the xylem, thus causing leaves to wilt. These diseases result from pathogen activity in the vessels or tracheids. Wilt pathogens are parasites that can move through the vascular tissue of trees. The pathogens can include fungi, nematodes, bacteria, or other micro-organisms.

The means of water disruption vary and are often not completely understood. In hardwoods, living parenchyma cells may balloon into vessels as a defense response to pathogens, resulting in vessel blockage. Pathogens may directly block water flow or cause air bubbles by damaging cell walls in the vessels or tracheids that disrupt water transport. Some wilt pathogens produce toxins that damage host cells or produce enzymes or other chemicals that disrupt flow. Pathogens can invade and kill the living parenchyma cells in the xylem tissues that are associated with water transport, thus disrupting flow. Any of these means can result in the typical wilt symptoms.

There are several wilt diseases in the Rocky Mountain Region (table 8). Some of the more damaging and/or common wilts in the Region are included in this table, but other wilt diseases may occur.

**Table 8.** Comparison of some more damaging and common wilts in the in the Rocky Mountain Region.

| Wilt | Pathogen | Type of organism | Common tree host |
|---|---|---|---|
| Dutch elm disease (Fig. 136) | *Ceratocystis ulmi* | Fungus | Elm; most damaging on native elm |
| Verticillium wilt (Fig. 137) | *Verticillium albo-atrum, V. dahliae* | Fungus | Several hardwood genera |
| Oak wilt (Fig. 138-139) | *Ceratocystis fagacearum* | Fungus | Oaks; most damaging on red oak group |
| Pine wilt disease (Fig. 140) | *Bursaphelenchus xylophilus* | Nematode | Pines and other conifers; most damaging on exotic pines |
| Black stain root disease (Fig. 141) | *Leptographium wageneri var. wageneri* | Fungus | Pinyon |

**Signs and Symptoms**—Signs are microscopic for many of the wilts. Therefore, microscopic examinations, isolation of the pathogen on agar media, or other laboratory tests are often the best ways to identify the causal organisms.

The common name for these diseases, "wilts," comes from the typical wilt symptoms that are attributed to drought stress, including drooping leaves and branches. The wilting leaves fade to yellow, then to brown, and then die.

Depending on the disease, leaves may be shed or remain on trees. In some cases, wilting and yellowing of individual branches and branch mortality may occur (figs. 136-141). Tree mortality is common with these diseases. Trees with wilt disease may appear healthy when soil moisture is high and then suddenly wilt and die during hot and dry periods. Vascular discoloration in the new sapwood is a common symptom of many wilt diseases.

General wilt symptoms can have many causes, including wilt disease. Root diseases and cankers often cause crown wilting. Prolonged periods of drought may cause wilt symptoms, especially on younger trees.

**Disease Cycle**—These diseases may spread by vectors, root grafts, and wind, or they may persist in the soil around infected trees. Vectors are motile

Figure 136. Crown wilt symptoms of Dutch elm disease. *Photo: Fred Baker, Utah State University, Bugwood.org.*

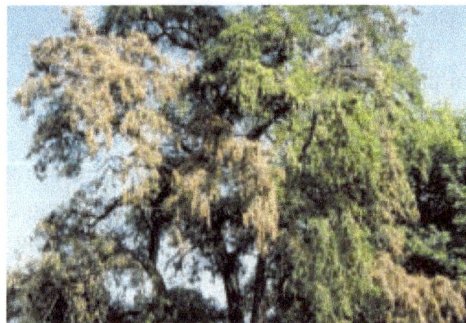

Figure 137. Crown wilt symptoms of Verticillium wilt. *Photo: USDA Forest Service Archive, Bugwood.org.*

Figure 138. Leaf wilt symptoms of oak wilt. *Photo: James T. Blodgett, USDA Forest Service.*

Figure 139. Crown wilt symptoms of oak wilt. *Photo: Joseph O'Brien, USDA Forest Service, Bugwood.org.*

organisms, often insects, that carry and help spread pathogens. Pathogens enter the xylem through wounds on the roots, stems, or branches. In some cases, wounds are caused by the associated insect vector. The pathogens then grow within the xylem, plugging vessels or tracheids and/or disrupting the integrity of the host cells resulting in the disruption of water transport.

**Figure 140.** Crown wilt symptoms of pine wilt disease. *Photo: James T. Blodgett, USDA Forest Service.*

**Impact**—Impacts include reduced growth, branch mortality, and tree mortality. Some infected trees can survive for many years, or mortality may be sudden. Mortality is common for many of the important wilt diseases in the region including Dutch elm disease, oak wilt, pine wilt disease, and black stain root disease.

**References: 140, 158**

**Figure 141.** Crown wilt symptoms of black stain root disease. *Photo: James T. Blodgett, USDA Forest Service.*

# Dutch Elm Disease

## A non-native invasive wilt

**Pathogen**—The fungus *Ophiostoma novo-ulmi* is the most aggressive pathogen involved, but other *Ophiostoma* species may occur within the Rocky Mountain Region. The pathogen was introduced into the United States in the 1930s and subsequently spread throughout the country.

**Vectors**—The native elm bark beetle (*Hylurgopinus rufipes*), smaller European elm bark beetle (*Scolytus multistriatus*), and potentially the banded elm bark beetle (*S. schevyrewi*) vector the pathogen. See the Elm Bark Beetles entry in this guide for more information.

**Hosts**—This disease affects trees in the elm family (Ulmaceae). Native elms are most susceptible.

**Signs and Symptoms—** Symptoms differ between trees that are infected through root grafts and those infected via beetles. Trees infected by root grafts wilt, their leaves turn brown, and the trees die rapidly, usually in the spring (fig. 142). When the disease is transmitted by bark beetles, the first symptoms are often yellowing and wilting of leaves on one or several branches. The leaves eventually turn brown and fall in mid to late summer (figs. 143-144). As the disease spreads to adjacent branches, additional branches die, and eventually tree mortality results. This often takes 1 or more years.

Slight symptom differences also occur among the beetle vectors. The smaller European elm bark beetle feeds in small twigs, usually high in the crown of mature trees. This results in the initial wilt symptoms on higher and smaller branches. In contrast, the native elm bark beetle bores into the bark of branches 2-4 inches (5-10 cm) in diameter to feed.

Brown streaks in the new sapwood are characteristic symptoms of infection (fig. 145). However, laboratory tests may be needed to confirm the presence of the pathogen because signs of the pathogen are usually microscopic. Sometimes the white spore-bearing bodies with tiny black stalks (synnemata, also known as coremia) of *Ophiostoma* may be visible in beetle galleries (fig. 146).

**Figure 142.** Trees killed by Dutch elm disease that spread through root grafts. *Photo: Edward L. Barnard, Florida Department of Agriculture and Consumer Services, Bugwood.org.*

**Figure 143.** Typical crown symptoms of Dutch elm disease associated with insect transmission. *Photo: Northeastern Area Archive, USDA Forest Service, Bugwood.org.*

**Disease Cycle—**The pathogen infects new trees when the bark beetle vectors feed on twigs or small branches of healthy trees; new trees are also infected throughout the growing season by root grafts. As the beetles feed, they introduce the pathogen spores attached to their body into the sapwood. The fungus quickly spreads in the vessels of the xylem and moves both up and, more importantly, down, eventually reaching the roots. As the disease spreads throughout the

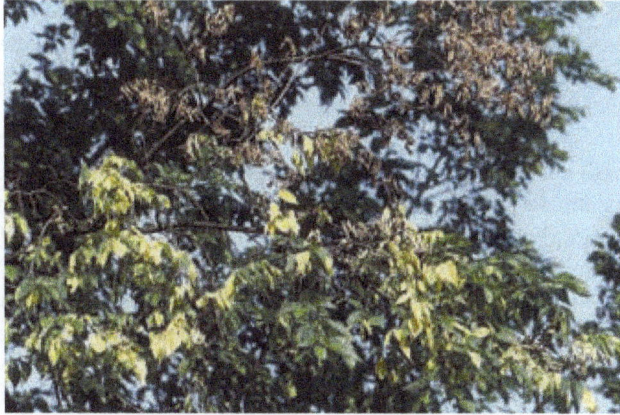

**Figure 144.** Close-up with yellow and brown leaves due to Dutch elm disease. *Photo: Joseph O'Brien, USDA Forest Service, Bugwood.org.*

tree, it disrupts water movement, causing trees to wilt. The fungus continues moving through the root system, infecting other adjacent trees through root grafts. As trees die, the beetles form breeding galleries in the recently dead or dying stems and branches of infected trees. The fungus sporulates in the galleries, and spores are picked up by the beetles in the galleries and are carried to a new, healthy host.

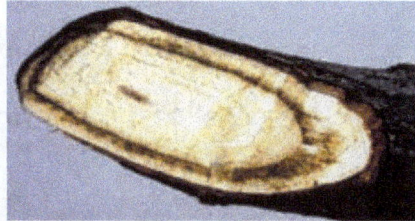

**Figure 145.** Brown streaks in the new sapwood caused by *Ophiostoma novo-ulmi*. *Photo: Petr Kapitola, State Phytosanitary Administration, Bugwood.org.*

**Impact**—Since its introduction in the 1930s, this disease has quickly spread throughout the range of elms and is now the most damaging disease of elm and the leading cause of elm mortality.

**Management**—Sanitation (removal of infected materials) can significantly reduce the disease spread of new infections. Sanitation efforts should focus on promptly removing infected trees and prohibiting the storage of elm wood with attached bark because beetles can infest cut wood and spread the disease. The fungus can persist in the sapwood for several years after trees die.

**Figure 146.** The pathogen *Ophiostoma novo-ulmi* sporulating along beetle gallery. *Photo: Joseph O'Brien, USDA Forest Service, Bugwood.org.*

Breaking root grafts is important to prevent root spread but may be difficult and expensive. Root graft disruption should be done before the infected trees are removed. Otherwise, the transpirational pull from the healthy trees will draw the pathogen from the diseased tree roots to the healthy tree.

A few varieties of American elm with some level of resistance are commercially available. However, many of the resistant elm varieties are susceptible to elm yellows. Planting resistant European or Asiatic elm species will reduce losses, but the growth forms are not the same as the American elm. Non-elm species adapted to the area are another option.

Preventive chemicals are available as tree injections that can protect healthy elms from the disease for up to 3 years. Because the chemicals are injected in the lower stems or upper roots and can only move upward in trees, these treatments are ineffective for root-graft infections. The treatments are also costly but might be appropriate for high-value trees. An appropriately registered fungicide for an individual state should be selected.

**References: 54, 107, 158**

# Oak Wilt

## Red oaks die quickly; white oaks may recover

**Pathogen**—Oak wilt is caused by the fungus *Ceratocystis fagacearum*. This disease occurs in Kansas, Nebraska, and South Dakota.

**Vectors**—Beetles in the Nitidulidae family and oak bark beetles (*Pseudopityophthorus* spp.) can transmit the fungus.

**Hosts**—Oak wilt is a disease of oak species. Species in the red oak group die more quickly than species in the white oak group.

**Signs and Symptoms**—The only visible signs are the black and gray fungal mats (fig. 147) that are common in infected red oaks but less common or less apparent in white oaks. The mats form beneath the bark soon after mortality. These mats occasionally raise and crack the bark (fig. 148). Other signs of the pathogen are microscopic. Laboratory tests and fungal culturing on agar media can be used to confirm the pathogen's presence.

For the red oak group, symptoms are often expressed in spring but can continue into the summer. Symptoms start from the tip and outer edges of leaves and move toward the midrib and base of leaves, often with a distinct margin (fig. 149). First, leaves turn dull green or bronze, can appear water-soaked, and wilt. Later, the leaves turn yellow and/or brown, curl around the midrib, and are shed at branch tips. Finally, both green and symptomatic leaves throughout the crown fall. Symptoms often develop quickly throughout the crown in red oak (fig. 150). Trees may die only 1 or 2 months after symptoms appear and seldom survive more than a year.

Disease symptoms are more variable for species in the white oak group. Symptoms can be the same as those of red oaks with quick mortality, but

Figure 147. Fungal mat of oak wilt disease. *Photo: Fred Baker, Utah State University, Bugwood.org.*

Figure 148. Cracked bark caused by fungal mat of oak wilt disease. *Photo: Joseph O'Brien, USDA Forest Service, Bugwood.org.*

Figure 149. Oak wilt symptoms on red oak leaves. *Photo: D. W. French, University of Minnesota, Bugwood.org.*

Figure 150. Typical crown symptoms of oak wilt on red oak. *Photo: Joseph O'Brien, USDA Forest Service, Bugwood.org.*

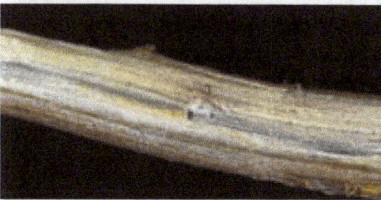

Figure 151. Black streaks in the new sapwood caused by oak wilt disease. *Photo: Fred Baker, Utah State University, Bugwood.org.*

typically, white oaks die slowly over several years, with only a few branches showing symptoms and dying per year. The leaves often remain attached with discoloration only at the margins. Other times, however, white oaks seem to recover. Brown to black discoloration commonly develops in the outer sapwood of infected white oaks (fig. 151). This symptom is less common in infected red oaks.

**Disease Cycle**—Root grafts and insect vectors spread the oak wilt fungus. Root grafts are an effective means of transmitting the fungus from infected to nearby healthy trees. These grafts are a major factor in local spread of the oak wilt pathogen, especially in areas with deep, sandy soils and where oaks grow close together (fig. 152).

**Figure 152.** Oak wilt spread by root grafts. *Photo: D. W. French, University of Minnesota, Bugwood.org.*

The fungal mats can enlarge and crack the bark (fig. 153). These aromatic mats attract insects such as sap-feeding Nitidulidae beetles and the fungal spores adhere to their bodies as they crawl over the mats (fig. 154). The beetles carry the spores to wounds on healthy oaks throughout the summer.

Oak bark beetles, *Pseudopityophthorus* spp., can also transmit the fungus (fig. 155). The beetles form breeding galleries in recently dead or dying oak stems and branches, including wilt-infected infected trees. The adults carry spores on their bodies when they emerge the following spring and infect healthy trees as they feed in spring and early summer.

**Figure 153.** Older fungal mat with bark beetle galleries. *Photo: C. E. Seliskar, North Carolina State University, Bugwood.org.*

**Figure 154.** Adult Nitidulidae beetle. *Photo: Pennsylvania Department of Conservation and Natural Resources, Forestry Archive, Bugwood.org.*

**Figure 155.** Adult oak bark beetle. *Photo: J. R. Baker & S. B. Bambara, North Carolina State University, Bugwood.org.*

This disease is a vascular wilt. Trees respond to infection by forming tyloses (where living parenchyma cells balloon into vessels as a defense response) that restrict water flow in the vessels. The pathogen spreads rapidly within vessels of red oaks and often more slowly in white oaks. The fungus overwinters as mycelium in infested trees and as fungal mats on dead trees.

**Impact**—The disease quickly kills red oaks and can cause reduced growth and mortality of white oaks. The disease range is expanding, and, in some areas, the incidence of disease is increasing.

**Management**—Sanitation (removal of infected materials) can significantly reduce new infections. If the mats or beetle galleries are present, trees should be cut, and the wood should be burned, buried, or chipped. The fungus can persist in cut wood with attached bark and in dead trees. After trees die, fungal mats continue to form and may attract beetles. Trees should not be pruned in spring or early summer when beetles are most active.

Mechanical and chemical barriers and severing root grafts between diseased and healthy trees are effective ways to prevent the spread of oak wilt through root grafts. New root grafts will not form between dead/dying trees and healthy trees.
**References: 5, 47, 136, 173**

# Pine Wilt

## Rapid wilting and death of pine

**Pathogen**—The pine wood nematode, *Bursaphelenchus xylophilus*, causes pine wilt disease. Nematodes are "roundworms" in the phylum Nematoda, which has over 80,000 described species. This disease can be a problem wherever non-native pines are planted but is most common in Kansas, Nebraska, and South Dakota.

**Vectors**—The pine sawyer beetles, *Monochamus* spp., transmit the nematode. Please see the Roundheaded Wood Borers (Longhorned Beetles) entry in this guide for more information.

**Hosts**—Scots, Austrian, and other non-native pines are often killed by this disease. Eastern white pine, a native pine, is also affected and may be killed by pine wilt disease. The nematode commonly infects other native pines and some native conifer species. However, most native species are resistant to the disease (e.g., native conifers may be infected and express little or no disease symptoms).

**Signs and Symptoms**—As with many wilts, signs are microscopic. The pine wood nematode is relatively large compared with other nematodes, but it cannot be seen with a hand-lens in infected wood. Laboratory tests are required to confirm its presence.

Pine wilt disease causes rapid wilting and death on non-native pines. Symptoms are often first expressed in early summer but can occur throughout the growing season. Symptoms may first appear on one or a few branches but

Figure 156. Symptoms of pine wilt disease on Austrian pine branch. *Photo: North Central Research Station Archive, USDA Forest Service, Bugwood.org.*

Figure 157. Crown symptoms of pine wilt disease on Scots pine. *Photo: James T. Blodgett, USDA Forest Service.*

often develop quickly throughout the crown, and trees may die only 1 or 2 months after symptoms appear. Trees seldom survive more than a year. Symptoms start with needle discoloration, which progresses rapidly from a grayish green to yellow and then to brown. Needles can but often do not show wilt-type symptoms, and needles are usually retained for a few months (figs. 156-157).

**Disease Cycle**—The nematode is introduced into pines as the pine sawyer beetles feed (fig. 158) or lay eggs. Eggs are laid in egg niches chewed by the females. When the beetle feeds on a healthy tree or chews the egg niches, the nematodes leave the beetle and enter the tree through the wounds. The nematodes reproduce rapidly in resin canals and go

Figure 158. Nematodes can be transmitted to pines when the adult pine sawyer beetle feeds on shoots. *Photo: L. D. Dwinell, USDA Forest Service, Bugwood.org.*

from egg to adult in 5 days with many eggs per nematode. They then spread rapidly within xylem tracheids and clog the water transport system of the pines, thus causing disruption of water movement throughout the tree (i.e., a vascular wilt disease). Nematodes can overwinter in dead or living pine.

Nematodes feed on fungi in the wood, including bluestain fungi (fig. 159) that are transmitted by engraver and other bark beetles. The nematodes can survive and reproduce with fungi in dead, stressed, and living pine. Therefore, this pathogen can be found on trees killed by other causes.

Nematodes can overwinter in a dormant state. As wood dries, they molt and enter a dormant phase high in lipids and do not feed. This phase is resistant to unfavorable environmental conditions.

The pine sawyer beetles lay eggs in dying or recently dead trees in spring. The eggs hatch, the beetle larvae bore into the wood to feed, and the beetle overwinters as larvae or pupae. Before emerging as young adults, the beetles acquire the nematodes. These beetles transmit nematodes to healthy trees.

**Impact**—The disease quickly kills exotic pines (especially when grown on poor sites with dry, shallow soils) but usually has little effect on native pines.

**Figure 159.** Bluestain fungal mycelium in pine wood is a source of food for the nematode. *Photo: North Central Research Station Archive, USDA Forest Service, Bugwood.org.*

**Management**—Sanitation (removal of infected materials) can reduce new infections. Infected trees should be cut and the wood should be burned, buried, or chipped. The nematode can survive for a time in cut wood that can attract beetles, which results in subsequent disease spread. Infected wood should not be kept for firewood without removing the bark. Anything that reduces pine sawyer beetle attraction and breeding success will reduce losses. This includes reducing stresses from other diseases, insects, or the environment.

Planting native tree species suitable to the site will reduce losses. Although they may be infected, native conifers seldom develop pine wilt disease symptoms. Even native pines planted off-site rarely develop this wilt disease. If exotic pines are desired, they should only be grown on favorable sites.

**References: 106, 117, 187**

# Verticillium Wilt

## Vascular wilt of hardwoods

**Pathogen**—Verticillium wilt is caused by two closely related species of fungi, *Verticillium albo-atrum* and *V. dahliae*.

**Hosts**—Verticillium wilt is a vascular wilt of hardwoods. Over 300 plant species are affected by Verticillium wilt. The disease is particularly destructive to trees in landscape plantings. Ash, catalpa, elm, sumac, and maple are the most common hosts in the Rocky Mountain Region.

**Signs and Symptoms**—Typical symptoms include chlorosis, wilting, marginal and interveinal necrosis of the foliage, branch dieback, and mortality (fig. 160). Symptoms vary among hosts and are not always completely diagnostic. Symptomatic wilting is most obvious on warm, sunny days. The disease often progresses from the lower crown upward, but it is not unusual for only a branch or portion of the tree to be impacted.

The sapwood may exhibit dark streaks or bands along the grain (fig. 161). The color of the streaking depends on the host and may be shades of green, red, brown, or black.

**Figure 160.** Verticillium wilt on silver maple. *Photo: William Jacobi, Colorado State University, Bugwood.org.*

**Figure 161.** Discolored streaks in the wood of a cherry branch infected by Verticillium wilt. *Photo: William Jacobi, Colorado State University, Bugwood.org.*

**Disease Cycle**—The fungi that cause Verticillium wilt are soil-borne and gain entry through roots or wounds near the ground. Once inside the host, they invade the xylem, which disrupts water transport and physiological function. Spread can occur throughout the plant by spores transported in the sap stream or by vegetative growth of mycelium. Plant-to-plant spread occurs by means of asexual spores (conidia) and by resting spore structures (microsclerotia) that allow the fungi to persist for long periods of time in the soil independent of any host material. Long-distance dispersal is a serious problem because microsclerotia can be carried in soil, bare roots, root balls, or equipment within nurseries to planting sites.

**Impact**—Small trees are very susceptible and may be killed rapidly by Verticillium wilt. Older trees generally deteriorate over time and can survive for several or many years with the disease. Damage is greatest in nurseries and landscape plantings. Stressed trees are more susceptible to Verticillium wilt and sustain more damage.

**Management**—The following control strategies can be used to reduce impacts of Verticillium wilt:
- Do not plant susceptible hosts in areas where the disease is present.
- Maintain tree vigor by fertilizing with "balanced" fertilizers (10-10-10 [N-P-K]) and watering. High-nitrogen fertilizers may increase damage.
- Prune back branches beyond any streaking in the wood. Destroy infected branches and sterilize pruning tools.
- Avoid damage to the roots and base of trees.
- Plant resistant or tolerant species.

**References: 140, 158**

# Introduction to Rust Diseases

## Specialized foliage and/or canker diseases often with alternate hosts

Rust diseases are grouped based on the taxonomic classification of the pathogens. They are fungi in the phylum Basidiomycota, class Pucciniomycetes, and order Pucciniales. Rusts are all obligate parasites that depend on the living cells of their hosts. Initially, they produce structures (haustoria) that grow into their hosts' living cells, from which they derive nutrients.

The diseases caused by this group are varied. Most of the rust fungi initially infect and cause diseases of foliage. However, some spread into branches and stems, colonizing the phloem and cambium. Rusts cause cankers (diseases of the bark), galls, brooms, foliage diseases, and/or cone diseases.

The majority of the rust diseases in this Region are native pathogens, with the exception of white pine blister rust. Table 9 provides examples of some of the most damaging and/or common rusts in the region (figs. 162-168). However, many other rusts occur within the Rocky Mountain Region.

**Signs and Symptoms**—The common name, "rusts," comes from the rust-like appearance of the various spore structures that are often rust-colored, yellow, and/or orange. These brightly colored signs help to distinguish rusts from other diseases. If signs are not present, symptoms may look like other foliage or canker diseases, with discolored foliage or branch flagging (faded and dying branches). Stem rusts often can be identified by the host and shape of the cankers or galls. The cankers usually have distinctive roughness on their margin that are the remains of earlier fruiting structures.

**Disease Cycle**—Rust disease cycles are complicated. Many require two different host species (table 9) that usually need to be less than a mile apart. The alternate hosts to the tree hosts can be other trees, shrubs, or herbaceous plants. The hosts are often specific species or groups within the same genus. Rust fungi may have as many as five different spore types or stages (table 9). White pine blister rust is an example of a five-stage (macrocyclic) rust with two hosts (fig. 162). There are rust diseases, such as western gall rust (*Peridermium harknessii*), that require only one host to complete their life cycle.

Spores are disseminated by wind and often infect hosts through needle stomata. However, some rusts can infect hosts through wounds. Prolonged periods of cool, wet weather or high humidity are necessary for infection and germination. Therefore, "wave years," or years with significant increases in new infections, occur during years of high rain and increased humidity.

Many rust pathogens have spore types called urediniospores. These spores permit the rusts to spread and only re-infect an alternate host group. This allows for increased spore levels (increased inoculum) during the summer.

**Impact**—Impacts include reduced growth, stem deformities, loss of wood quality, dead tops and branches, and tree mortality. Stem cankers, especially

**Table 9.** Comparison of the more damaging and common rusts in the Rocky Mountain Region.

| Rust | Scientific name | Common tree hosts | Common alternate hosts | Number of spore types | Main damages on tree host |
|------|-----------------|-------------------|------------------------|-----------------------|---------------------------|
| White pine blister rust (Fig. 163) | *Cronartium ribicola* | Five-needle pines | Currants and gooseberries | 5 | Stem and branch cankers |
| Comandra blister rust (Fig. 164) | *C. comandrae* | Lodgepole and ponderosa pine | Bastard toadflax | 5 | Stem and branch cankers |
| Stalactiform rust | *C. coleosporioides* | Lodgepole pine | Indian paintbrush | 5 | Stem and branch cankers |
| Western gall rust (Fig. 165) | *Peridermium harknessii* | Ponderosa, lodgepole, Scots, and pine | None | 1 | Branch galls and stem cankers |
| Spruce broom rust (Fig. 166) | *Chrysomyxa arctostaphyli* | Engelmann and Colorado blue spruce | Bearberry | 4 | Branch brooms |
| Fir broom rust (Fig. 167) | *Melampsorella caryophyllacearum* | White and subalpine fir | Chickweed | 5 | Branch brooms |
| Melampsora leaf rust | *Melampsora medusae*, other *Melampsora* spp. | Aspen, cottonwood, and willow | Douglas-fir and ponderosa and lodgepole pine | 5 | Foliage disease |
| Gymnosporangium (Fig. 168) | *Gymnosporangium* spp. | Junipers | Species in family Rosaceae | 4 | Branch galls, stem swelling, and brooms |

## Pine | Ribes

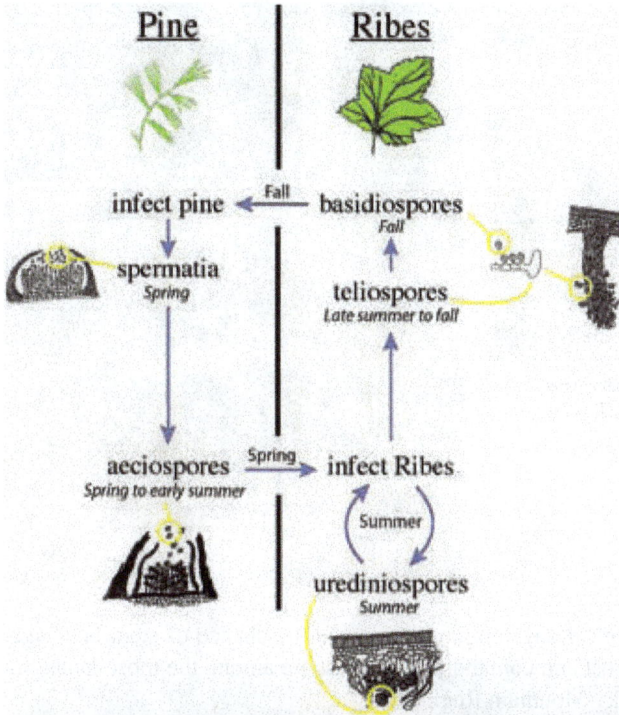

Pine

Ribes

infect pine ←—Fall— basidiospores
*Fall*

spermatia
*Spring*

teliospores
*Late summer to fall*

aeciospores —Spring→ infect Ribes
*Spring to early summer*

Summer

urediniospores
*Summer*

**Figure 162.** White pine blister rust disease cycle as an example of a five-cycle rust. *Image: Jim Worrall, USDA Forest Service.*

**Figure 163.** White pine blister rust. *Photo: James T. Blodgett, USDA Forest Service.*

**Figure 164.** Comandra blister rust. *Photo: James T. Blodgett, USDA Forest Service.*

**Figure 165.** Western gall rust. *Photo: James T. Blodgett, USDA Forest Service.*

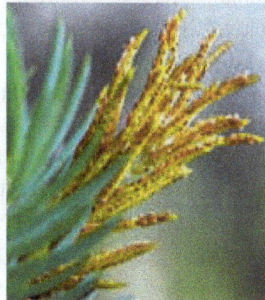

**Figure 166.** Close-up view of spruce broom rust fruiting on needles. *Photo: James T. Blodgett, USDA Forest Service.*

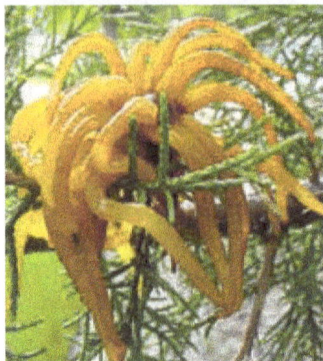

**Figure 167.** Gymnosporangium rust. *Photo: James T. Blodgett, USDA Forest Service.*

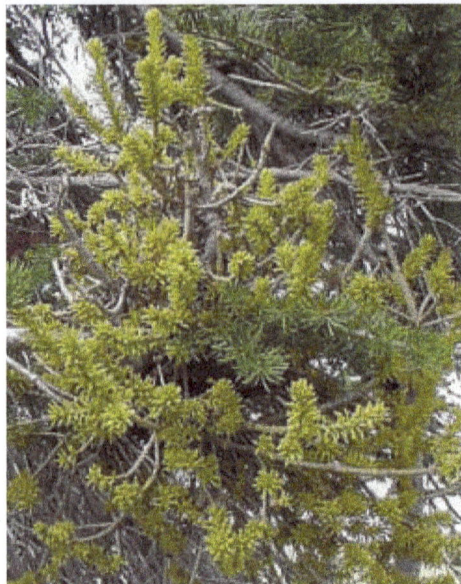

**Figure 168.** Fir broom rust. *Photo: James T. Blodgett, USDA Forest Service.*

when associated with decay fungi, can increase the chance of stem breakage. White pine blister rust and comandra blister rust are among the most damaging diseases in the Rocky Mountain Region.

**References: 1, 6, 158, 201**

# Broom Rusts of Spruce and Fir

## Dense, pale, or dead-looking brooms

**Pathogen**—Fir broom rust is caused by the fungus *Melampsorella caryophyllacearum*.

Spruce broom rust is caused by the fungus *Chrysomyxa arctostaphyli*.

**Hosts**—Many true fir species are susceptible to fir broom rust, including white fir and subalpine fir in the Rocky Mountain Region. Alternate hosts are chickweeds (*Cerastium* and *Stellaria* spp.) (fig. 169).

Spruce broom rust primarily affects Engelmann spruce and Colorado blue spruce in the Rocky Mountain Region. The primary alternate host is bearberry or kinnikinnick (*Arctostaphylos uva-ursi*) (fig. 170), but manzanitas (*Arctostaphylos* spp.) are occasional

**Figure 169.** Chickweed is the alternate host for fir broom rust. *Photo: Karan A. Rawlins, University of Georgia, Bugwood.org.*

**Figure 170.** The alternate host for spruce broom rust is kinnikinnick. *Photo: Dave Powell, USDA Forest Service, Bugwood.org.*

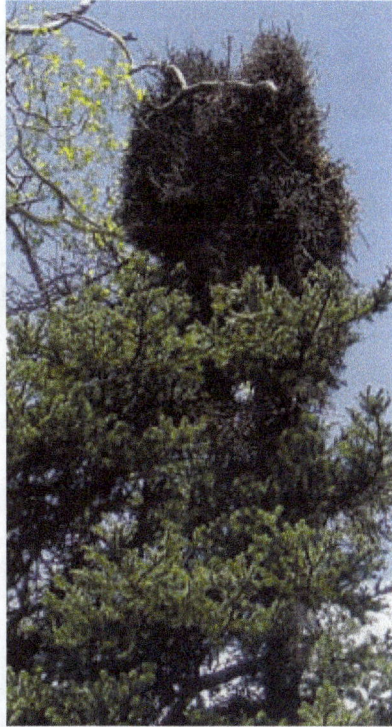

**Figure 171.** Rust broom on Engelmann spruce after infected needles produced that year died. Note that the tree top has failed. *Photo: Kelly S. Burns, USDA Forest Service.*

alternate hosts on the Uncompahgre National Forest.

**Signs and Symptoms**—These diseases are easily identified by the dense proliferation of branches (witches' brooms) on spruce and fir (fig. 171). Infected needles are chlorotic (pale), short, and thick and are shed each fall, making the broom appear dead over the winter, but new chlorotic foliage is produced within the broom each spring. Spermogonia form on the underside of needles between the cuticle and epidermis in the spring and have a strong, characteristic odor. Aecia appear in the summer. Brooms are most obvious at this time because the chlorotic foliage is covered with bright orange aecia, which contrast to adjacent healthy foliage (figs. 172-173). Witches' brooms commonly lead to cankers, dead tops and branches, broken tops and branches, and mortality. Rust

**Figure 172.** Close-up of infected spruce needles sporulating during summer. *Photo: William M. Ciesla, Forest Health Management International, Bugwood.org.*

brooms are sometimes confused with dwarf mistletoe brooms. However, rust brooms are denser, yellow, and lack mistletoe shoots. Dwarf mistletoes do not occur in true firs and spruce in the Rocky Mountain Region. Both pathogens cause a leaf spot and shoot blight on the alternate hosts.

**Disease Cycle**—Both pathogens require an alternate host to complete their life cycles. Their life cycles are similar to one another and to other macrocyclic rust fungi, except *C. arctostaphyli* does not produce uredinia. Wind-blown spores (basidiospores) produced on the alternate host are needed to start new infections on trees. Infection

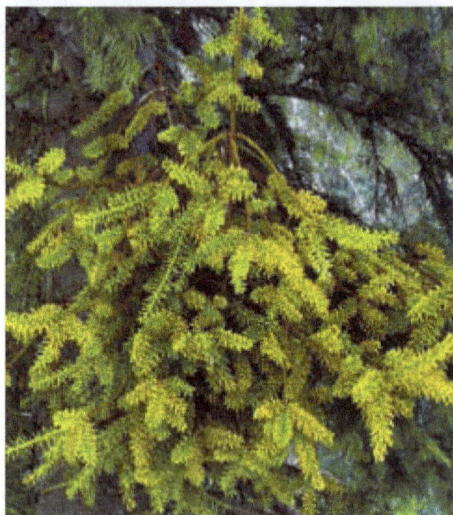

**Figure 173.** Rust broom on subalpine fir sporulating in midsummer. *Photo: Susan K. Hagle, USDA Forest Service, Bugwood.org.*

occurs on newly emerging twig and bud tissue. Once a tree is infected, the fungus stimulates bud formation, leading to broom development. Spermogonia form on the underside of needles in the spring; their strong, fetid odor attracts insects that cross-fertilize the fungus. Bright yellow-orange pustules (aecia) of spores (aeciospores) rupture through the leaf surface in early summer. Aeciospores disseminate in the wind to infect the alternate host. Moist conditions are conducive to infection.

**Impact**—Broom rusts may cause stem cankers and deformations, growth loss, top-kill, and tree mortality. Trees weakened by broom rust may be more susceptible to other insects and diseases. A few brooms on a tree can result in more than 20% reduction in diameter and height growth. Rust brooms also serve as infection courts for decay fungi such as *Porodaedalea* (= *Phellinus*) *pini.*

Rust brooms are especially damaging when they occur near the stem. Stem breakage may occur at the point of infection, creating hazards in recreation areas. Portions of the San Juan and Rio Grande National Forests have high levels of infection (up to 23-29%) of spruce. Fir broom rust is also common on the San Juan National Forest, with some sites having 37-42% infection. However, average incidence in Colorado is 4.2% for spruce and 2.3% for fir.

The incidence of broom rust is associated with the distribution and abundance of hosts, microclimatic conditions, and host susceptibility. Mature stands are likely more susceptible simply because there is more susceptible target area, but damage is generally greater on small trees because their needles are so close to the main stem.

**Management**—Trees with stem cankers or brooms can be selectively removed during stand treatments. Pruning out broom rust infections may reduce the risk of stem breakage and maintain tree vigor. Trees with dead tops and adjacent dead brooms are often heavily decayed and should be considered hazardous in recreation areas. Consider the following management recommendations (ref. 91).

Tree removal priorities for commercial and precommercial entries:
- Remove trees with spike tops or broken tops while maintaining sufficient numbers of snags.
- Remove all infected trees that have symptoms of bark beetle attack, decay, and/or root disease.
- Remove trees with one or more bole cankers that girdle more than 33% of the stem circumference.
- Remove trees with one or more dead brooms and two or more live brooms within 1 ft (30 cm) of the main stem.
- Remove all infected trees while maintaining adequate stocking.

On developed sites:
- Remove all dead trees, hazardous trees, and hazardous branches.
- Maintain as many live trees as possible.

**References: 8, 91, 132, 201**

# Comandra and Stalactiform Blister Rusts

## Branch and stem rusts of lodgepole and ponderosa pines

**Pathogen**—The rust fungi *Cronartium comandrae* and *C. coleosporioides* cause comandra and stalactiform blister rusts, respectively.

**Hosts**—Lodgepole and ponderosa pine are hosts of both rusts, but stalactiform is uncommon on ponderosa pine. The alternate hosts also differ (figs. 174-175). The alternate hosts for comandra blister rust are bastard toadflax (*Comandra*

**Figure 174.** The alternate host for comandra blister rust is bastard toadflax. *Photo: William Jacobi, Colorado State University, Bugwood.org.*

**Figure 175.** Principal alternate hosts for stalactiform blister rust are species of Indian paintbrush. *Photo: Andrew Kratz, USDA Forest Service.*

*umbellata*) and northern comandra (*C. livida*). The alternate hosts for stalactiform rust include several broadleaved plants but mainly species of Indian paintbrush (*Castilleja* spp.). Comandra rust is especially severe in Wyoming and areas of northern Colorado. Stalactiform rust is less common in the Region.

**Signs and Symptoms—** Symptoms on pines include branch flagging, rough branch cankers, elongate stem cankers, top-kill (spike tops), and tree mortality (figs. 176-177). Rodents may feed on branch and stem canker margins, and damage is often accompanied with heavy resin flow. Top-kill is much more common with comandra rust.

**Figure 176.** Comandra blister rust canker on an older stem. *Photo: James T. Blodgett, USDA Forest Service.*

Stem cankers initially have rough bark and heavy resin flow. With time, stem cankers slough the dead bark, revealing perennial stem cankers with concentric ridges of resinous sapwood and dead cambium that result from the annual canker growth. Stem cankers caused by the rusts are similar in appearance, but comandra rust cankers tend to be one and one-half to four times longer than wide, while stalactiform rust cankers tend to be eight to many times longer than wide.

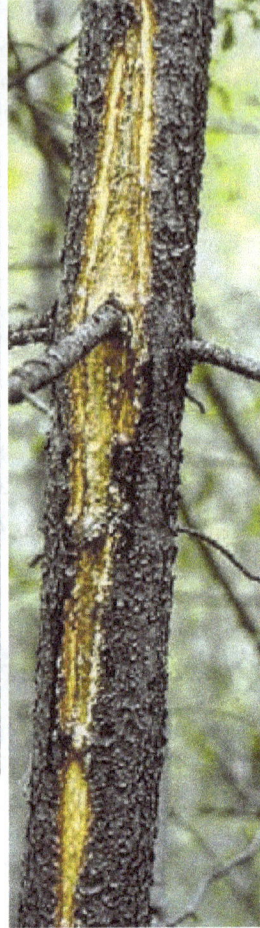

**Figure 177.** Stalactiform blister rust canker on an older stem. *Photo: John Guyon, USDA Forest Service, Bugwood.org.*

For pines, the signs include light yellow to orange pustules (aecia) of spores (aeciospores) produced at canker margins in late spring and early summer (figs. 178-179). Cankers caused by comandra blister rust tend to be yellow-colored and cause top-kill after girdling the stem. The fungi are best differentiated by microscopic examination of spores.

Symptoms on the alternate hosts can include tan to brown leaf spots. However, the signs may be present with few symptoms. Signs, visible on the underside of leaves, include orange-colored fruiting bodies (uredinia) with spores (urediniospores), and later in the year, rust-brown, hair-like structures (telia) form.

USDA Forest Service RMRS-GTR-241. 2010.

**Disease Cycle—** Both rusts require two different hosts to complete their life cycle. The incidences of these diseases are correlated with presence of their alternate hosts. Studies have shown a high incidence of comandra blister rust in pine stands near sagebrush. Bastard toadflax is an obligate parasite of sagebrush and many other woody perennial plants.

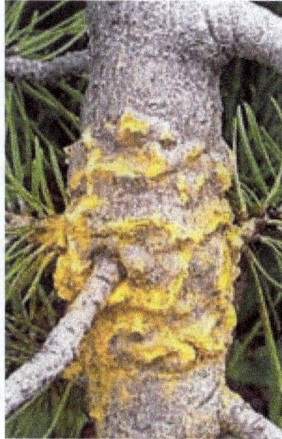

**Figure 178.** Comandra blister rust sporulating on a young stem. *Photo: James T. Blodgett, USDA Forest Service.*

**Figure 179.** Stalactiform blister rust sporulating on a young stem. *Photo: USDA Forest Service Archive, Bugwood.org.*

Both rusts have five spore stages. Three spore stages (uredinia, telia, and basidia) develop on the alternate hosts, and two spore stages (spermogonia and aecia) develop on the pines.

Basidiospores are wind-blown from the alternate hosts and infect pine needles and new shoots. The basidiospores cannot infect the alternate hosts. Fungal hyphae grow through the phloem and cambium of branches, forming perennial cankers that can spread to the stem. Spermogonia develop within the canker in the spring and early summer and aecia are produced in the same tissue the following spring and early summer. The aeciospores are wind-disseminated. These spores can only infect the alternate host's leaves. The urediniospores produced on the underside of leaves re-infect the alternate host leaves during the summer but cannot infect pines. Basidiospores develop on the telia in late summer or fall for comandra rust or late spring for stalactiform rust and infect the pines. Long periods of high humidity are required for infection of the pine hosts. Therefore, rust epidemics often involve a burst of infections during a year with a long, moist season.

**Impact—**Comandra rust is one of the more important diseases of lodgepole pine in the region. Both rusts can cause stem deformities, growth reduction, and cankers that girdle branches or stems, resulting in top-kill or tree mortality. Stem girdling and top-kill are far more frequent with comandra rust. Trees may survive several decades with spiked tops, and at times, a lower branch will assume terminal dominance and produce a new top. However, the cankers can continue to grow down the stem, killing the new top. Infection is occasionally heavy, causing high volume losses in stands.

These rusts affect the form, lumber quality, and growth rate of trees. Although these fungi can kill individual trees, they do not kill whole stands. On large trees, stem cankers often result in non-merchantability of infected logs. Infected seedlings and young trees are frequently killed. Because fruiting bodies are sparse on old stem cankers and rodents (especially porcupines and squirrels) chew canker margins, the damage is often attributed entirely to other causes.

**Management**—Removing infected trees during partial cuts, using disease-free trees as leave trees during seed cuts, pruning infected branches, and planting non-host species are ways to reduce losses.
**References: 1, 72, 95, 175, 201**

# Gymnosporangium Rusts

## Commonly called juniper rust

**Pathogen**—Rust fungi in the genus *Gymnosporangium* cause these diseases, commonly called juniper rust. There are at least nine (and likely more) *Gymnosporangium* species in the Rocky Mountain Region.

Cedar-apple rust, *G. juniperi-virginianae*, is often incorrectly credited for all rusts in this group and is only common in the eastern part of the Rocky Mountain Region, in the Great Plains.

**Hosts**—The Gymnosporangium rusts require two hosts to complete their disease cycle. In this Region, one spore stage (telial stage) occurs on juniper species, and the other spore stage (aecial stage) occurs on rosaceous species. Different Gymnosporangium rusts can have different hosts.

Various juniper species are affected by these diseases, including Rocky Mountain juniper, common juniper, and eastern redcedar. *Rosaceous* hosts in the Region include apple, crabapple, hawthorn, Juneberry, and serviceberry. Cedar-apple rust is found on apple and crabapple along with juniper host species.

**Signs and Symptoms**—On junipers, the telial stage looks like exploding, orange "Jell-O" masses on branches (figs. 180-181), stems, or needles. These telial horns are finger-like, bright orange, gelatinous, and emerge from the areas where symptoms occur. On rosaceous species, the aecial stage is the most visible sign. During this stage,

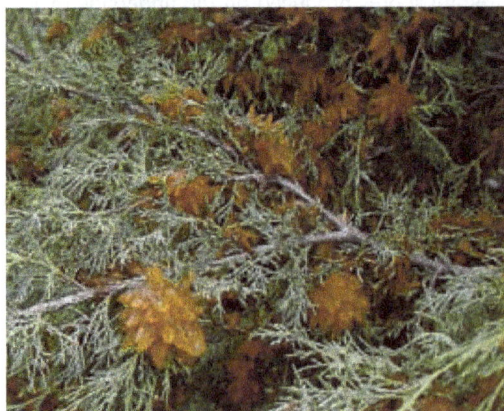

**Figure 180.** Sign of Gymnosporangium rust: fresh telial horns on branches. *Photo: Rocky Mountain Region, USDA Forest Service.*

USDA Forest Service RMRS-GTR-241. 2010.

yellow to orange fungal tissue can be seen within the lesions and spots (fig. 182). These areas first contain small, black, pimple-like fruiting bodies (spermogonia). Later, in the same area, tube-like, orange fruiting structures (aceia) form.

**Figure 182.** Signs of Gymnosporangium rust: pycnia and aecia on leaves. *Photo: Clemson University, USDA Cooperative Extension, Bugwood.org.*

On junipers, symptoms include branch galls, branch knots, brooms, stem swelling, and small needle lesions

**Figure 181.** Close-up of dried telial horns. *Photo: Petr Kapitola, State Phytosanitary Administration, Bugwood.org.*

(figs. 183-184). The type of symptom depends on the rust species involved. On rosaceous species, symptoms are lesions and circular spots on leaves, fruit, petioles, or young twigs.

**Disease Cycle**—These rusts have four spore types or spore stages, and two hosts are required for the pathogens to complete their disease cycle.

Usually, after a heavy spring rain, telial horns extrude from the branch galls, branch knots, brooms, stem swelling, or needle lesions on the juniper host. Teliospores in the gelatinous horns produce basidiospores that are wind-blown

**Figure 183.** Symptom of Gymnosporangium rust: branch galls. *Photo: James T. Blodgett, USDA Forest Service.*

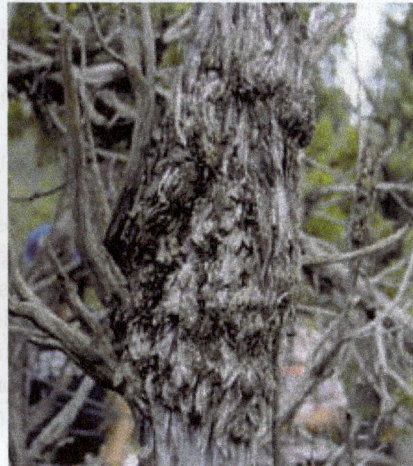

**Figure 184.** Symptom of Gymnosporangium rust: stem swelling. *Photo: William Jacobi, Colorado State University, Bugwood.org.*

to rosaceous hosts. These basidiospores do not survive long, so the distance of spread is generally short. Telial horns dry and re-wet rapidly during spring.

If infection is successful in the rosaceous host, spermogonia develop in orange lesions or leaf spots, often on the upper surface of a leaf. Aecia also develop in orange lesions or leaf spots, but often on the lower surface of the leaf. These two rust stages can also develop on fruit, petioles, or young twigs; the locations where fruiting occurs vary by the rust species and hosts. Aeciospores (spores produced in the aecia) are wind-blown to the juniper host the same year from early summer to fall, depending on the rust species.

If infection is successful in the juniper host, galls, witches' brooms, or stem swellings form. The telial horns grow from these symptomatic tissues the next spring or a year after the next spring.

**Impact**—The biggest impacts from these rusts are on the rosaceous hosts. Damages include growth loss and degraded fruit quantity and quality. Numerous infections, which can be common in wet years, can reduce rosaceous host vitality and result in attacks by other diseases or insects.

Although the strange appearance and bright color of the telial horns cause much concern, damage is usually minor. The aesthetic value is only impacted for about a week and soon fades. Rarely, numerous infections may reduce juniper host vigor, resulting in attacks by other diseases or insects. The *Gymnosporangium* species that cause stem swelling and occasionally the ones that cause branch knots can kill the tree host. The rust species that cause brooms can deform branches but seldom stems.

**Management**—It may be possible to reduce new infections in landscape settings by eliminating one of the hosts from an area. Rust infections will be minimized with a 2-mile separation of the hosts. Because basidiospores (produced on the juniper host) are more delicate and short-lived, eliminating juniper at a shorter distance will reduce new infections on rosaceous hosts.

Some varieties of both host groups are resistant to some of these rusts and can be used in areas where these rusts are a problem. Gymnosporangium rusts may also be host-specialized. For example, planting apple near juniper may not be a problem if the dominant rust in the area has hawthorn or Juneberry as its alternate host. However, host resistance has not been determined for some Gymnosporangium rusts.

Pruning the affected parts of the juniper can reduce losses, especially for rust species that cause stem swelling, branch knots, or brooms. Preventive fungicides are registered to control some of these rusts on both hosts. However, these treatments are costly and are usually not needed. If used, an appropriately registered fungicide for the individual state should be selected. Depending on the rust species, fungicides should be applied in spring to the rosaceous host and to the juniper host when aecia form on their rosaceous host.

**References: 74, 104**

# Western Gall Rust

## Pine-to-pine rust of branches and stems

**Pathogen**—Western gall rust is caused by the fungus *Peridermium* (*Endocronartium*) *harknessii*.

**Hosts**—Western gall rust is a pine-to-pine rust with no alternate host. Most two and three-needle pines in the West are susceptible, but the disease is especially common on lodgepole and ponderosa pine.

**Signs and Symptoms**—The disease causes round to pear-shaped woody swellings (galls) on branches or stems (fig. 185). Galls and cankers are most obvious in the spring and early summer when sporulation occurs. Flared target cankers, called "hip" cankers, are very common on larger lodgepole pine stems in the Rocky Mountain Region (fig. 186).

**Figure 185.** Branch gall sporulating on lodgepole pine. *Photo: James T. Blodgett, USDA Forest Service, Bugwood.org.*

**Disease Cycle**—*Peridermium harknessii* does not require an alternate host to complete its life cycle but spreads directly from pine to pine. White to orange pustules (aecia) full of yellow-orange spores (aeciospores) form in bark cracks on galls or, less commonly, at the edges of stem (hip) cankers in spring. Aecia rupture during moist weather and release spores that disperse in the wind. Most infections occur on the current year's shoots or needles, which are highly susceptible until they reach 90% elongation. Infection rarely occurs through wounds on older shoots. Trunk infection occurs through smaller side branches or from branch galls near the main stem (figs. 187-188). Galls are produced at the point of infection and sporulation typically occurs the second or third year. Host cambial cells are stimulated by the pathogen to divide rapidly, causing spherical gall formation.

**Figure 186.** Hip canker on lodgepole pine. *Photo: William Jacobi, Colorado State University, Bugwood.org.*

**Impact**—Western gall rust affects trees of all ages, causing growth loss, branch death, and deformity. Mortality is most common in seedlings and saplings because galls can quickly girdle the small stem. Branch galls typically only live a few years until the branch and the gall die. Mortality may result when numerous branch galls occur throughout the crown.

Hip cankers can severely deform larger trees, and wind snapping is common. They may persist for many years (100-200 years), but mortality is rare because they usually expand faster in a vertical direction. Stem cankers have very little impact on growth but can greatly reduce merchantable volume.

Mass infection tends to occur in wave years when conditions are particularly favorable. In the Rocky Mountain Region, wave years are somewhat rare, occuring approximately every 5-15 years. Vigorous trees are more prone to infection during this time because they have a larger proportion of susceptible foliage.

**Figure 187.** Stem canker that was initiated by a branch gall severely deforms the stem at the point of the canker. *Photo: James T. Blodgett, USDA Forest Service.*

**Figure 188.** Stem gall on ponderosa pine. *Photo: James T. Blodgett, USDA Forest Service.*

**Management**—Management of western gall rust is complicated because of the lag time between infection and symptom development. The following options may be useful for reducing disease impacts, but complete sanitation is difficult.

• *Sanitation.* This option involves removing all trees with stem infections and carefully selecting leave trees that are disease-free or only have branch galls or stem cankers high in the crown. Trees with stem cankers can be hazardous in recreation areas and should be given priority for removal.

• *Pruning.* Pruning infected branches provides little benefit to the tree because branches with galls usually die anyway. However, pruning may reduce inoculum levels in some areas.

• *Prepare for disease losses.* Regenerate stands at increased stocking levels to compensate for future rust-caused mortality.

• *Manage species.* Plant non-host species that are adapted to the site.

• *Destroy and regenerate.* It may be necessary to start over in areas where infection is severe and managing species is not an option.

**References: 122, 130, 133, 177**

# White Pine Blister Rust

## Non-native, invasive rust of five-needle pines

**Pathogen**—White pine blister rust is caused by *Cronartium ribicola*, an Asian fungus that was introduced into North America from Europe in the early 1900s. The disease continues to spread to five-needle pines throughout North America.

**Hosts**—All North American white pines (members of subgenus *Strobus*) are susceptible. In the Rocky Mountain Region, hosts include limber pine, whitebark pine, Rocky Mountain bristlecone pine, and southwestern white pine. Alternate hosts include currants and gooseberries in the genus *Ribes* and, occasionally, species of *Pedicularis* and *Castilleja*.

**Signs and Symptoms**—Signs of white pine blister rust are visible on *Ribes* spp. in the summer and fall (uredinia and later telia) on the undersurface of leaves. Symptoms such as leaf spots and premature defoliation occur on *Ribes* spp. but otherwise, the disease causes little damage.

Bright red, recently killed "flagged branches" are the most obvious symptom of white pine blister rust from a distance (fig. 189). However, other agents, such as dwarf mistletoe and twig beetles, can cause flagging. The first detectable symptoms on pines are yellow needle spots. Diamond-shaped stem cankers are often swollen and resinous and sometimes have an orange margin. Cankers are most obvious in spring and early summer when pustules (aecia) full of orange aeciospores rupture through the bark.

**Figure 189.** Flagged branches on limber pine. *Photo: William Jacobi, Colorado State University.*

The cankered bark becomes roughened and dark as it dies following sporulation, but the fungus continues to expand into adjacent healthy tissue. Rodents often gnaw the bark off around cankers (fig. 190)

**Disease Cycle**—White pine blister rust cannot spread from pine to pine but is transmitted to pines from basidiospores produced on infected *Ribes* spp. leaves. Basidiospores are short-lived and primarily disperse short distances (usually less than 1,000 ft [300 m] but possibly a few miles). Pines are infected through needle stomata in the late summer and early fall. Following infection, the fungus grows down the needle and into the bark where a perennial canker forms (fig. 191). In spring to early summer 2-4 years later, spermogonia (pycnia) form within the canker (fig. 192), and aecia are produced in the same tissue the following year. Aeciospores can travel long distances (potentially hundreds of miles) in the wind to infect susceptible *Ribes* spp. Urediniospores are produced in orange pustules (uredinia) on the underside of infected *Ribes* spp. leaves throughout the summer. These spores re-infect other *Ribes* spp. leaves; they cannot infect pines. In late summer, hair-like columns (telia) with teliospores are produced on infected *Ribes* spp. leaves (fig. 193). Teliospores form basidiospores that later infect pines, completing the cycle.

**Figure 190.** Canker with orange canker margin and rodent feeding. *Photo: William Jacobi, Colorado State University.*

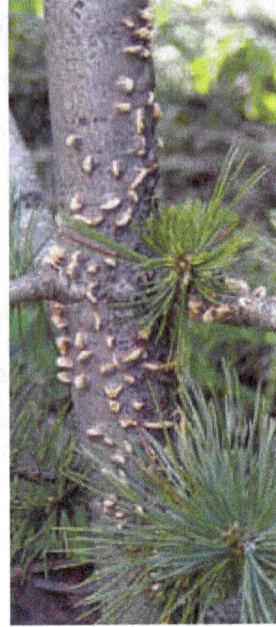

**Figure 191.** White pine blister rust stem canker sporulating on limber pine. *Photo: Kelly S. Burns, USDA Forest Service.*

**Impact**—White pine blister rust affects trees of all ages and sizes and can essentially eliminate white pines from certain ecosystems. Branch and stem cankers eventually lead to branch death, top-kill, or whole tree mortality. The probability of branch infections reaching the bole declines with distance, and branch infections more than 2 ft (61 cm) from the trunk will usually kill the

**Figure 192.** Infected limber pine seedling with needle spots and spermogonia. *Photo: Isabelle Lebouce, Dorena Genetic Resource Center, USDA Forest Service.*

**Figure 193.** *Ribes* spp. leaf with infection that is producing telia. *Photo: Kelly S. Burns, USDA Forest Service.*

branch before reaching the main stem. Small trees are especially susceptible because most infections occur close to the main stem, quickly girdling the tree. In the Rocky Mountain Region, mortality is common on larger trees without a stem infection when numerous branch infections occur throughout the crown. White pine blister rust may severely impact reproductive potential by weakening and killing cone-bearing branches. Infected trees may become susceptible to other damaging agents such as bark beetles.

**Management**—The most promising strategy for managing white pine blister rust is to increase the frequency of rust-resistant individuals across the landscape. The conventional approach is to develop a rust resistance breeding program in which seed is collected from putatively resistant trees, progeny are screened for resistance, and a seed orchard is established to supply seed for restoration and reforestation. An alternate strategy for areas not yet challenged by rust is to increase age class diversity by establishing young cohorts within mature stands. Upon invasion, natural selection would occur more quickly in the young cohort than in the mature cohort, ultimately accelerating the development of a more resistant stand.

Management strategies such as pruning and *Ribes* spp. removal to decrease inoculum potential may be used to reduce infections and prolong the life of existing trees. *Ribes* spp. removal may be feasible in certain high-value areas (for example, campgrounds) where all plants within 0.6 miles (1 km) of the area to be protected can be located and completely removed. Sanitation pruning is used to remove the infection before it reaches the main stem.

Burns and others (ref. 21) present detailed management options for white pine blister rust in the Rocky Mountain Region.

**References: 21, 25, 87, 116, 119, 135, 144, 151, 152**

# Introduction to Foliage Diseases

## Spots, blights, and casts

Diseases of tree foliage can be broken roughly into three groups: hardwood foliage diseases, needle casts, and needle blights (figs. 194-197). Hardwood foliage diseases are diverse.

**General Features**—General features of foliage diseases:

**Figure 194.** Example of a foliage disease: early stage of inkspot of aspen. *Photo: Jim Worrall, USDA Forest Service.*

- Foliage pathogens penetrate either directly through the cuticle and epidermis or enter through stomates. Wounds are not normally involved.
- Hardwood foliage diseases usually have minor impact unless they cause repeated defoliation.

- Conifer foliage diseases (table 10) can significantly affect growth. Conifers cannot refoliate like hardwoods; conifers depend on several years of foliage. A defoliated hardwood will have a full complement of leaves the next year if not the same year.
- Foliage pathogens are usually damaging only under certain circumstances such as in trees that are off-site or out of native range; pure, dense stands; small trees, usually seedling to pole size; and particular weather patterns.

**Figure 195.** Marssonina blight of aspen. *Photo: Jim Worrall, USDA Forest Service.*

- Most overwinter in foliage, either on the tree (conifers) or on the ground (mostly hardwoods). Some hardwood foliage pathogens also grow into the twig, causing twig cankers, and can overwinter there.
- Although leaf wetness is required for spore production, germination, and infection, the required duration is only a few days or less. So foliage diseases can be successful even during dry years if sufficient moisture occurs at the right time. Some (e.g., Davisomycella needle cast) are even more damaging during hot, dry conditions, if infection occurred earlier in a wet period.

**Figure 196.** Discoloration and reduced needle retention due to needle cast of ponderosa pine. *Photo: Jim Worrall, USDA Forest Service.*

**Figure 197.** Unusually severe damage and mortality due to red band needle blight of lodgepole pine in British Columbia. *Photo: Jim Worrall, USDA Forest Service.*

**Table 10.** Conifer foliage pathogens (other than rusts) and their hosts documented in Region 2 and nearby areas. Most diseases do not have well-established names, so they are not provided. Alternate names are provided where confusion is likely. From refs. 9, 44, 164, and 165. DF = Douglas-fir, F = true firs, LP = limber pine, LPP = lodgepole pine, P = pine, PiP = pinyon, PP = ponderosa pine, S = spruces, SAF = subalpine fir, WF = white fir, WP = white pines.

| Pathogen | Host |
|---|---|
| Bifusella linearis | LP |
| Bifusella pini | PiP |
| Bifusella saccata | PiP, LP |
| Cyclaneusma niveum | LPP, PP |
| Davisomycella medusa | LPP, PP |
| Davisomycella ponderosae | PP |
| Dothistroma pini | P |
| Dothistroma septosporum (Mycosphaerella pini) | P |
| Elytroderma deformans | PiP, PP |
| Hemiphacidium planum | LP, PP |
| Herpotrichia juniperi (H. nigra) | P, S, F |
| Isthmiella (Bifusella) abietis | SAF |
| Isthmiella (Bifusella) crepidiformis | F, S |
| Lecanosticta (Coryneum) cinerea | LPP, PP |
| Lirula abietis-concoloris | F |
| Lirula macrospora (Lophodermium filiforme) | S |
| Lophodermella (Hypodermella) arcuata | LP |
| Lophodermella cerina | PP |
| Lophodermella concolor | LPP |
| Lophodermella montivaga | LPP |
| Lophomerum (Lophodermium) autumnale | WF |
| Lophodermium decorum | WF |
| Lophodermium juniperinum | Rocky Mountain juniper |
| Lophodermium nitens | WP |
| Lophodermium piceae | S |
| Neopeckia coulteri | P, S |
| Phaeocryptopus gaeumannii | DF |
| Phaeocryptopus nudus (Adelopus balsamicola) | SAF |
| Rhabdocline pseudotsugae | DF |
| Rhizosphaera kalkhoffii | S |
| Virgella (Hypoderma) robusta | WF |

**Hardwood Diseases**—Hardwood foliage diseases typically overwinter in foliage on the ground. The primary inoculum in spring is ascospores that are shot into the air from the foliage. Some produce secondary inoculum from lesions on current-year leaves for further infections during the year. Others grow down the petiole, invade twigs, and cause cankers there in which they can overwinter.

**Needle Casts**—Characteristics of needle casts:
- Needles are often lost, or cast, prematurely. However, there are some needle casts (for instance, on larch) where the needles are kept longer than normal.
- Needle casts have only one infection period per year.
- Most infections are caused by a group of fungi in the family Rhytismataceae, order Rhytismatales, but some pathogens are in other groups of the phylum Ascomycota.

- Needle-cast pathogens in the family Rhytismataceae usually have specialized fruiting structures called hysterothecia. Hysterothecia may be elongated and have a covering that develops a longitudinal slit in the middle. Special cells at the outer edges absorb water under wet conditions and force the slit open to expose the spore-producing surface. When the weather is dry, they close again, functioning as automatic doors.

- Ascospores are forcibly shot into the air. They are usually long and narrow, which may increase the likelihood of hitting a needle. They have a sticky sheath that helps them stick to needles.

- Most needle casts infect young, current-year needles, and their sporulation is synchronized with needle elongation. Some infect mostly older needles, but they are less serious diseases and verge on the saprobic species.

- Symptoms usually do not appear during the year of infection. If they do, it is not until autumn.

**Needle Blights**—The term "blight" is used various ways, and needle blights vary considerably. In general, however, they can infect multiple times during the year, whenever temperature and moisture are favorable. They usually have two kinds of spores: sexual spores that cause the initial infection and asexual spores that can propagate the epidemic during suitable weather. Dead needles may remain attached to twigs instead of being prematurely cast.

**References: 9, 44, 164, 165**

# Anthracnose

## Common foliage disease of deciduous trees

**Pathogen**—Anthracnose diseases are caused by a group of morphologically similar fungi that produce cushion-shaped fruiting structures called acervuli (fig. 198). Many of the fungi that cause anthracnose diseases are known for their asexual stage (conidial), but most also have sexual stages. Taxonomy is continually being updated, so scientific names can be confusing. A list of common anthracnose diseases in the Rocky Mountain Region and their hosts is provided in table 11.

**Hosts**—A variety of deciduous trees are susceptible to anthracnose diseases, including ash, basswood, elm, maple, oak, sycamore, and walnut. These diseases are common on shade trees. Marssonina blight of aspen (see the Marssonina Leaf Blight entry in this guide for more information) is an anthracnose-type disease. The fungi

**Figure 198.** *Apiognomonia quercina* acervuli on the mid-vein of an oak leaf. *Photo: Great Plains Agriculture Council.*

**Table 11.** Common anthracnose pathogens in the Region by host and part of tree impacted (ref. 163).

| Host | Pathogen | Part of tree impacted |
| --- | --- | --- |
| Ash (especially green) | *Apiognomonia errabunda* conidial state = *Discula* spp. | Leaves and twigs |
| Basswood | *Apiognomonia tiliae* | Leaves and twigs |
| Elm | *Stegophora ulmea* conidial state = *Gloeosporium ulmicolum* | Leaves |
| Maple | *Kabatiella apocrypta* conidial state unknown | Leaves |
| Oak (especially white) | *Apiognomonia quercina* conidial state = *Discula quercina* | Leaves, twigs, shoots, and buds |
| Sycamore and London plane-tree | *Apiognomonia veneta* conidial state = *Discula* spp. | Leaves, twigs, shoots, and buds |
| Walnut | *Gnomonia leptostyla* conidial state = *Marssoniella juglandis* | Leaves, twigs, and nuts |

that cause anthracnose diseases are host-specific such that one particular fungus can generally only parasitize one host genus. For example, *Apiognomonia errabunda* causes anthracnose only on species of ash, and *A. quercina* causes anthracnose only on oaks.

**Signs and Symptoms—** Symptoms of anthracnose vary considerably from host to host. Most anthracnose fungi cause blotchy, necrotic spots on leaves associated with veins and sometimes cause leaf distortion and premature defoliation (figs. 199-201). On oak and sycamore, the disease also impacts twigs, shoots, buds, and fruits and occasionally causes stem cankers and brooming. When leaves are infected in early spring, they often turn black and may be confused with frost-damaged leaves.

**Disease Cycle—**These fungi overwinter on infected leaves and branches on the tree or the ground.

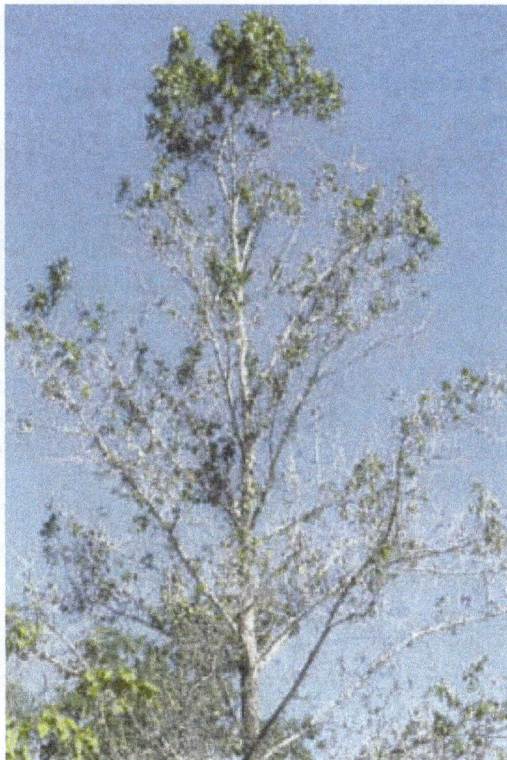

**Figure 199.** Crown symptoms caused by *Apiognomonia veneta* on sycamore. *Photo: William Jacobi, Colorado State University, Bugwood.org.*

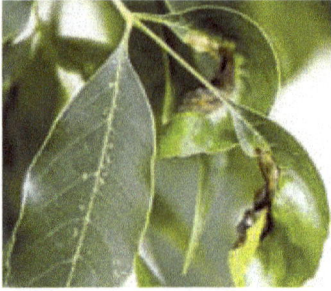

**Figure 200.** Ash leaves infected with *Apiognomonia errabunda* have irregularly-shaped blotches and are distorted. *Photo: Joseph O'Brien, USDA Forest Service, Bugwood.org.*

**Figure 201.** *Apiognomonia quercina*-infected oak leaves with irregular necrotic area associated with veins. *Photo: Joseph O'Brien, USDA Forest Service, Bugwood.org.*

In spring, spores are produced during periods of cool, moist weather. These may be conidia or ascospores, but conidia appear to play the most important role in dissemination and infection. Spores are spread by wind and splashing rains and infect buds, twigs, and newly emerging leaves if conditions are conducive. Lesions eventually form and symptoms become visible in mid to late summer. In some species (e.g., ash and walnut), secondary spores (called "summer spores") are produced on primary lesions during periods of cool, wet weather, allowing the diseases to intensify throughout the summer.

**Impact**—Disease severity appears to be associated with prevailing weather. Periods of cool, wet weather are conducive to spore production, spread, infection, and disease development, and warm temperatures (above 55 °F) inhibit the fungus. In the Rocky Mountain Region, damage is usually minor and vigorous trees can recover fairly rapidly. Trees weakened by other factors such as environmental stress, nutritional imbalance, or insects and diseases may be more susceptible to the diseases. Anthracnose can also stress trees, making them more susceptible to other diseases or insects.

**Management**—Anthracnose damage is generally insignificant in the Region, but management may be necessary in certain high-value areas. The following are control strategies that can be used to reduce disease severity:
- Plant resistant or tolerant species in areas where the disease is a problem or where air circulation is poor. Species vary in their susceptibility to anthracnose diseases. For example, true London plane is more resistant than American sycamore, and the red oaks tend to be more tolerant than the white oaks.
- Maintain proper spacing between trees to increase air circulation and improve tree vigor.
- Proper fertilization may increase disease tolerance.
- Raking infected leaves and litter in the fall may help reduce infections in the spring.
- Effective protective fungicides labeled for anthracnose control are available.

**References: 15, 159, 163**

# Brown Felt Blight

## Snow mold of conifers (also called bear wipe)

**Pathogen**—Two fungi cause brown felt blight: *Neopeckia coulteri* and *Herpotrichia juniperi*. These fungi are commonly referred to as snow molds.

**Hosts**—Almost all high-elevation conifers are susceptible to brown felt blight. *Neopeckia coulteri* generally infects only pines. *Herpotrichia juniperi* infects a variety of conifers, including Engelmann spruce, Colorado blue spruce, subalpine fir, Douglas-fir, lodgepole pine, and whitebark pine.

**Signs and Symptoms**—The most obvious sign of snow mold is the felt-like mat of mycelium that grows on infected twigs and branches (figs. 202-203). Under careful inspection, black, globose fruiting bodies may be visible on the mycelium. The two fungi can be differentiated based on host preference or by microscopic examination of spores. The latter is necessary for specific identification in pines.

**Disease Cycle**—The fungi develop beneath the snow and produce thick, gray, felt-like mats of mycelium that smother needles, branches, and twigs. *Neopeckia coulteri* is able to penetrate living needles, while *Herpotrichia juniperi* enters needles and twigs after they are dead. As snow melts, mycelium turns gray to black, and fungal growth ceases. Spherical, black fruiting bodies (perithecia) may be visible protruding from mycelium during the second summer. Spread occurs when needles come into contact with infected litter or spores under snow. The disease is favored by deep, long-lasting snow packs.

**Impact**—Damage from brown felt blight may result in branch death and growth loss. Severe infections

Figure 202. Close-up of brown felt blight on a lodgepole pine branch. *Photo: Kelly Burns, USDA Forest Service.*

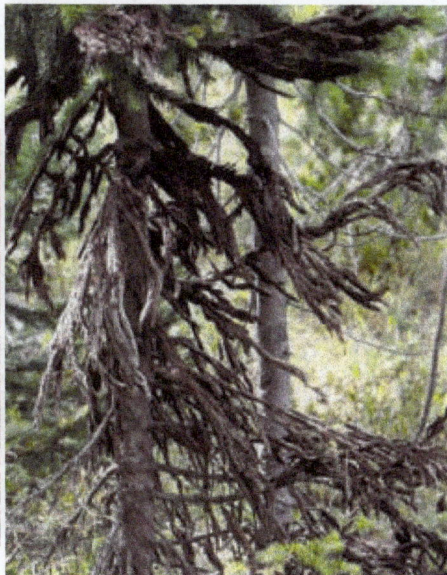

Figure 203. Snow mold on Engelmann spruce. *Photo: Whitney Cranshaw, Colorado State University, Bugwood.org.*

on seedlings and saplings may lead to mortality, but the disease has little effect on larger trees.

**Management**—Brown felt blight usually does not warrant management. Possible management options for high-value areas include installing snow breaks to divert snow pack from target areas and removing infected branches to reduce inoculum.

**References: 71, 155, 157, 159**

# Cercospora Blight of Junipers

## Serious disease in shelterbelts

**Pathogen**—Cercospora blight of junipers is caused by the fungus *Pseudocercospora juniperi* (*Cercospora sequoiae* var. *juniperi*).

**Hosts**—Many members of the Cupressaceae family are, but Rocky Mountain juniper and eastern redcedar are the most severely impacted in the Great Plains.

**Signs and Symptoms**—Foliage of the inner branches of the lower crown becomes bronzed, then necrotic, and eventually sheds, leaving the inner crown devoid of foliage (fig. 204). The extremities remain green. Small, gray, fuzzy asexual fruiting bodies (sporodochia) may be visible with a hand lens on dead needles (fig. 205). Defoliation progresses from the lower portion of inner crown outward and upward.

**Disease Cycle**—Sporodochia are produced on infected foliage in September. Sporulation occurs continuously from April to October during periods of warm, wet weather. Conidiospores disperse short distances (<6 ft [1.8 m]) by means of rain splash. Free moisture is required for sporulation and infection. Infection occurs through stomata or directly through the cuticle. Infection is initiated on current and previous years' juvenile needles (sharply pointed young needles) and previous years' spur leaves (scale-like mature leaves) of the inner crown and eventually spreads to the entire branchlet. Whip foliage (elongated shoot tips) appears to be resistant. Infection usually occurs in June and July, and symptoms develop about 2-3 weeks after infection. The fungus overwinters in infected needles on living trees. The sexual state of *P. juniperi* is not known.

**Impact**—Cercospora needle blight can cause devastation and mortality to juniper in shelterbelts and other established plantings, particularly with repeated years of infection. Impacts are more severe on Rocky Mountain juniper than on eastern redcedar and in areas or years with significant moisture and humidity.

**Management**—Several management strategies may be used to control Cercospora needle blight:
- Do not plant Rocky Mountain juniper and eastern redcedar in areas where the disease is a problem.
- Maintain wide spacing between trees to promote air circulation and reduce humidity.

**Figure 204.** Inner branches of the lower crown of Rocky Mountain junipers infected with Cercospora needle blight are defoliated while the extremities remain green. *Photo: Ned Tisserat, Colorado State University.*

**Figure 205.** Small, gray, fuzzy fruiting bodies of Cercospora on infected leaves may be visible with a hand lens. *Photo: Ned Tisserat, Colorado State University.*

- Protective fungicides labeled for Cercospora control (e.g., Mancozeb and Bordeaux mixture) can be used to protect susceptible trees. Two applications are usually necessary, depending on the type of foliage present (juvenile, spur, or both).

**References: 159, 172**

# Davisomycella and Lophodermella Needle Casts

## Occasionally epidemic diseases of pines

**Pathogen—***Davisomycella* is a genus of needle-cast pathogens in the family Rhytismataceae, order Rhytismatales. In the Rocky Mountain Region, *D. (Lophodermium) ponderosae* infects ponderosa pine; *D. (Hypodermella) medusa* infects both ponderosa and lodgepole pines.

The related genus *Lophodermella* also contains several pathogens that are occasionally important in this Region. They were first described in the genus *Hypodermella* and occur in older literature under that name. *Lophodermella arcuata* occurs on limber pine, *L. cerina* occurs on ponderosa pine, and *L. concolor* and *L. montivaga* occur on lodgepole pine.

**Hosts**—Ponderosa, lodgepole, and limber pines can be infected by different species and combinations of species of *Davisomycella* and *Lophodermella* (see "Pathogen" above).

**Signs and Symptoms**—These fungi cause browning and premature casting of needles. Symptoms of *Davisomycella* species may appear on needles of any age class except current-year needles (fig. 206). Whole needles or portions may be affected. After symptoms appear, fruiting structures (hysterothecia) may begin to form. In *D. ponderosae*, they begin to form in spring and mature in summer. Hysterothecia are dark brown and elliptical to linear; they may be partly fused end-to-end, forming long, sometimes sinuous or forked lines (figs. 207-209). They open by a longitudinal slit and release spores into the air during wet weather. Hysterothecia of *D. medusa* are also dark and raised but are considerably shorter. They occur in greenish straw-colored areas of the needle. As older needles are lost and only current needles are present, twigs may take on a lion's tail appearance.

With *Lophodermella concolor*, needles infected the previous year die and turn reddish brown by spring, becoming straw-colored during the summer. Hysterothecia are colorless (the specific epithet means that they blend with their background). They are usually evident as shallow, elliptical depressions up to 1/25 inch (1 mm) long.

**Disease Cycle**—The disease cycles of Davisomycella needle casts are somewhat unusual. Primarily current-year needles are infected after hysterothecia mature in late summer. As with most foliage diseases, wet weather is required for spore dispersal and infection. Symptoms may not appear the next year; in fact, the needles may appear healthy and may be retained for the normal time. This is a latent infection. Evidence suggests that symptom development is triggered by drought stress. When drought occurs, any age classes of needles that were infected (i.e., had suitable weather conditions and inoculum during their first year) may develop symptoms.

Disease cycles of Lophodermella needle casts vary somewhat (table 12). *Lophodermella concolor* and *L. montivaga* have a 1-year cycle. Infection occurs in current-year needles in early or late summer, respectively. No symptoms develop until late winter. By spring, infected needles are dead (figs. 210-211). Hysterothecia develop in them and mature in early or late summer, after which needles are cast.

**Figure 206.** *Davisomycella ponderosae*: discoloration and needle loss to large, open-grown ponderosa pines. *Photo: Jim Worrall, USDA Forest Service.*

Figure 207. *Davisomycella ponderosae*: immature hysterothecium in late May. *Photo: Jim Worrall, USDA Forest Service.*

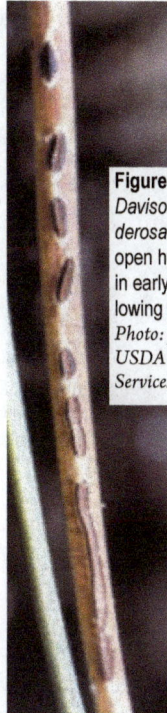

Figure 208. *Davisomycella ponderosae*: mature, open hysterothecia in early October following wet weather. *Photo: Jim Worrall, USDA Forest Service.*

Figure 209. *Davisomycella ponderosae*: another view of mature, open hysterothecia in early October following wet weather. *Photo: Jim Worrall, USDA Forest Service.*

Table 12. *Davisomycella* and *Lophodermella* species that cause needle cast in the Region (includes hosts and phenology).

| Pathogen | Host | Dispersal period | Symptoms appear |
|---|---|---|---|
| *Davisomycella medusa* | Lodgepole, ponderosa | July-September | Following year to not at all |
| *Davisomycella ponderosae* | Ponderosa | July-October | Following year to not at all |
| *Lophodermella arcuata* | Limber | July-August | Following spring |
| *Lophodermella cerina* | Ponderosa | Spring | Beginning that autumn |
| *Lophodermella concolor* | Lodgepole | June-July | Following spring |
| *Lophodermella montivaga* | Lodgepole | July-September | Beginning that autumn |

**Impact**—Epidemics of needle casts are sporadic. Damage may be severe and widespread in some years, while infections may be hard to find in others. Nearby trees may contrast greatly in infection levels, suggesting genetic variation in resistance. Often, damage is most severe on smaller trees, in the lower crown of larger trees, and in dense stands. However, damage is sometimes severe throughout the crown of large, open trees. Growth losses have been documented for needle casts and can be significant with successive years of infection. During an epidemic of *Davisomycella medusa* associated with drought, 10-year growth

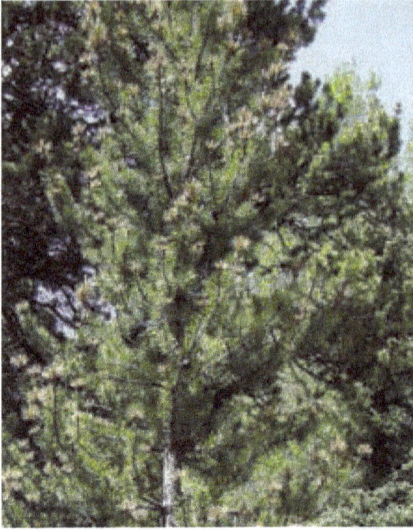

Figure 211. *Lophodermella montivaga*: single shoot showing symptoms and fruiting on last year's needles in late June. *Photo: Leanne Egeland, USDA Forest Service.*

Figure 210. *Lophodermella montivaga*: discoloration of sapling lodgepole pine. *Photo: Leanne Egeland, USDA Forest Service.*

of infected trees was reduced by 69%, while growth of uninfected trees was reduced by only 29%. As noted above, *Davisomycella* species may be most damaging when drought follows a year that is favorable for infection.

**Management**—Direct management approaches are not known for needle casts and are generally not needed. Growth impacts are usually localized and temporary. Thinning may be helpful where epidemics recur frequently enough to cause concern in high-value forests. Thinning would speed drying of the foliage, thereby reducing the periods available for infection. Also, growth increases associated with thinning may compensate for reductions caused by the disease.

**References: 158, 166, 178, 198**

# Elytroderma Needle Cast

## Perennial, compact brooms in pines

**Pathogen**—Elytroderma needle cast is caused by the fungus *Elytroderma deformans*.

**Hosts**—The disease is most severe on ponderosa pine, but lodgepole pine, jack pine, and pinyon pine are also susceptible.

**Signs and Symptoms**—Perennial infections on vigorous branches often cause small to large compact brooms with upward turning branches and many dead, straw-colored needles. This is the only needle cast that is perennial in host twigs.

The fungus invades twigs and branches, causing stunting of needles, reddened foliage, defoliation, and conspicuous brooming (figs. 212-213). Brown,

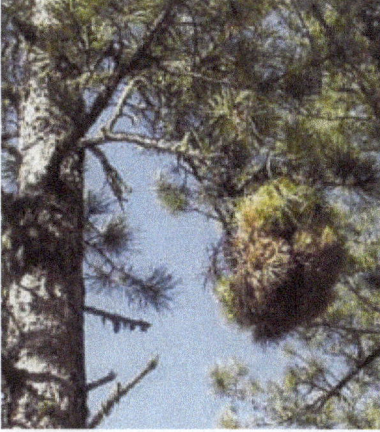

**Figure 212.** Large, compact witches' broom typical of Elytroderma disease on ponderosa pine. *Photo: Thomas R. Erikson, USDA Forest Service.*

**Figure 213.** A perennial infection on ponderosa pine with upward-turning branches. *Photo: USDA Forest Service Archive, Bugwood.org.*

resinous lesions often develop in the phloem of infected twigs (fig. 214). Infected needles are red-brown, becoming pale over the summer, but the basal portion (1/4-3/8 inch [6-10 mm]) remains green. Fruiting bodies appear near the base of dead and dying needles in the spring and are thin, elongated (approximately 1/2 inch [13 mm] long), and dull black (fig. 215). Brooms are compact, globose, and contain discolored needles, most of which are shed by fall.

Dwarf mistletoe brooms on pines sometimes look similar to Elytroderma brooms. However, dwarf mistletoe plants or remnant basal cups should be visible on dwarf mistletoe infections and Elytroderma brooms are more compact and globose.

**Disease Cycle**—Airborne spores infect needles in late summer and fall and occasionally in spring, following rainstorms. When free moisture is available,

**Figure 214.** Necrotic, resinous flecks are often visible in the inner bark of infected twigs. *Photo: John Schwandt, USDA Forest Service, Bugwood.org.*

**Figure 215.** Black, elongated fruiting bodies appear in the spring near the base of needles. *Photo: John Schwandt, USDA Forest Service, Bugwood.org.*

spores germinate and penetrate young needles. The fungus grows within the needle tissue and eventually invades twigs and becomes systemic within the host, infecting new shoots and needles each year. The following spring, infected needles turn reddish brown, and by early summer, fungal fruiting bodies develop on dead needles if conditions are conducive. These fruiting bodies, which appear as black, elongated lines or slits, split open during wet weather and release spores.

**Impact**—Impacts are greatest in areas with moist environmental conditions such as drainages. Dry spring weather tends to suppress spore production and infection. Infection is favored by dense stand conditions and injury is greatest on poor crowns. Elytroderma needle cast is more common and damaging in the northern Rockies than in the southern and central portions of the range. The disease is occasionally found in the Colorado Front Range, but damage in the Rocky Mountain Region is greatest on ponderosa pine in the Black Hills of South Dakota. Infection results in loss of the previous year's needles and death of branch cambium. Stem infections can severely damage and deform small trees. When infection is severe, growth loss occurs and entire tree tops may be deformed. Severely infected trees may be weakened, predisposing them to attack by bark beetles or other damages, or killed outright.

**Management**—In young stands, damage can be reduced by maintaining good spacing through thinning. Selectively remove moderately and severely infected trees and trees with infections high in the crown. Accelerated logging in heavily infested mature stands not only serves to salvage valuable timber but also prevents the establishment of secondary pests. Remove all trees with more than one-quarter of their twigs killed.

**References: 26, 159**

# Ink Spot

## Dark, raised blotches or shotholes in aspen leaves

**Pathogen**—Ink spot is caused by the fungus *Ciborinia whetzelii*.

**Hosts**—Aspen is the predominant host, although other poplars are also susceptible.

**Signs and Symptoms**—Ink spot is characterized by brown to black spots on blighted leaves (figs. 216-218). Early summer symptoms of ink spot can look like leafminer insect damage with concentric zones that are light and dark. Reddish brown blotches become visible on infected leaves within a few weeks, and ink spots (sclerotia) develop several weeks later. Infected leaves quickly turn brown, and sclerotia begin to drop out, leaving shotholes in the dead leaves. *Ciborinia whetzelii* forms well-defined, circular, stalked fruiting structures called apothecia that produce and release spores.

**Disease Cycle**—*Ciborinia whetzelii* overwinters as sclerotia (hardened masses of fungal hyphae) in forest litter. In the spring, sclerotia produce fruiting bodies

**Figure 216.** Early season ink spot on aspen leaf. *Photo: William Jacobi, Colorado State University, Bugwood.org.*

(apothecia) that release wind- and rain-disseminated ascospores that infect developing leaves. Infected leaves develop sclerotia (ink spots) that eventually drop out. Infected leaves begin to die midsummer, but defoliation may not take place until autumn. Cool, moist weather in the spring is conducive to disease spread.

**Figure 217.** Brown, immature sclerotia on blighted leaves. *Photo: Jim Worrall, USDA Forest Service.*

**Impact**—Ink spot is more severe on small trees and in the lower crowns of larger trees. The incidence of the disease is greatest in dense stands. Occasionally, this disease is responsible for 25-100% defoliation in localized areas. This disease rarely causes long-term damage or mortality. As with many aspen diseases, some clones appear to be more susceptible than others.

**Figure 218.** Black ink spots (sclerotia) on aspen leaf. *Photo: Jim Worrall, USDA Forest Service.*

**Management**—Spring infections may be reduced if infected leaves are raked up and removed the previous fall. Increasing the spacing between trees may create a less favorable microclimate for disease spread and infection. Fungicides can be used to prevent infection, but they must be applied before infection occurs.

**References: 79, 159**

# Marssonina Leaf Blight

## Midsummer spots and defoliation of aspen

**Pathogen**—Marssonina leaf blight is caused by at least two fungi in our area: *Marssonina brunnea* (sexual stage *Drepanopeziza tremulae*) and *Marssonina* *populi* (*Drepanopeziza populorum*). The disease is also commonly referred to as black leaf spot.

**Hosts**—Aspen is the predominant host, but narrowleaf cottonwood and other poplars are also susceptible.

**Signs and Symptoms**—Small, brownish spots appear on infected leaves as early as the end of leaf expansion but are especially

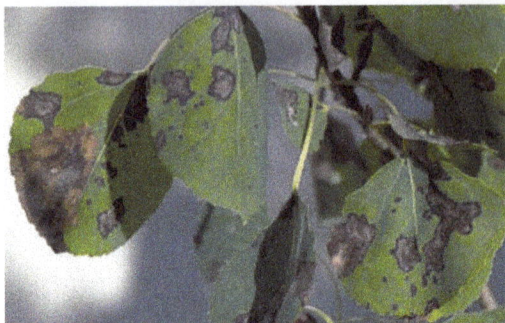

**Figure 219.** Marssonina blight on aspen. *Photo: William Jacobi, Colorado State University, Bugwood.org.*

noticeable in July and August (figs. 219-222). *Marssonina brunnea* often causes small spots, initially about 1/25 inch (1 mm) in diameter, while *M. populi* causes larger spots. The spots later enlarge to various sizes and shapes, turn brownish black, and often have a yellow margin and a white center. Spots may coalesce to form vein-limited, necrotic blotches. Acervuli (asexual fruitbodies) appear as tiny, ring-like blisters in the center of spots when conditions are moist. Infected leaves often fall during the summer. This may be more common when infections are on or near the petiole, even if the rest of the leaf is relatively healthy.

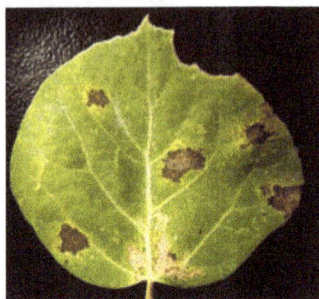

**Figure 220.** Close-up of an infected aspen leaf, showing irregularly-shaped spots with yellow margins, probably *Marssonina populi*. *Photo: Jim Worrall, USDA Forest Service.*

Symptoms intensify throughout the summer and progress upward in the crown as the season advances. From a distance, canopies of heavily infested stands often have a bronze or brown hue. Crowns may appear thin from defoliation, and infected leaves are often seen on the forest floor and roads (figs. 223-224). Fall colors are usually less intense in such stands.

**Disease Cycle**—Primary infection occurs in spring when spores produced on twig lesions or infected fallen leaves infect developing leaves. Second-ary infection occurs throughout the summer when

**Figure 221.** Early stage of infection by *Marssonina brunnea*. *Photo: Jim Worrall, USDA Forest Service.*

spores from fruiting bodies on primary lesions are released with wet weather. Disease incidence is closely associated with warm, wet conditions. The pathogens overwinter as tiny stromata in fallen leaves and in twig lesions.

**Impact**—Marssonina leaf blight is the most common leaf disease of aspen in the Rocky Mountain Region. Because of the dependence of the disease on weather,

Figure 222. Close-up of leaf spots on aspen, showing vein-limited margins and tiny blisters (acervuli) in the center. *Photo: Jim Worrall, USDA Forest Service.*

Figure 223. Defoliation caused by *Marssonina brunnea* in early July. *Photo: Suzanne Bethers, USDA Forest Service.*

Figure 224. Fresh leaves on the ground in mid-July due to infection by *Marssonina brunnea. Photo: Jim Worrall, USDA Forest Service.*

it may be severe one year and difficult to find the next. Infected leaves are often stunted and shed prematurely. Defoliation can be severe, with nearly complete defoliation by early August. Trees may refoliate in late summer and early fall, but twig dieback often occurs the following winter because late shoots lack sufficient cold hardiness. Mortality is rare unless the outbreak occurs over consecutive years or is combined with other stresses. However, damaged trees may be more susceptible to other damages, and wood production is reduced for 1 or more years following an epidemic. Clonal variation in suceptibility may be visible during severe epidemics.

**Management**—Because the disease rarely leads to mortality and management approaches are often impractical, management of Marssonina leaf blight is not often contemplated in forests managed for multiple objectives. In high-value sites, the most effective way to manage the disease is by planting or managing for resistant or tolerant clones. Removing and destroying diseased material from trees and the ground may help reduce infections. Increasing space between trees may create a less favorable microclimate for spread and infection. Fungicides can be used to prevent infection, but they must be applied at bud break before infection occurs.

**References: 140, 158**

# Melampsora Rusts

## Common leaf rusts of poplars and willows

**Pathogen**—Several *Melampsora* species occur in the Rocky Mountain Region, but the most common and important is *M. medusae* (= *M. albertensis*), the fungus that causes conifer-aspen leaf rust. Less common species include *M. occidentalis*, which causes conifer-cottonwood rust, and *M. epitea*, which causes willow rust.

**Hosts**—Aspen, cottonwood, and willow are hardwood (telial) hosts. Aecial hosts include Douglas-fir, lodgepole pine, ponderosa pines, true firs, *Saxifraga* species, and *Ribes* species. Host relationships are described in table 13.

**Signs and Symptoms**—Melampsora rusts are the common leaf rusts of poplar and willow. The most obvious indicators are the yellow leaf spots, which eventually become necrotic, and the orange powdery pustules on the underside of poplar and willow leaves. On conifers, infected needles shrivel and die soon after sporulation, but yellow aecia are sometimes visible on cones.

**Disease Cycle**—The life cycles of *Melampsora* species are complex and generally require two unrelated host plants and five different spore stages. Teliospores overwinter on dead poplar or willow leaves on the ground. In the spring, teliospores germinate, producing wind-disseminated basidiospores that infect current year's conifer needles and cones. Spermogonia (pycnia) and aecia (yellow pustules) develop within several weeks on the underside of needles. Orange-yellow aeciospores are dispersed in the wind to infect hardwood hosts. Conifer needles become necrotic, shrivel, and are shed shortly after sporulation. Infected poplar and willow leaves produce urediniospores that infect other leaves of the same species throughout the summer, intensifying the disease (figs. 225-227). In the fall, brown, crust-like telia form in place of uredinia. Mild temperatures (65-70 °F) and moist conditions (continuous moisture on leaf surface for 2-24 hours) are required for infection.

**Impact**—Impacts are more severe on hardwoods than on conifers. The disease causes discoloration, shriveling, and premature dropping of poplar and willow leaves. Severe defoliation may reduce growth and increase susceptibility to other diseases or insects. Impacts to conifer hosts are minimal.

**Table 13.** Hosts and alternate hosts for Melampsora rusts that occur in the Rocky Mountain Region.

| Fungus | Common name | Host (aecial) | Alternate host (telial) |
|---|---|---|---|
| *Melampsora medusae* | Conifer-aspen rust | Douglas-fir, lodgepole, and ponderosa pine | Aspen (most common) and cottonwood |
| *M. occidentalis* | Conifer-cottonwood rust | Douglas-fir and possibly lodgepole and ponderosa pines | Cottonwood and other Poplars |
| *M. epitea* | Fir-willow rust and *Ribes*-willow rust | True firs and *Ribes* spp. | Willow |

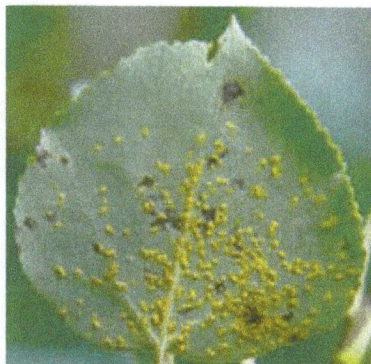

**Figure 225.** Orange, powdery pustules (uredinia) on the underside of an aspen leaf. *Photo: Mike Schomaker, Colorado State Forest Service, Bugwood.org.*

**Figure 226.** Close-up of uredinial pustules on the underside of an aspen leaf. *Photo: William Jacobi, Colorado State University, Bugwood.org.*

**Management**—Plant resistant or tolerant clones, where available. Removing and destroying diseased leaves from the ground may help reduce infections. Wide spacing between trees may create a less favorable microclimate for spread and infection. Chemical controls are ususaly not warranted.

**References: 140, 159, 201**

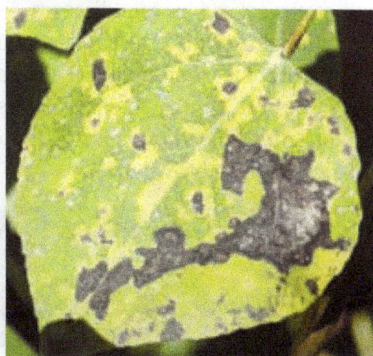

**Figure 227.** Yellow and necrotic spots on the upper surface of infected aspen leaves. *Photo: Whitney Cranshaw, Colorado State University, Bugwood.org.*

# Red Band and Brown Spot Needle Blights

## Discolored lower crowns of pines

**Pathogen**—Red band needle blight (also called Dothistroma needle blight) is caused primarily by *Dothistroma septosporum* and, to a lesser extent, by *D. pini* (= *Scirrhia pini*). The sexual form of *D. septosporum* is *Mycosphaerella pini*, while *Dothistroma pini* has no known sexual stage.

Brown spot needle blight is caused by *Scirrhia acicola* (= *Systremma acicola*). The sexual form is *Mycosphaerella acicola*.

**Hosts**—Most pines are hosts, but the diseases are common on ponderosa, lodgepole, and Austrian pine in the Region. Brown spot needle blight is also common on Scots pine. Needle blights can occur throughout the range of the hosts but are most common in areas that trap high humidity, in dense stands, and in off-site plantings.

**Signs and Symptoms**—With both diseases, symptoms are usually most severe in the lower crowns (especially on the north side) and on small trees and saplings (fig. 228). Symptoms first appear on older needles as yellow or tan bands or spots (figs. 229-230). Red band needle blight develops distinctive red transverse bands (fig. 231). Brown spot needle blight produces brown spots (fig. 232) but can form bands. Both diseases produce clearly defined margins on the needles. Needles progressively turn light green, yellow, tan, and brown from the tips back. With both diseases, black dots (fruiting bodies of the fungi) can occasionally be seen in the bands or spots throughout the year. Unlike needle casts, previous-year needles (dead needles) are usually retained in needle blights. The needles can have a drooping appearance. Needle blights look similar to early needle cast diseases, winter desiccation, drought, chemical damage, air pollutions, and symptoms caused by root diseases.

Symptoms of these diseases are quite similar. They can only be distinguished by microscopic examination of the spores (asexual spores called conidia). Even the spores are similar in shape and size, and both have cross-walls. However, conidia of *Dothistroma* species are clear, and conidia of *Scirrhia acicola* are usually greenish brown.

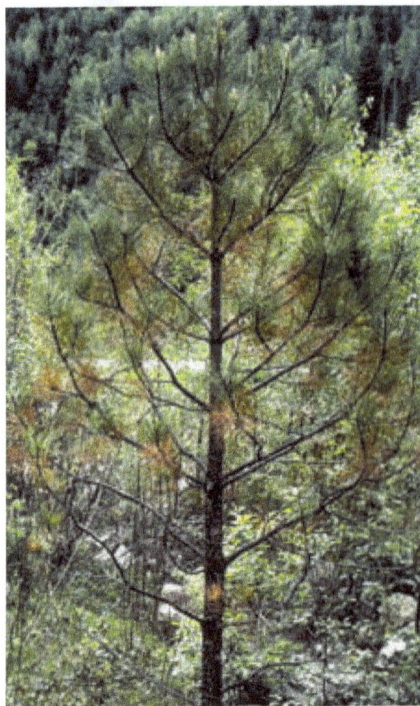

**Disease Cycle**—These fungi can infect multiple times during the year (from April to November) when temperatures are favorable during moist periods. Most have both sexual and asexual spores, but only the asexual forms are common in this Region. Under moist conditions, the asexual spores are dispersed by rain splash and, to a much lesser degree, by wind. The pathogens enter

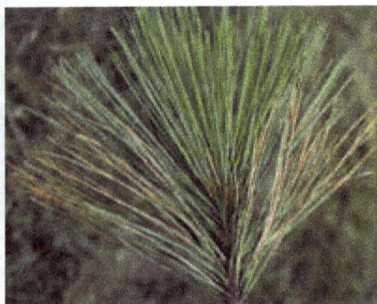

**Figure 228.** Needle blight on a ponderosa pine sapling. *Photo: Robert L. James, USDA Forest Service, Bugwood. org.*

**Figure 229.** Infected ponderosa pine branch. *Photo: Susan K. Hagle, USDA Forest Service, Bugwood.org.*

needles through stomata and colonize the needles. These pathogens overwinter in the dead or infected foliage that often remains attached to the branches for a year or two. Fruiting bodies develop below the epidermis, split the epidermis, and erupt through the needle surface in the bands or spots.

**Impact**—Impacts include growth reduction and, in small trees with extensive infections, mortality. Occasional outbreaks appear dramatic, but damage is almost always minor and trees recover.

**Management**—Management is often not needed. Growth losses are usually localized and only occur in moist years. In dense stands, thinning can increase airflow, which will promote foliage drying and reduce infections. Thinning can also be used to remove the most damaged trees and the most susceptible species. In high-value sites, applying an appropriately registered fungicide to pines can reduce symptoms. Given the occasional nature of outbreaks and the high costs involved, fungicide treatments are often inappropriate in forest situations.

**References: 9, 49, 129, 145, 159**

**Figure 230.** Infected lodgepole pine branch. *Photo: Jane E. Taylor, USDA Forest Service, Bugwood.org.*

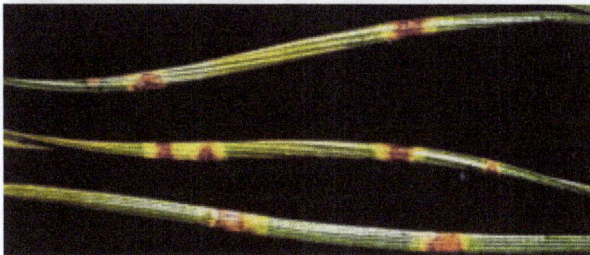

**Figure 232.** Typical spots caused by brown spot needle blight on Scots pine. *Photo: Darroll D. Skilling, USDA Forest Service, Bugwood.org.*

**Figure 231.** Distinctive red, transverse bands of red band needle blight on ponderosa pine. *Photo: Robert L. James, USDA Forest Service, Bugwood.org.*

# Septoria Leaf Spot and Canker

## Premature defoliation in windbreaks and other plantings

**Pathogen**—Septoria leaf spot is caused by fungi in the genus *Mycosphaerella* (*M. populorum* and *M. populicola*). The asexual stages are *Septoria musiva* and *S. populicola*, respectively.

**Hosts**—Septoria diseases are most common in planted windbreaks in the Great Plains. All North American poplars and aspens are susceptible to leaf spots (fig. 233) caused by *Mycosphaerella populorum*, but the fungus also causes cankers on cottonwood, hybrid poplars, and introduced poplars (fig. 234). Hybrid poplars and some introduced poplars are most susceptible.

Hosts for *M. populicola* include black and eastern cottonwoods, balsam poplar, and narrowleaf cottonwood.

**Figure 233.** Close-up of an irregularly-shaped Septoria leaf spot lesion with fruiting bodies (pycnidia) in the center. *Photo: Michael Kangas, NDSU-North Dakota Forest Service, Bugwood.org.*

**Signs and Symptoms**—The appearance of foliar lesions varies within and among hosts. Typically, lesions appear as small, circular to angular, white, grayish, tan, brown, or purplish spots with a dark border. Spots may coalesce to form blotches but remain small on more resistant species. Dark specks (pycnidia) are scattered in older lesions. Pink tendrils of conidia may exude from pycnidia during moist weather. Microscopic examination of conidia is necessary to positively identify the species.

Cankers only occur on trees with leaf spots, and their severity is proportional to leaf spot severity. Cankers usually occur

**Figure 234.** Septoria cankers on young poplar branches. *Photo: T.H. Filer, Jr.; USDA Forest Service; Bugwood.org.*

on the lower portions of the tree within 5 ft (1.5 m) of the ground and originate at wounds, lenticels, stipules, or leaf bases. Cankered bark initially darkens but becomes tan in the center. Pycnidia may be visible on young cankers but are rare on older cankers.

**Disease Cycle**—The pathogens overwinter on fallen infected leaves and in infected branches and stems. Primary infections occur in the spring by means of ascospores produced in fruiting bodies (pseudothecia) on fallen leaves or in infected branches and stems. Lesions develop about 1-2 weeks after infection, and asexual fruiting bodies (pycnidia) are produced in about 3-4 weeks. Secondary infections may occur throughout the summer during warm, moist conditions when conidia are released from fruiting bodies (pycnidia) and are spread by wind or rain splash to infect new leaves and stems.

**Impact**—*Mycosphaerella populicola* is less virulent than *M. populorum*. The diseases cause little to no damage in natural stands, but *M. populorum* can cause severe damage in plantings and windbreaks. Premature defoliation may occur in more susceptible species, and cankers can girdle stems. *Cytospora chrysosperma* and other canker fungi may invade the edges of Septoria cankers.

**Management**—Control strategies for Septoria diseases include: planting only resistant or tolerant clones; increasing spacing between trees to increase air circulation and reduce humidity; raking and destroying infected overwintering leaves and stems; and applying fungicides to protect propagation beds and landscape trees.

**References: 108, 159**

# Pine Shoot Blight and Canker

## Branch tip flagging and black dots on cone scales

**Pathogen**—The fungus *Diplodia pinea* (= *Sphaeropsis sapinea*) causes pine shoot blight and canker. This disease is also known as Diplodia or Sphaeropsis shoot blight. A second species, *Diplodia scrobiculata*, occurs in the United States and may be present in this region.

**Hosts**—Pine shoot blight is an important disease of pines and other conifers. Seedlings to fully mature pines are affected, and damage occurs in nurseries, plantations, Christmas tree and ornamental plantings, and natural stands. This disease is widely distributed and locally severe in much of South Dakota, Nebraska, and Kansas where it affects mostly ponderosa, Austrian, and Scots pine.

**Signs and Symptoms**—*Diplodia pinea* can cause shoot blight, canker, crown wilt, collar rot, and root disease. New shoots are killed rapidly by the fungus (fig. 235). Damage may be confined to the new shoots, particularly on trees with shoots that are infected for the first time. Infected needles become discolored (tan to gray) while still encased in fascicle sheaths. Dead needles usually remain attached to the twig. Cankers in new shoots can cause stunting or crooking of the shoots. Repeated infections result in dead branches and tree mortality (fig. 236). Resinous cankers with brown phloem and cambial tissue develop on branches and stems (fig. 237).

Figure 235. Diplodia infected shoot-tip. *Photo: James T. Blodgett, USDA Forest Service.*

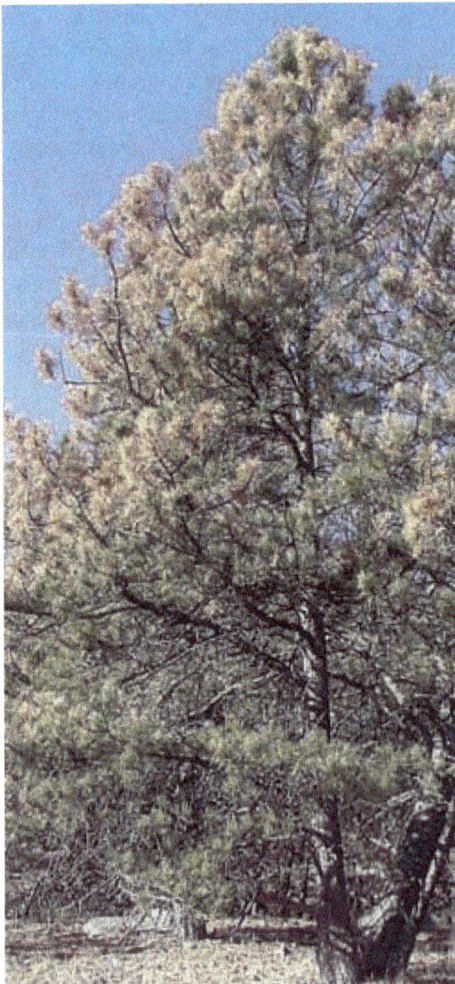

Figure 236. Infected ponderosa pine with hail damage. *Photo: James T. Blodgett, USDA Forest Service.*

Figure 237. Branch with cankered tissue. *Photo: James T. Blodgett, USDA Forest Service.*

Small, black fruiting bodies (pycnidia) are produced abundantly on needles and cones (figs. 238-239). The fruiting bodies can be seen with a hand lens. However, microscopic examination of the spores is necessary for accurate identification.

**Disease Cycle**—Spores develop in the fruiting bodies on needles, needle fascicle sheaths, scales of second-year cones, and bark. These black fruiting bodies, which erupt through the epidermis, are often numerous at the base of needles and on scales of second-year seed cones.

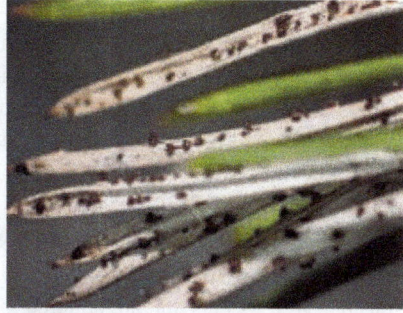

**Figure 238.** Pycnidia of Diplodia on needles. *Photo: James T. Blodgett, USDA Forest Service.*

Spores are dispersed from spring to fall, with large numbers of spores being dispersed during rain and high relative humidity. Rain splash is a common means of spread, but spores can be carried by high wind or by insects.

Moderate temperatures and very moist conditions are needed for spore germination and infection. Young shoots are most susceptible to infection. Spores can penetrate needles and young shoots, or *D. pinea* can infect

**Figure 239.** Pycnidia on cone scales. *Photo: James T. Blodgett, USDA Forest Service.*

both current-year and older tissues through wounds. Because this pathogen can persist in stems and branches of pines with no symptoms, wounding might act as a trigger, allowing the fungus to rapidly infect. This could explain the rapid disease development often associated with hail events. Both planting stress and water stress can also result in rapid disease-associated mortality of already infected trees.

The large numbers of spores produced on cones can contribute to infection and branch mortality of older trees or infection of seedlings and young trees growing near the older infected trees. The fungus commonly kills the entire new shoot by the end of summer. Infections may continue to develop into year-old tissues.

Increased disease development has been associated with water deficits; competing vegetation; high soil nutrition; poor site; and damage by hail, snow, and insects. Seasonal variations and climate differences also have been related to differences in the amount of disease caused by this pathogen.

**Impact**—Pine shoot blight kills current-year shoots and major branches. Severe infections may lead to death of trees of all sizes. The disease can predispose trees to attack by other diseases and insects, including the pine engraver beetles. This pathogen often severely affects trees wounded by hail, insects, or other damage. Pruning and shearing wounds may become infected and are susceptible to infection for several days.

**Management**—Reducing water stress is an important factor in controlling pine shoot blight. Techniques to reduce disease losses include managing competing vegetation, stand thinning, using planting techniques that reduce drought, and selecting tree species that are adapted to the site.

Pruning or shearing should be avoided during periods when conditions are highly favorable for infection because of the danger of infection through wounds. Pruning tools should be disinfected.

Seedlings and small trees can become infected when they are located close to infected cone-bearing pines. Avoid planting adjacent to infected pines, or remove infected trees.

Fertilization of pines at levels recommended for ornamental and shade trees can result in increased disease and should be avoided.

Infection of new shoots can be reduced significantly by applying an appropriately registered fungicide to pines during the 2-week period when shoots are highly susceptible to infection. This period begins with bud opening, usually from late April to mid-May. Fungicide applications during late April and mid-May will protect new shoots but will not prevent infection of cones.

**References: 20, 48, 126, 127, 128, 175**

# Shepherd's Crook

## Leaf and shoot blight of young aspen

**Pathogen**—Shepherd's crook, also known as aspen shoot blight, is caused by the fungus *Venturia tremulae* var. *grandidentatae* (= *V. moreletii*). The asexual stage (anamorph) of the fungus is *Fusicladium radiosum* var. *lethiferum* (= *Pollacia radiosa*). A similar disease is caused by *Venturia populina* but usually on other *Populus* species. Several other less common *Venturia* species cause similar diseases in other parts of North America.

**Hosts**—*Venturia tremulae* infects *Populus tremuloides*, *P. alba*, *P. grandidentata*, and hybrids. It has also been reported on *P. angustifolia* and occasionally infects other *Populus* species. The pathogen occurs across North America and in Europe.

**Signs and Symptoms**—Dark brown to black lesions first appear in the spring on leaf blades, petioles, or current-season shoots. Affected leaves and shoots often droop and curl, leading to the name of the disease (figs. 240-241). Early symptoms can look similar to frost damage, but patterns in the landscape differentiate the two. Infections can progress from leaves to petioles to shoots. Dead

**Figure 240.** Shepherd's crook on young as- **Figure 241.** A group of aspen suckers, all with blighted
pen, showing the typical "crook" (drooping)   terminals. *Photos: Jim Worrall, USDA Forest Service.*
of the young, dead apical shoots and lesions
on lower, living foliage. *Photo: Jim Worrall,*
*USDA Forest Service.*

tissues eventually become dry and brittle and break off. During wet weather, lesions develop a fine, olive-colored fuzz due to production of asexual spores on the surface.

**Disease Cycle**—The pathogen can survive the winter as mycelium in the stubs of blighted shoots. It also overwinters in blighted foliage on the ground, where pseudothecia (sexual fruiting bodies) develop during the dormant season. In the spring, sexual spores are released into the air and can cause initial infections. The mycelium in stubs of blighted shoots may also produce infective asexual spores that are dispersed by rain splash. If the weather remains wet, subsequent cycles of infection are caused by asexual spores produced on diseased tissues. Lateral shoots cease growth early in the season and become resistant, so most infections later in the season are on terminals that continue to grow and remain succulent.

**Impact**—During very wet years, particularly early in the growing season when tissues are succulent, the disease can kill nearly all terminal shoots of suckers and small saplings. Growth rate is reduced and the stem may become crooked as lateral shoots assume dominance. If infection is repeated for years, growth may take on a shrubby form as the terminals are repeatedly killed back. Smaller plants may be killed when disease is severe. The disease is most severe on moist sites.

A quantitative study using wounding to simulate the disease concluded that it significantly reduces height growth (ref. 16). However, the effect was temporary; differences between treated and control plants diminished over several years.

There is a high degree of genetic variation in resistance to shepherd's crook. Resistant genotypes have higher concentrations of condensed tannins in their tissues.

**Management**—No practical management approaches are available for this disease in a forest environment. Because the effects are usually temporary and aspen populations recover, the disease is usually not a major management concern.

**References: 16, 84, 158**

# Burls, Galls, and Tumors

## Abnormal swellings on stems and branches

**Cause**—Burls, galls, and tumors may be caused by bacteria, fungi, insects, environmental stress, or genetic predisposition. In the Rocky Mountain Region, insect and pathogen-induced burls and galls are common on hardwoods, but the cause of conifer stem galls is often unknown. Galls on ponderosa and lodgepole pines, caused by western gall rust, are discussed separately in this guide.

**Hosts**—Many conifer and hardwood species may be hosts. In the Rocky Mountain Region, incidence is especially common in lodgepole pine, Douglas-fir, Engelmann spruce, subalpine fir, and aspen.

**Signs and Symptoms**—Burls, galls, and tumors are abnormal swellings that typically occur on stems and branches (figs. 242-244). They often form in succession along the stem. Swellings vary in size and may be smooth or rough with rounded or flared edges. Adventitious buds or sprouts sometimes protrude from the surface. Trees of all ages and sizes may be affected, and trees may occur singly or in groups. Incidence is especially frequent in high-elevation sites of the Rocky Mountain Region.

**Impact**—Growth impacts are minimal, but burls and other abnormal swellings may cause significant defect and deformation, which lowers timber value. On the other hand, researchers found less decay in aspen trees that had black stem galls in Alberta, Canada, suggesting that galled trees may be more valuable from a timber perspective in certain situations.

**Figure 242.** Several burls on an Engelmann spruce stem, cause unknown. *Photo: Kelly S. Burns, USDA Forest Service.*

**Figure 243.** Black stem gall on quaking aspen, cause unknown. *Photo: William Jacobi, Colorado State University.*

**Figure 244.** Galls on a quaking aspen caused by the poplar vagabond aphid (*Mordwilkoja vagabunda*). *Photo: Whitney Cranshaw, Colorado State University, Bugwood.org.*

**Figure 245.** A sculpture created from a burled spruce tree. *Photo: Jim Worrall, USDA Forest Service.*

The distinctive wood is also treasured by artists who use it for sculptures, furniture, and other forms of woodworking (fig. 245).

**Management**—Control strategies have not been developed for these swellings, and management is not generally warranted. However, avoiding wounding and removing damaged trees during intermediate stand treatments and maintaining host vigor may reduce impacts.

**References: 31, 131, 159**

# Animal Damage

## Various animal species cause physical damage

**Cause**—A small number of animal species cause significant damage to trees. The animals that primarily damage trees in the Rocky Mountain Region include deer, elk, porcupines, beaver, mice, squirrels, gophers, rabbits, and birds.

**Hosts**—All tree species can be damaged by animals, and damage occurs throughout the Region. The type and extent of damage varies by animal and tree species because animals differ in their preference for tree species.

**Signs and Symptoms**—In some situations, many trees in an area are affected, and in other situations, only individual trees are damaged. Damages include removal of bark, wood, foliage, and twigs. Effects of the damage include tree girdling, scarring, deforming, brooming, stunting, and callus ridge formation. Callus ridges are the result of natural healing of trees as new bark forms around damaged bark.

**Figure 247.** Repeated browsing can cause the production of multiple stems or brooming. *Photo: Susan K. Hagle, USDA Forest Service, Bugwood.org.*

**Figure 248.** Aspen sucker browsed by elk. *Photo: James T. Blodgett, USDA Forest Service.*

**Figure 246.** Antler rubbing damage with shredded bark and callus ridges. *Photo: Susan K. Hagle, USDA Forest Service, Bugwood.org.*

Common damages and their causes in the Region include: antler rubbing of tree bark by deer and elk (fig. 246); browsing of foliage, buds, and young shoots by ungulates (figs. 247-248); bark feeding by elk (fig. 249) or bear; girdling by various mammals (fig. 250), including porcupines (fig. 251) and beavers; rabbit and gopher chewing; and debarking or scratching by bears (fig. 252). Porcupines chew through the outer bark of branches and boles of conifers, especially in the winter. Bears may rip wide strips of bark from trees and scrape bark with their claws. The sapsuckers, large birds in the genus *Sphyrapicus*, make uniform horizontal rows of holes in the bark of conifer and hardwood species (fig. 253). Sapsuckers feed on sap that oozes from holes they make on the stem and on insects attracted to the sap. Tooth, claw, or antler scrape marks are usually visible in the sapwood of damaged trees. Rubbing by deer and elk often produces shredded bark attached to the damaged areas. Mechanical damage can be difficult to differentiate from animal damage unless the scrape marks or shredded bark are evident. Circumstances can provide clues to the cause of the damage.

Squirrels, porcupines, and mice often chew the bark around rust-induced cankers. However, they also feed in the top or mid-stems of healthy trees. Squirrels also clip small

**Figure 249.** Elk feeding damage on aspen bark, Salida Ranger District, San Isabel National Forest. *Photo: Dave Powell, USDA Forest Service, Bugwood.org.*

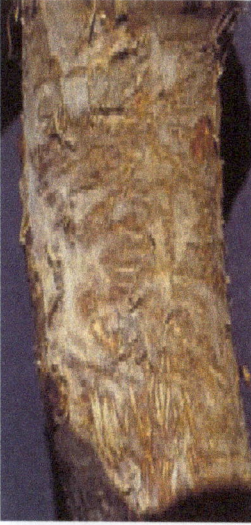

**Figure 250.** Tooth marks from vole feeding on a root. *Photo: Paul Bachi, University of Kentucky Research & Education Center, Bugwood.org.*

**Figure 251.** Tooth marks from porcupine feeding within the margins of a sporulating rust canker. *Photo: Brian W. Geils, USDA Forest Service, Bugwood. org.*

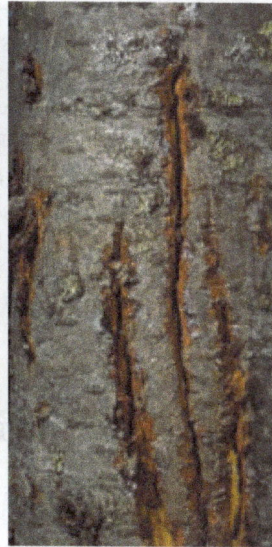

**Figure 252.** Bear claw marks are distinct when fresh. *Photo: USDA Forest Service, Bugwood.org.*

branches from trees, leaving green branches on the ground. Rabbits and ground-dwelling rodents feed on the bark of young trees near the ground and can scar or girdle trees. Beavers cut down larger trees for construction and/or collect branches from trees and saplings to eat. This results in damage near the ground, including girdling and felling of trees. Underground rodents such as pocket gophers feed on roots.

Seeds and cones can be damaged by rodents and birds and occasionally larger mammals such as bears.

**Impact**—The overall losses from wild animal damage are usually far less than losses caused by diseases, insects, fire, or even domestic animals. However, large losses can occur in limited areas, especially to aspen regeneration. The wounds allow entry of diseases (especially canker and wood decay fungi) and may attract insects.

If stems or branches are girdled, mortality of the tree part above the girdle results. Extensive damage without girdling can result in growth loss or stunting. In most cases, callus will form and trees will recover if not attacked by diseases or insects.

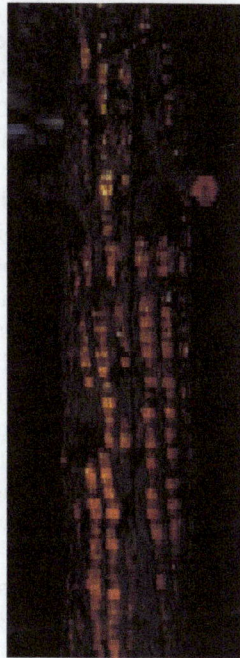

**Figure 253.** Sapsuckers make distinctive holes on the stem that are regularly spaced in rows. *Photo: James T. Blodgett, USDA Forest Service.*

Regeneration can also be impacted. Browsing can cause brooming, mortality, and vegetation removal, resulting in severe loss of forest regeneration. This is especially common with aspen and other hardwood species. Rodents can girdle seedlings and saplings. Rodents and birds eat and disperse seeds, which can both negatively and positively affect regeneration.

**Management**—Numerous management methods and tools are available, including chemical deterrents/repellents, exclusion with fencing or other means, devices to frighten animals, habitat modification, relocation, traps, or lethal control. The application of any of these methods depends on the pest involved, the setting, and local laws and regulations.

**References: 61, 112**

# Abiotic Foliage Damage

## Often affects many host species in an area

**Cause**—Common causes of injury or damage to foliage by non-living agents include fire, drought, frost, and chemicals.

**Hosts**—All tree species throughout the Rocky Mountain Region can be damaged.

**Signs and Symptoms**—Identification of abiotic foliage damage often requires learning the history of the site, comparing injury between trees and other plants of the same and different species, looking for patterns across the site, and ruling out damage from other agents such as diseases and insects. The factors causing these damages often operate for a brief period, but some can persist for much of the growing season as with drought.

Common damages in the region:

- Fire injury to foliage can be from direct burning or from radiant heat. Mortality is common, especially if stems are affected. See the Abiotic Stem Damage entry in this guide. The damage can look similar to some foliage diseases or other abiotic foliage damage, but charred or burned vegetation around the trees is often evident.

- Drought symptoms are usually expressed in the foliage, but drought affects the whole tree (figs. 254-255). Direct drought mortality is not common in older, well-established trees, but damage to young trees and especially regeneration is common. Drought stress is a normal phenomenon in trees when water loss by transpiration exceeds the rate of absorption from soil. In this Region, drought stress occurs frequently in mid to late summer. Direct mortality from drought only occurs when a tree's water potential falls below the permanent wilting point. Drought stress at levels that do not cause mortality can predispose trees to some diseases (especially cankers and root diseases), insects (especially bark beetles), and fire.

**Figure 254.** This tree shed older needles first due to drought. *Photo: Susan K. Hagle, USDA Forest Service, Bugwood.org.*

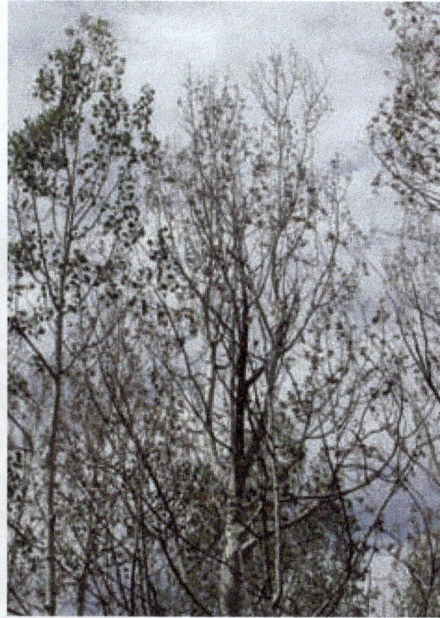

**Figure 255.** Aspen after chronic drought. *Photo: Southwestern Region, USDA Forest Service.*

Symptoms of drought include reduced growth, wilting, foliage discoloration and damage, premature leaf fall, and/or early dormancy. With extreme drought, new foliage is shed, shoots die back to lateral buds, and tree mortality may occur. In conifers, common symptoms are premature browning and shedding of older needles, stunted needles, and brown tips of new foliage. In hardwoods, common symptoms include wilting, yellowing of foliage, reddish brown to brown leaf tip and/or margins (scorch), and premature leaf drop. Injury is most severe on south and southwest slopes, shallow or sandy soils, and during periods of heat and unusually low rainfall. Symptoms are similar to those of trees suffering from root, wilt, or canker diseases. With root diseases, roots die before the foliage and a root disease pathogen can be found in the roots. Trees with wilt diseases may have vascular discoloration and a wilt disease pathogen can be found in the outer xylem tissues. Canker diseases cause the same crown symptoms when a canker pathogen girdles stems and/or branches.

- Winter desiccation injury occurs when solar warming of south and southwest tree crowns causes leaves to transpire when roots are frozen and unable to replace moisture (fig. 256). Foliar damage may be predominantly on the southwest aspect of the crown, but occasionally, damage is located more on the side of prevailing winds. The portions of crowns covered by snow are not damaged.

Winter desiccation can damage or kill conifer foliage but seldom damages the buds and branches. Needles damaged or killed often remain green when temperatures are cold, turn yellow or red-brown when temperatures increase in spring, and are shed in early summer. New foliage usually emerges in late spring or early summer. Because buds and branches usually survive, winter desiccation typically results in minor growth loss and aesthetic problems of conifers. If buds and/or branches die, multiple stems often form, resulting in a bushy tree form. Although not common, top-kill or mortality has been reported. Symptoms occur across the landscape, but individual trees vary in symptom expression.

**Figure 256.** Winter desiccation damages the foliage exposed above the snow. *Photo: Scott Tunnock, USDA Forest Service, Bugwood.org.*

- Red belt is a type of winter desiccation (figs. 257-258). This damage affects distinct patches of conifers, often in well-defined horizontal bands (belts). South and west slopes are usually affected more than north and east slopes. This is a weather-related winter phenomenon on slopes where rapid temperature changes cause desiccation of foliage but rarely cause bud mortality. Red belt is believed to be caused by air inversions in valleys. A thin layer of warm air is trapped and cannot mix downward, so it continues across the slopes.

Symptoms are expressed in the spring when needles turn red-brown along distinct patches or horizontal bands. Needles can become entirely red-brown

**Figure 257.** Red belt damage on hillside in early spring. *Photo: USDA Forest Service, Bugwood.org.*

**Figure 258.** Close-up of red belt damage. The outer foliage is usually the most damaged. *Photo: Oscar Dooling, USDA Forest Service, Bugwood.org.*

**Figure 259.** Frost damage to buds of Douglas-fir. *Photo: USDA Forest Service, Bugwood.org.*

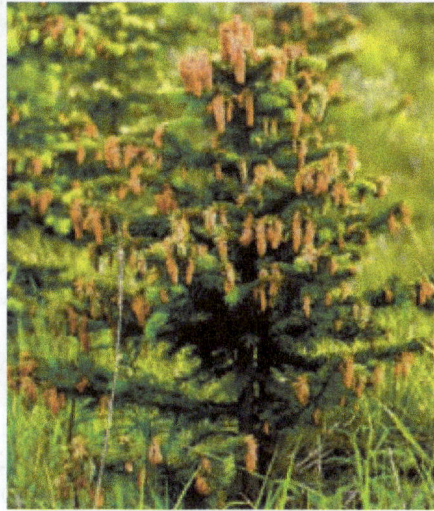

**Figure 261.** Frost damage with branch droop. *Photo: Petr Kapitola, State Phytosanitary Administration, Bugwood.org.*

**Figure 260.** Frost damage to red oak shoots. *Photo: Andrew J. Boone, South Carolina Forestry Commission, Bugwood. org.*

or, in less severe cases, discolored only at their tips. Upper portions of trees and the outermost foliage are most affected. Trees usually recover during the growing season with little residual effect.

• Frost or freeze injury occurs in the spring when susceptible new foliage or young shoots are exposed to sudden freezing temperatures (figs. 259-261). Greatest vulnerability is from bud break to final shoot and leaf elongation. The extent of damage depends on the stage of development at the time of frost. New leaves and shoots are often killed, especially during earlier expansion. Damage frequently occurs on several species in an area.

**Figure 262.** Salt damage is evident only near the road. *Photo: William M. Ciesla, Forest Health Management International, Bugwood.org.*

Within a day or two of freeze injury, foliage and shoots become limp and begin to fade to yellow. Damaged tissues at first appear water soaked. After a week or more, tissues become shriveled; foliage turns reddish brown, dark brown, or black, depending on the species, and can droop; and young shoots turn brown and can droop. Dead buds become dark brown in the interior. Dead leaves and shoots eventually break off or abscise during the next few weeks. New shoots and leaves begin to grow almost immediately.

- Road salt is a common type of chemical damage (fig. 262). Salt can affect the foliage indirectly through root absorption or directly when salt spray settles on trees near high-speed roads. Damage can occur wherever chlorides (sodium, calcium, or magnesium) are applied for deicing in the winter or dust abatement in the summer. Most of the time, symptoms are insignificant and injury is very close to a road. Symptoms may be severe along highways where large quantities of salt are used. Symptoms can also occur in low areas where salts accumulate after runoff. Plant tissue can be analyzed in a laboratory to determine the amount of salt present. Symptoms can differ between the two entry methods.

Root uptake damage is evident after spring thaw in conifers in second-year needles and soon after budbreak in hardwoods. Burn of needle tips, with the bases of needles remaining green, is common in conifers, and browning of leaf margins is common in hardwoods. With higher salt concentrations, buds will break with reduced foliage, more pronounced tip burn or brown leaf margins develop, and trees have reduced growth. Symptoms can intensify through the season with 1-year-old needles showing symptoms mid-winter and premature leaf drop of the 2- and 3-year-old needles in conifers. Symptoms of severe damage include delayed or lack of bud break and branch or tree mortality.

**Figure 263.** Herbicide damage on pine. *Photo: Susan K. Hagle, USDA Forest Service, Bugwood. org.*

**Figure 264.** Distorted shoots due to herbicide damage on pine. *Photo: Minnesota Department of Natural Resources, Bugwood.org.*

On conifers, tip browning from salt spray is common, and browning can progress toward the base of needles throughout the growing season. Salt spray in hardwoods often results in mortality of terminal buds. This mortality can result in bud break of several lateral buds, causing brooms. Unlike damage caused by root uptake, marginal browning rarely occurs with salt spray on hardwoods.

• Herbicide damage is another type of chemical damage (figs. 263-265). Herbicides can affect foliage directly or indirectly through root uptake. Herbicide damage often follows a pattern consistent with application of the chemical, such as a swath about the width of a spray boom where weeds were sprayed. Weed treatment areas include: along roads, rights-of-way, fuelbreaks, or around dwellings. In many situations, plant samples can be submitted to a laboratory to determine the amount and type of chemical in the tissue.

**Figure 265.** Herbicide damage on hardwood species. *Photo: Joseph O'Brien, USDA Forest Service, Bugwood.org.*

Symptoms vary by chemical. Preemergence: photosynthetic inhibitors (simazine) can cause leaf yellowing. Postemergence: growth regulators (Glyphosate) can cause abnormal leaf development, tip yellowing, and dieback; Picloram (2,4-D) symptoms are similar to glyphosate and also include twisted petioles and shoots. Contact (Paraquat) injury results from drift, causing small dead spots on foliage. High concentrations of any of these herbicides can cause mortality.

**Impact**—Losses in the Region from abiotic foliage damage are usually minor compared to those caused by diseases and insects, with the exception of fire. Most biotic diseases and insects affect a single tree species or group of species. Abiotic agents often affect various tree, shrub, and herbaceous species in an area. However, some tree species are more resilient to some of the damage agents.

Extensive damage from any of these agents can result in growth loss or stunting. However, abiotic foliage damage rarely causes direct mortality, with the exception of fire and, in extreme cases, herbicide damage. Stress associated with these damages can make trees more susceptible to some diseases (especially cankers and root diseases) and insects (such as wood borers and bark beetles), which can result in mortality.

**Management**—There are few management options for many of the abiotic foliage damages, with the exception of fire, drought, salt, and herbicide damage.

Fire management should include elements of fire prevention and fire suppression. Because many fires are caused by people, educating the public is the best option. Methods that reduce the spread of fires such as reducing fuel loads, thinning stands, and creating firebreaks will reduce the number of trees damaged. After fire, prompt salvage cuts can be used to remove trees of value before they are colonized by decay fungi and wood borers, if consistent with management objectives.

For drought, management of competing vegetation, stand thinning, planting techniques that reduce drought, and selection of a tree species compatible with a site will reduce losses.

For road salt, prudent use will reduce losses. To reduce over-application, machines should be calibrated to apply the optimum level of salt for reducing surface ice or dust in order to not apply excessive levels. Mortality can be high when salts are applied for dust abatement before heavy rainstorms, so this should be avoided. In areas where trees are planted along roads, salt-tolerant species could be used. The amount of salt damage is directly related to the distance from a road, with the exception of salt runoff in limited downhill areas. Therefore, planting trees farther from the road or cutting trees near a road will reduce salt damage and also improve visibility.

To reduce herbicide damage, read the label and apply the chemical as directed. Also, check and properly calibrate equipment. Select herbicides that will do the job but pose the least risk to trees in the area. Tree species vary in tolerance to some herbicides.

**References: 17, 18, 19, 38, 40, 83, 121, 159**

# Abiotic Stem Damage

## Injury or damage from non-living agents

**Cause**—Common causes of injury or damage to stems by non-living agents include mechanical abrasion, physical constriction, fire, lightning strikes, hot and cold extremes, and high winds.

**Hosts**—All tree species throughout the Rocky Mountain Region can be damaged. Species with thin bark and most young trees are more susceptible to mechanical damage, fire, freeze injury/sunscald, solar heat injury, frost cracks, and hail.

**Signs and Symptoms**—Common damages in the Region:

• Mechanical damage is caused by impacts or rubbing from various sources such as vehicles (skidders, cars, etc.), falling trees or branches hitting other trees, people chopping with axes and hatchets, or other contact sources (fig. 266). Mechanical injury may lead to resin production in conifers, and callus ridges eventually form at the edges of the damage. Initially, scarring or deformation of the tree results. Depending on the wound size, the callus can form new bark over time and the wound can heal. The wounds allow entry of wood-decay fungi. If trees are damaged completely around the stem (girdled), the part above the girdle dies.

**Figure 266.** A mechanical wound caused during logging. *Photo: Susan K. Hagle, USDA Forest Service, Bugwood.org.*

• Fire injury can be from direct burning or from radiant heat (figs. 267-268). Initially, fire injury is obvious and mortality is common. Fire scars can be superficial or deep and are often resinous on conifers that are not killed. Callus ridges eventually form at the edges of the damage on trees that survive the initial fire. Over time, some evidence of char usually can be found on the bark and/or on exposed wood at the base of trees. Trees that survive the initial fire injury are often

**Figure 267.** Old fire scar with intact bark covering most of the injury. *Photo: Susan K. Hagle, USDA Forest Service.*

**Figure 268.** Charred bark and resin flow are common symptoms of fire damage. *Photo: Kurt K. Allen, USDA Forest Service.*

**Figure 271.** Sunscald on the south side of thin-barked trees. *Photo: Susan K. Hagle, USDA Forest Service, Bugwood.org.*

**Figure 270.** Lightning injury to an elm. *Photo: William Jacobi, Colorado State University, Bugwood.org.*

**Figure 269.** Lightning-struck trees are often attacked by bark beetles or decay fungi. *Photo: James T. Blodgett, USDA Forest Service.*

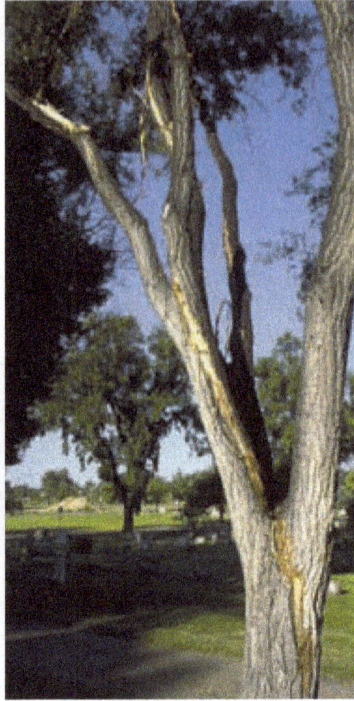

attacked by diseases such as root, butt, or stem decays and/or by insects such as bark beetles.

• Lightning can result in direct mortality. However, in most cases, trees are only slightly damaged (figs. 269-270). Trees can be blown apart at impact, or lightning can cause fires that result in fire injury. More often, damage is in the form of a long, narrow furrow in the bark with thin layers of wood blown out. The furrow often extends down to the soil and may spiral to some extent. Bark beetles occasionally colonize trees soon after lightning strikes, and the wounds can allow entry of diseases, especially wood decays.

• Freeze injury/sunscald is caused by a sudden drop in temperature in the winter (i.e., sudden freezing temperatures) that damages sun-warmed and, therefore, non-hardened cambium (fig. 271). Symptoms are seen on the south to southwest side of stems. At first, the bark is discolored; if the injury is severe enough the bark will become rough and dead bark will flake. If the cambium is killed, the scar extends to the sapwood and bark is eventually sloughed off. Little or no resinosis results. A ridge of callus will form between live and dead cambium.

• Solar heat injury looks similar to freeze injury/sunscald but occurs when bark is suddenly exposed to intense sun (heat), which can result from pruning or stand thinning. This does not occur if the bark developed in the sun or when the bark is slowly exposed to the sun, and it only occurs in thin-barked species. As with freeze injury/sunscald, solar heat injury occurs on the south to southwest side of stems and produces the same symptoms. Damage to stems from solar heat is not well documented and is occasionally referred to as sunscald.

• Frost cracks are caused by an extreme drop in temperature during the winter when trees are dormant (fig. 272). Shrinking of the bark or outer wood as it cools quickly, compared to warmer inner wood, causes the bark or wood to separate or split. This can cause short or long cracks that are usually vertical, but long cracks can spiral somewhat on the stem. Unlike lightning damage, frost cracks usually have little bark loss. Cracks can form deep into the wood. Frost cracks seldom result in mortality. However, this damage often leads to extensive heartrot when cracks extend deep into the stem. Callus ridges form on both sides of the crack.

• Mechanical girdling is caused most often by people wrapping wire, rope, or other objects partially or completely around trees (figs. 273-274). Vines or other

**Figure 272.** Long, vertical frost crack. *Photo: James Solomon, USDA Forest Service, Bugwood.org.*

**Figure 273.** Girdling by barbed wire. *Photo: Robert L. Anderson, USDA Forest Service, Bug-wood.org.*

**Figure 274.** Girdling by a wire. *Photo: Francis Gwyn Jones, Bugwood.org.*

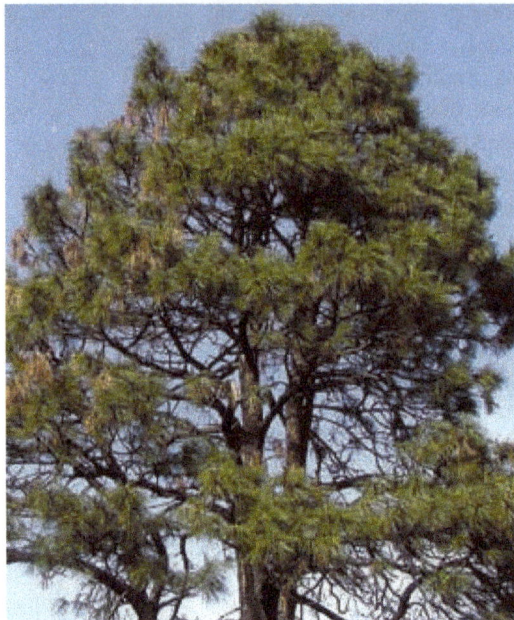

Figure 275. Directional damage caused by hail. *Photo: James T. Blodgett, USDA Forest Service.*

Figure 276. Close-up of hail damage on an aspen. *Photo: Steven Katovich, USDA Forest Service, Bugwood.org.*

natural objects occasionally cause mechanical girdling. Because trees continually grow in diameter, mechanical girdling can eventually result in tree mortality or mortality of the part of the tree exposed to the girdling object.

• <u>Hail</u> damage is a type of mechanical damage caused by hail impacts (figs. 275-276). This damage often affects trees in a larger patchy area, is directional (on the upper side of horizontal branches), and does not girdle stems or branches. Therefore, stem and branch mortality does not occur unless the stems or branches are snapped by the impacts. Initial scarring results, but if trees are not infected by diseases such as pine shoot blight or other canker diseases, callus ridges eventually form and wounds heal.

• <u>Wind</u> damage from very strong winds can result in snapping of the main stem (fig. 277). This type of damage is not

Figure 277. Damage from strong winds. *Photo: Gil Wojciech, Polish Forest Research Institute, Bugwood.org.*

common with healthy stems in the Rocky Mountain Region but has been reported. Snapping or breaking of stems is common when associated with decay. Snapping associated with stem decay or root and butt rot can occur in conditions from no wind to high winds. Snapping of solid (non-decayed) stems requires very high winds and usually results in many damaged trees in an area in a short time. Wind-broken wood has a more jagged appearance than breaks associated with decay. Winds can also shake stems, causing separation of annual rings. Resin deposition in wood at the damage site and/or wood separation can result in significant log defects.

**Impact**—Direct losses from abiotic stem damage are usually minor compared to those caused by diseases and insects, with the exception of fire. In some situations, many trees and/or species in an area are affected (such is the case from fire, high wind, and hail), and in other situations, only individual trees are damaged. Extensive damage from any of these agents can result in growth loss, stunting, or mortality. Wounds are important as points of entry for diseases (especially cankers and wood decays) and insects. Callus ridges are the result of natural healing of trees as new bark forms around damaged bark. In most cases, callus will form and trees will recover if not attacked by diseases or insects.

**Management**—There are few management options for most of the abiotic stem damages, with the exception of mechanical damage and mechanical girdling caused by people and fire damage. Care should be exercised in pruning and thinning to avoid sudden exposure that could lead to freeze injury/sunscald, solar heat injury, or frost cracks.

For mechanical damage and mechanical girdling caused by people, public education is the best option. For mechanical damage associated with logging, options include: minimize stand entry during management, keep vehicles away from leave trees, use bumper trees, and include fines or value reductions in contracts for the number of trees damaged.

Fire management should include elements of fire prevention and fire suppression. Because many fires are caused by people, public education is the best option. Methods that reduce the spread of fires such as reducing fuel loads, thinning stands, and creating firebreaks will reduce the number of trees damaged. After fire, prompt salvage cuts can be used to remove trees of value before they are colonized by decay fungi and wood borers. Sanitation cuts can be used to remove damaged trees, if consistent with management options.

**References: 19, 159**

# Wetwood

## Watersoaked, discolored, and often smelly wood in living trees

**Cause**—Wetwood is not a disease. It results from a physiological process that occurs when the living cells in the wood die. Although it is sometimes called

**Figure 278.** Wetwood in cross section of narrowleaf cottonwood. *Photo: Jim Worrall, USDA Forest Service.*

**Figure 279.** Wetwood in cross section of white fir. *Photo: Jim Worrall, USDA Forest Service.*

"bacterial" wetwood, this is a misnomer because it is not caused by bacteria. However, various bacteria colonize wetwood and, in some cases, cause additional damaging effects.

For many years, wetwood was thought to be a bacterial disease. However, later experiments showed

**Figure 280.** Wetwood in radial section of white fir. Note the narrow, dry transition zone between sapwood on left and wetwood. *Photo: Jim Worrall, USDA Forest Service.*

that wetwood can be formed following wounding under conditions that preclude bacterial growth. During death of the parenchyma in wood, calcium, potassium, and magnesium salts are mobilized into the area, lowering the osmotic potential. Moisture accumulates in response to the osmotic gradient, and a drier transition zone with living parenchyma separates sapwood from wetwood.

**Hosts**—Wetwood occurs in many hardwoods and in some conifers. In this Region, wetwood often occurs in elm, maple, ash, cottonwood, aspen, and white fir.

**Signs and Symptoms**—Wetwood is wood that does not conduct water but has a watersoaked, dark appearance (figs. 278-280) and a fetid, fermentative odor. It is the normal condition of the heartwood in some species but may occur uncommonly in others. It may also appear in sapwood in response to wounds or biological attacks.

Because of bacterial fermentation, the wetwood liquid may be under pressure. In some areas where white fir is harvested, squirting of the foul liquid under pressure onto loggers has led to the appellation "piss-fir." Externally, especially in elm, the fermented liquid may be exuded through cracks or branch stubs, leaving streaks of bleached or discolored bark (figs. 281-282). The exudate is often colonized by additional micro-organisms when it reaches the surface, leading to a thickened consistency and additional odors. Such liquid is called wetwood slime and the condition is called slime flux.

**Impact**—Wetwood inhibits wood-decay fungi. Wood that is wet has very low $O_2$ availability, and the bacteria present in wetwood further reduce the $O_2$ content to a level too low to support fungal growth. Bacteria also produce organic acids that strongly inhibit fungal growth. Thus, the bacteria generally seem to function more as beneficial symbionts, aiding the tree's defense, rather than as pathogens.

In some species, most notably elm, wetwood fluid may develop pressure due to gasses produced by the bacteria. When this happens, the liquid may be forced into and kill adjacent sapwood and cambium. In severe cases foliar symptoms may result, including scorch, wilt, yellowing, and defoliation. The fermented liquid may seep out through branch stubs and cracks and discolor the bark surface as it flows down.

Wetwood is associated with a variety of problems when making wood products:

**Figure 281.** Bark bleaching and discoloration from wetwood fluid flowing from branch stub in elm. *Photo: William Jacobi, Colorado State University, Bugwood.org.*

**Figure 282.** Bark bleaching and discoloration from wetwood fluid flowing from crack in elm. *Photo: Mike Schomaker, Colorado State Forest Service, Bugwood.org.*

- Wood is more difficult and requires more energy to dry.
- Wood dries unevenly and may warp and twist.
- During kiln drying, vapors of the volatile organic acids (acetic, propionic, butyric) cause kiln corrosion.
- Wetwood is associated with ring shake and honeycomb, two lumber defects. Ring shake in elm has led to the term "onion elm" in the lumber trade.

**References: 30, 158, 195, 196, 197**

# Insects

# Introduction to Bark Beetles

Bark beetles are the most destructive insects in western coniferous forests. It has been estimated that 90% of insect-caused tree mortality and more than 60% of the total insect-caused loss of wood growth in the United States is due to bark beetles.

The Rocky Mountain Region has a large complex of bark beetles composed of many genera and species. Frequently, several species are found attacking the same host tree, and therefore, it may be difficult to discern what species initiated the attack. Although species of *Dendroctonus* are the most significant tree killers in the western United States, other bark beetle species also play important roles in the conifer forests of the Rocky Mountain Region (fig. 283).

a. Dendroctonus

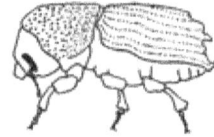

b. Ips

c. Scolytus

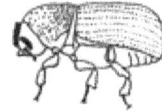

d. Dryocoetes

**Figure 283.** Adult bark beetles (from Hagle and others 2003).

**General Features**—Bark beetles derive their name because they live and develop in the bark and wood of trees and shrubs. Adults excavate egg galleries in living bark (phloem). All bark beetle life stages are spent in the phloem, inner bark, and bark, except when adults leave the tree they developed in to fly to new host trees. Bark beetles feed on the phloem of their host trees during adult and larval stages.

Crowns of successfully attacked trees turn from green to yellow to reddish brown. This color change, an indication of a dying tree, may occur from a month to more than 2 years after successful attack, depending on the temperature, moisture conditions, and density of beetles in the tree. Close inspection of infested tree trunks will show either small globules of resin (pitch tubes), small holes through the bark, or reddish boring dust in bark crevices and around the tree base. The removal of bark from infested trees will reveal two types of galleries: egg and larval (fig. 284). Egg galleries constructed by adult beetles are rather uniform in width. Larval galleries depart at right angles from egg galleries and increase in size as the young grow.

**Life Cycle**—Bark beetles have four stages of development during their life: egg, larva, pupa, and adult. All stages are found under the bark. The small, white eggs are in niches along the sides of the egg gallery. Larvae are small, white grubs with distinct heads and no legs (fig. 285a). They are C-shaped and are found in the feeding galleries off the sides of the egg gallery. The pupae are found in small chambers at the end of the feeding galleries.

**Figure 284.** Bark beetle gallery patterns (from Hagle and others 2003).

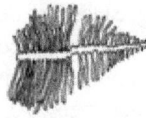

a. Fir engraver

b. Western pine beetle

c. Western balsam bark beetle

d. Mountain pine beetle, Jeffrey pine beetle

e. Spruce beetle

f. Douglas-fir beetle

g. Pine engraver, Piñon engraver beetles

h. Red turpentine beetle

Many bark beetles prefer weakened host trees, however, during environmental conditions favorable for beetle development, populations may build up rapidly and successfully attack healthy trees. Most bark beetles have a symbiotic relationship with blue-stain fungi. The blue stain fungi can completely penetrate the sapwood within a year. The fungi invade the living tissues in sapwood—the

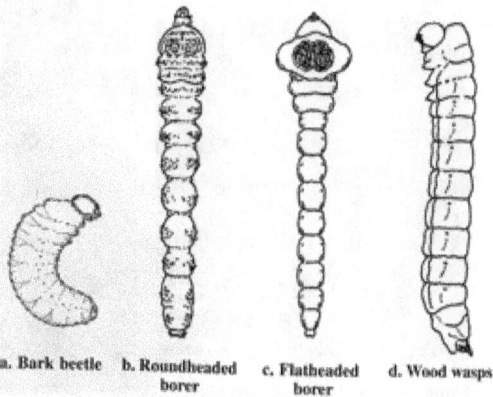

a. Bark beetle  b. Roundheaded borer  c. Flatheaded borer  d. Wood wasps

**Figure 285.** Larval bark beetle compared to larvae of other common boring insects (from Hagle and others 2003).

ray parenchyma and epithelial cells of resin canals—and cause the death of the sapwood. This action, plus the bark beetle feeding, causes the death of a host tree.

Many bark beetles produce chemical compounds called pheromones that are used to communicate with other beetles. Aggregation pheromones cause beetles to congregate in certain areas and mass-attack trees. Anti-aggregation pheromones cause beetles to disperse to neighboring trees or other areas.

**Reference: 60**

# Ash Bark Beetles

## Causes branch dieback on ash

**Name and Description**—*Hylesinus* spp. [Coleoptera: Curculionidae: Scolytinae]

Larvae of ash bark beetles are white, C-shaped grubs. Adults are small, gray beetles about $1/13$-$1/6$ inch (2-4 mm) long with black markings (fig. 286).

**Hosts**—Ash

**Life Cycle**—Adults overwinter in tunnels under the bark of infested branches. In the spring, females begin constructing egg galleries in trees that are typically recently felled or weakened host trees. The egg galleries run perpendicular to the branch or trunk (fig. 287). Eggs are laid along the sides of the galleries. Larvae tunnel between the bark and the wood throughout the summer, feeding away from the egg gallery. There can be one to three generations per year, depending on location.

**Figure 286.** Adult ash bark beetle. *Photo: David Cappaert, Michigan State University, Bugwood.org.*

**Figure 287.** Ash bark beetle galleries. *Photo: James Solomon, USDA Forest Service, Bugwood.org.*

**Figure 288.** Ash bark beetle damage. *Photo: James Solomon, USDA Forest Service, Bugwood.org.*

**Damage**—Generally, the favored breeding material is recently cut or broken trees. Living trees weakened by mechanical damage or disease may also be attacked. Entrance, exit, and breathing holes can be found on the outside of infested trees (fig. 288). In July or August, the leaves on branches that have been girdled will turn yellow and then brown as the branch dies.

**Management**—Management is generally not warranted. As these beetles generally attack much-stressed trees, maintaining tree vigor will reduce impact. Infested branches can be removed prior to beetle emergence if necessary.

**References: 39, 174**

# Aspen bark beetles

## Attack stressed aspen

**Name and Description**—*Trypophloeus populi* Hopkins and *Procryphalus mucronatus* (LeConte) [Coleoptera: Curculionidae: Scolytinae]

Two closely related scolytid beetles (fig. 289) attack stressed aspen, and both are members of the tribe Cryphalini. Difficulty in distinguishing the two arises from their close taxonomic relationship and from their very small size. Both species of beetles are less than 1/10 inch (1.7-2.2 mm) long. While a number of characteristics can help determine probable species identity in the field and aid in identification when suitable specimens are unavailable (i.e., post-emergence), the only certain means of identification is collecting specimens for later examination in the laboratory.

**Host**—Aspen

**Figure 289.** Adult *Trypophloeus populi*. Size and appearance of *Procryphalus mucronatus* is similar; however, note the distinct basal margin of the pronotum on this insect. The basal pronotum margin in *P. mucronatus* is rounded. Also, there is a lack of setae on the basal half of the pronotum of *Trypophloeus populi*, while *Procryphalus mucronatus* has setae covering the entire pronotum. *Photo: Tom Eager, USDA Forest Service.*

**Life Cycle**—The natural history of these beetles has been described in detail in ref. 134. This description includes information regarding the insects' mating behaviors, gallery construction methods, and phenologies. For both of these species, the female initiates gallery construction and is soon joined by a male. Both beetles are monogamous species, and the male goes to some lengths to exclude rival males from the gallery. Following gallery construction and mating, a series of eggs are deposited in the gallery. Larvae feed upon phloem tissues in galleries extending from the egg galleries. The sudden appearance of the male at female initiated galleries indicates that some sort of aggregating pheromone may be utilized by these beetles. The exact nature of the pheromones and whether they serve as further isolating mechanisms is

**Table 14.** Comparison of aspen bark beetle phenology.

| | Overwinter | Late May to mid-June | Mid-June to Autumn |
|---|---|---|---|
| Procryphalus mucronatus | Larvae, pupae, and pre-emergent adults | Emergence, attack of new host, and oviposition | Egg hatch and larval to pupal stages with some pre-emergent adults |
| Trypophloeus populi | Larvae | Completion of metamorphosis to pre-emergent adults | Adult emergence, oviposition, and egg hatch to early larval instars |

unknown. Despite the many similarities in the life history of these beetles, they do differ in phenology. These differences can be seen in table 14.

**Damage**—As indicated above, these two species of beetles attack aspens that are under stress. When discussing insects that attack trees, it is common to refer to insects that attack otherwise healthy trees as primary insects. Insects that specialize in attacking trees under stress or that are close to death are referred to as secondaries. Insects that feed upon completely dead trees can be considered tertiary insects. However, it has been noted that *Trypophloeus populi* attacks trees that still have a large component of "green bark," while *Procryphalus mucronatus* is found in trees on which the bark is almost entirely dead. Thus, even though both of these species of insects are considered secondaries, there is still a degree of host selection discrimination based on tree vigor.

Other aspects of the beetles' biologies fit into this construct that *Trypophloeus populi* attacks earlier in the process of host decline than does *Procryphalus mucronatus*. The galleries of *Trypophloeus populi* are typically stained black (fig. 290), which indicates that they carry a symbiotic fungus that helps them overcome the host tree defense mechanisms. The eggs of *T. populi* are laid in clusters, and the larvae hatch out to feed upon the fungal-infused host material. In contrast, *Procryphalus mucronatus* larvae are more spread out in the presumably nutrient deficient dead host tissues. A comparison of several of the key aspects of qualitative differences in the two species of beetles can be seen in table 15.

In addition to the characteristics that can be observed in the field, a number of morphological characteristics can be observed under a high-powered microscope. These distinguishing traits are summarized in table 16.

**Figure 290.** Dark, fungal staining associated with *Trypophloeus populi* galleries. *Photo: Jim Worrall, USDA Forest Service.*

**Management**—The status of these two beetles as secondary insects indicates that the best strategy to reduce their impact is to maintain host vigor. The sudden prominence of these two previously obscure beetles is indicative of the relationship between recent drought and aspen health.

**References: 134, 190, 191**

**Table 15.** Qualitative characteristics to distinguish two species of aspen bark beetles.

|  | Gallery staining | Egg distribution | Gallery construction | Host condition | Post-hatching bark surface |
|---|---|---|---|---|---|
| Procryphalus mucronatus | No | Singly along length of gallery | Generally deeper in wood; entrance nearly perpendicular to bark surface | Tree either dead or very close to death | Bark remains intact above galleries |
| Trypophloeus populi | Black staining surrounding galleries | In clusters in egg niches | Closer to surface of bark; entrance angled into bark | Stressed tree but maintaining green bark | Bark tends to crack open above galleries |

**Table 16.** Morphological characteristics used to determine aspen bark beetle species in laboratory.

|  | Funicle structure (segments of antennae below club) | Antennal club structure | Pronotal setae | Pronotal margins |
|---|---|---|---|---|
| Procryphalus mucronatus | Four segmented funicle | Pointed club | Pronounced setate only on anterior portion of pronotum | No distinct, raised line on basal and lateral margins of pronotum |
| Trypophloeus populi | Five segmented funicle | Elongate, oval club | Pronounced setae on entire surface of pronotum | Distinct, raised line on basal and lateral margins of pronotum |

# Blue Spruce Engraver

## An important pest of ornamental blue spruce

**Name and Description**—*Ips hunteri* Swaine [Coleoptera: Curculionidae: Scolytinae]

The blue spruce engraver is a common bark beetle of Colorado blue spruce. The insect has become a problem in ornamental plantings of Colorado blue spruce along the Front Range of Colorado and southeastern Wyoming. In some forested situations, the blue spruce engraver can be found infesting the tops of Colorado blue spruce trees killed by the spruce beetle, *Dendroctonus rufipennis*. The beetle is approximately 1/8-1/4 inch (3-4 mm) long. The larva is a C-shaped, legless grub.

**Host**—Colorado blue spruce, *Picea pungens*, appears to be the only confirmed host for *Ips hunteri*, although Engelmann spruce, *Picea engelmannii*, is also listed as a host. This bark beetle has been recorded in Arizona, Colorado, Utah, and Wyoming.

**Life Cycle**—The blue spruce engraver has at least two generations per year. The biology of the blue spruce engraver is not known in any detail. Beetle flight is thought to begin in early spring along Colorado's Front Range.

**Damage**—The blue spruce engraver attacks weakened trees and windthrown trees. In urban plantings, however, the beetle infests the tops of drought stressed trees and progressively kills the trees over several years and multiple generations (fig. 291). The egg gallery (fig. 292) generally consists of two branches extending in opposite directions, and it may be diagonal, transverse, longitudinal, or curved. The beetle is occasionally seen killing Colorado blue spruce in forested situations (fig. 293).

**Management**—Homeowners should maintain adequate soil moisture levels for their Colorado blue spruce throughout the entire year with a regular watering schedule. Injury and additional stresses, such as construction activities that damage the root systems of the trees, should be avoided.

Figure 291. Blue spruce that has been top-killed by the blue spruce engraver, *Ips hunteri*. *Photo: William M. Ciesla, Forest Health Management International, Bugwood.org.*

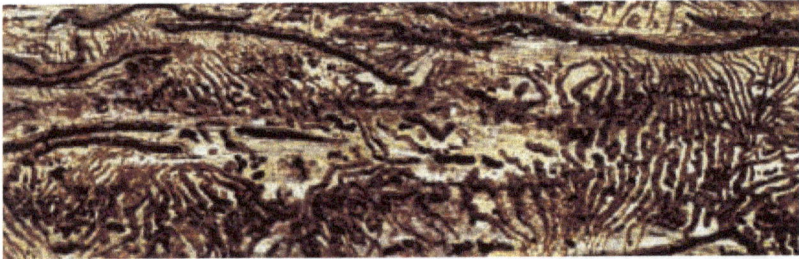

Figure 292. Egg galleries of the blue spruce engraver. *Photo: Whitney Cranshaw, Colorado State University, Bugwood.org.*

Figure 293. Aerial view of blue spruce engraver mortality. *Photo: William M. Ciesla, Forest Health Management International, Bugwood.org.*

Ornamental plantings of Colorado blue spruce that have been top-killed by the blue spruce engraver should be removed because the trees generally decline quickly and contribute to the abundance of beetles in the area. In addition, stand sanitation and removal of windthrown branches can circumvent an increase in beetle populations. Protective insecticide sprays have been used successfully to prevent infestation; treatment should be completed by April along Colorado's Front Range.

**References: 35, 190**

# Cedar Bark Beetles

## Can kill twigs, branches, or entire trees

**Name and Description**—*Phloeosinus* spp. [Coleoptera: Curculionidae: Scolytinae]

Cedar bark beetles are small, reddish brown to black beetles that are approximately 1/8 inch (3 mm) long. Larvae are small, cream-colored, legless grubs with brown head capsules similar to other bark beetle larvae. Cedar bark beetles are not known as aggressive tree killers but can kill twigs, branches, or entire trees.

**Hosts**—Rocky Mountain, Utah, and one-seed junipers, as well as ornamental and windbreak plantings of eastern red cedar.

**Life Cycle**—There is one generation per year. Beetles overwinter beneath the bark as larvae. Adults mature and fly to colonize new hosts from July through September. Timing depends on the particular beetle species and local weather conditions.

**Damage**—Cedar bark beetles typically colonize broken branches and trees stressed by drought, soil compaction, stem breakage, animal damage, and other similar factors. Beetles feed on twigs prior to brood production. Twigs are hollowed out and their tips die as a result. Dead twig/branch tips (often called flagging) may be scattered throughout a tree's crown. This impact may or may not be significant, depending on factors such as number of flagged branches, tree size, and overall tree health. Cedar bark beetles' most significant impact is when they colonize a tree's branches and trunk to produce brood beneath the bark. Death of major branches or of the entire tree may result.

**Figure 294.** Galleries of cedar bark beetles that are etched into the face of sapwood. *Photo: William M. Ciesla, Forest Health Management International, Bugwood.org.*

Evidence of beetle colonization, though often difficult to see, includes the presence of very fine boring dust in bark crevices and around the base of the trunk. If boring dust is present, distinct galleries can likely be detected by removing a small section of bark. The gallery pattern will be etched into the face of the sapwood (fig. 294). Crowns of infested trees will fade from green to yellow and eventually to red. By the time the entire tree crown is red, beetles have likely matured and are gone, having flown to other host trees. If so, many tiny, dust-free holes should be evident on the surface of the bark.

**Management**—Management of cedar bark beetles is seldom necessary in general forest settings. Preventive insecticide sprays, applied in late spring or early summer, have been used to protect individual high-value trees in recreational and historic settings and within windbreaks.

**References: 50, 113**

# Douglas-Fir Beetle

## Attacks and kills Douglas-fir trees

**Name and Description**—*Dendroctonus pseudotsugae* Hopkins [Coleoptera: Curculionidae: Scolytinae]

The Douglas-fir beetle is a common bark beetle that kills Douglas-fir trees. Adult beetles are cylindrically-shaped and about a 1/4 inch (6 mm) long. The head and thorax are black, and wing covers are reddish brown. Eggs are white and very small (1/25 inch [1 mm] long). Larvae are legless and white with light brown heads. Larvae can grow up to 1/4 inch (6 mm) long. Pupae are white, and some adult features are often present.

**Hosts**—Douglas-fir

**Life Cycle**—Douglas-fir beetles have a 1-year life cycle and overwinter as adults or larvae. Beetles usually emerge mid to late spring, when the temperature is 60° F and above. However, a small portion of beetles emerge later in midsummer. Some adults that make early spring attacks can reemerge and make a second attack from late June to August. Distinctive vertical egg galleries (5-12 inches [13-30 cm] long) are constructed by the female in the phloem layer (fig. 295). Eggs are laid in groups, alternating along opposite sides of the gallery. Eggs hatch in 1-3 weeks, and newly hatched larvae mine out at right angles from the egg gallery. Mature larvae construct a pupal chamber at the end of their mines.

Figure 295. Douglas-fir beetle egg and larval galleries. *Photo: Kenneth Gibson, USDA Forest Service, Bugwood.org.*

**Damage**—The larvae feed under the bark in the phloem layer, introducing fungi, yeasts, and other organisms, and lead to tree death. The first sign of attack is reddish orange frass in bark crevices that is expelled by attacking beetles (fig. 296). However, frass can wash away and attacks may be above eye-level, making it difficult to locate attacked trees. Pitch-tubes are not usually present, but many trees will have pitch streaming (clear resin) down the tree bole from the top of the beetle-colonized area. Tree foliage discolors several months to a year later, transitioning from green to reddish brown in that time.

Douglas-fir beetles prefer to attack trees that are injured by fire scorch, defoliation, windthrown, or root disease. When low beetle populations are present, individual or small groups of trees will be attacked. Once populations build up, large outbreaks can occur that kill thousands of trees. Stand conditions and weather can also strongly influence Douglas-fir beetle populations.

**Figure 296.** Reddish orange frass from Douglas-fir beetle attack. *Photo: Sandy Kegley, USDA Forest Service, Bugwood. org.*

**Management**—The best management is to promote stand vigor by thinning. Prompt removal of windthrown, severely fire-damaged trees or trees damaged by other stand disturbances is also recommended. Because Douglas-fir beetles preferentially attack burned trees, removing fuels from beneath large-diameter Douglas-fir trees before a prescribed burn can reduce tree scorch and, consequently, the tree's susceptibility to attack by Douglas-fir beetle. Attacks are most severe in unmanaged stands, on trees that are largest in diameter, and in dense stands. If direct control is deemed necessary, trees can be protected using the anti-aggregation pheromone methylcyclohexanone (MCH), which disrupts beetle aggregation. Combining MCH with salvage of infested trees has been successful at reducing subsequent tree mortality. However, under condition of intense or long-lived outbreaks, even MCH has sometimes failed to protect trees. Direct control is usually implemented in small, high-value areas.

**References: 50, 150**

# Douglas-Fir Pole and Engraver Beetles

## Attack small Douglas-fir trees

**Name and Description**—Douglas-fir pole beetle—*Pseudohylesinus nebulosus* (LeConte)

Douglas-fir engraver beetle—*Scolytus unispinosus* LeConte

An engraver beetle—*Scolytus monticolae* (Swaine) (= *S. tsugae* [Swaine])
[Coleoptera: Curculionidae: Scolytinae]

Douglas-fir pole and engraver beetles attack small-diameter Douglas-fir trees and tops of larger trees. They are commonly active during droughty periods. The Douglas-fir pole beetle adults are brown, slender, about 1/8 inch (3 mm) long, appear dull due to the dense covering of scales, and have a round posterior. The engraver beetles average less than 1/8 inch (3 mm) long and have a "sawed-off" posterior.

**Hosts**—Douglas-fir is the principal host.

**Life Cycle**—Depending on the location, Douglas-fir pole beetles and engraver beetles have one to two generations per year. Beetles usually emerge and attack in the spring. A short (1-3 inches [2.5-7.6 cm]), longitudinal egg gallery is constructed in the cambium layer, often with two branches—one up and one down the trunk—originating from the central entrance tunnel. The Douglas-fir engraver beetle gallery can be unbranched, extending in one direction from an enlarged chamber or notch. The galleries of the Douglas-fir engraver beetles (fig. 297) can be distinguished from the galleries of Douglas-fir pole beetle (fig. 298) by the well-defined nuptial chamber made by engraver beetles. Larval galleries tend to turn upward and downward depending on if they are above or below the notch. Douglas-fir pole beetle adults overwinter in niches cut into the bark. Douglas-fir engraver beetles overwinter as larvae.

**Figure 297.** Galleries of Douglas-fir engraver beetle, *Scolytus unispinosus*. *Photo: Wayne Brewer, Auburn University, Bugwood.org.*

**Damage**—These beetles cause mortality in smaller trees and top-kill or branch-kill in larger trees, with occasional mortality in larger trees. They also commonly attack thin-barked portions of logging slash. The larvae feed under the bark in the phloem layer. They can be one of several agents that kill a tree. The size of the emergence holes and distinctive galleries distinguish the species from other beetles such as Douglas-fir beetle.

Douglas-fir pole beetles and engraver beetles prefer to attack trees that are injured by fire scorch, defoliation, blowdown, or root disease. Stand conditions and weather can also strongly influence beetle populations. Under drought conditions, they have been known to attack and kill Douglas-fir as large as 12 inches (30 cm) in diameter.

**Figure 298.** Galleries of Douglas-fir pole beetle, *Pseudohylesinus nebulosus*. *Photo: Karen Ripley, Washington Department of Natural Resources.*

**Management**—Because Douglas-fir pole beetles and engraver beetles are secondary insects associated with trees under stress, enhancing tree/stand quality will help to prevent attacks. The best management approach is to promote stand vigor by thinning and promptly removing windthrown trees or trees damaged by other stand disturbances.

**References: 35, 50**

# Elm Bark Beetles

## Native and introduced bark beetles of elm

**Name and Description**—Native elm bark beetle—*Hylurgopinus rufipes* Eichhoff

Smaller European elm bark beetle—*Scolytus multistriatus* (Marsham)

Banded elm bark beetle—*S. schevyrewi* Semenov [Coleoptera: Curculionidae: Scolytinae]

Three species of bark beetles are associated with elms in the United States: (1) the native elm bark beetle (fig. 299) occurs in Canada and south through the Lake States to Alabama and Mississippi, including Kansas and Nebraska; (2) the introduced smaller European elm bark beetle (fig. 300) occurs throughout the United States; and (3) the introduced banded elm bark beetle (fig. 301) is common in western states and is spreading into states east of the Mississippi River. Both the smaller European elm bark beetle and the banded elm bark beetle were introduced into the United States from Europe and Asia, respectively. *Hylurgopinus rufipes* adults are approximately 1/12-1/10 inch (2.2-2.5 mm) long; *Scolytus multistriatus* adults are approximately 1/13-1/8 inch (1.9-3.1 mm) long; and *S. schevyrewi* adults are approximately 1/8-1/6 inch (3-4 mm) long. The larvae are white, legless grubs.

**Figure 299.** Native elm bark beetle. *Photo: J.R. Baker and S.B. Bambara, North Carolina State University, Bugwood.org.*

**Figure 300.** Smaller European elm bark beetle. *Photo: J.R. Baker and S.B. Bambara, North Carolina State University, Bugwood.org.*

**Figure 301.** Banded elm bark beetle. *Photo: Jim LaBonte, Oregon Department of Agriculture.*

**Hosts**—Hosts for the native elm bark beetle include the various native elm species in the United States and Canada, while the introduced elm bark beetles also infest introduced species of elms, such as English, Japanese, and Siberian elms. American elm is the primary host tree for the native elm bark beetle. Siberian elm is the native host tree species for the banded elm bark beetle in Asia. Siberian elm is a host for the smaller European elm bark beetle in the United States but not in Europe because the tree species' range does not overlap the beetle's native range in Europe.

**Life Cycle**—The native elm bark beetle has a variable life cycle depending on latitude, with two generations in the southern portion of its range and a 1-year life cycle in the northern portion of its range. The egg gallery for this species has two branches and barely etches the wood (fig. 302). It is not known whether the native elm bark beetle utilizes aggregation pheromones.

The smaller European elm bark beetle and the banded elm bark beetle have two generations per year in most locations and may have a third generation in some warmer locations. Beetles fly in the spring and infest dying elms. The egg galleries have a single branch and generally score the wood surface (figs. 303-304). Larvae develop through the summer months and overwinter as larvae in brood trees. The smaller European elm bark beetle produces an aggregating pheromone, but the banded elm bark beetle apparently does not produce one. Host odors play an important role in both of these introduced species.

**Damage**—The first sign of attack is the accumulation of boring dust around the base of the tree. Trees infected with Dutch elm disease, caused by the

**Figure 302.** Native elm bark beetle egg gallery. *Photo: Roland J. Stipes, Virginia Polytechnic Institute and State University, Bugwood.org.*

**Figure 303.** Smaller European elm bark beetle egg gallery. *Photo: William M. Ciesla, Forest Health Management International, Bugwood.org.*

**Figure 304.** Banded elm bark beetle egg gallery. *Photo: Jose Negron, USDA Forest Service.*

exotic fungal pathogen, *Ophiostoma novo-ulmi*, are highly attractive to elm bark beetles (fig. 305). The native elm bark beetle undergoes a period of feeding on healthy elms before seeking a dying tree for gallery construction and brood production, which provides an opportunity for adults to transmit the Dutch elm disease pathogen to uninfected elms. The smaller European elm bark beetle is an efficient vector of Dutch elm disease in American elms because newly emerged adults undergo a period of maturation feeding in tunnels on branches of healthy elm canopies prior to dispersing to dying elms for gallery construction, thereby

**Figure 305.** American elm dying from Dutch elm disease. *Photo: Jose Negron, USDA Forest Service.*

creating an opportunity to transmit the fungal pathogen to healthy trees. The banded elm bark beetle appears to have the same behavior and is known to carry the fungal pathogen when brood beetles leave diseased trees, but it has not yet been confirmed as a vector of Dutch elm disease.

**Management**—The native elm bark beetle is not an aggressive species and was not a major pest problem prior to the introduction of Dutch elm disease. The smaller European elm bark beetle and the banded elm bark beetle are serious pests of native and introduced elms. The banded elm bark beetle is more aggressive than the smaller European elm bark beetle and appears to be displacing this species in western states. Drought stress predisposes exotic elm species to infestation by these two introduced elm bark beetle species. In urban settings, rapid removal of Dutch elm disease-infected American elms followed by destruction of the host material is the key to successfully managing these two species.

**Reference: 190**

# Fir Engraver

## Horizontal egg galleries

**Name and Description**—*Scolytus ventralis* LeConte [Coleoptera: Curculionidae: Scolytinae]

The fir engraver is an important bark beetle of true firs. In the central Rocky Mountains, this insect is most frequently observed in white fir, *Abies concolor*. Epidemics of fir engraver are observed most frequently in the Pacific Northwest and California and are often associated with periods of drought. The adults are $1/10$-$1/7$ inch (2.5-3.8 mm) long (fig. 306). The larvae are small, white, legless grubs.

Egg galleries are transverse, often with a visible nuptial chamber at the center of the two arms of the gallery. The female deposits eggs in niches on both sides of the gallery. The larval galleries are longitudinal, and both the egg and larval galleries score the wood deeply (fig. 307). Pupation occurs in the inner bark at the end of the larval galleries.

**Figure 306.** Fir engraver adult. *Photo: Don Owen, California Department of Forestry and Fire Protection, Bugwood.org.*

**Hosts**—In the southern Rocky Mountains, this insect is most commonly found in white fir. Throughout the West, the fir engraver may be found in a variety of true fir species, including white fir, grand fir, and California red fir and may occasionally be found infesting other species, including Douglas-fir. The range of this bark beetle includes British Columbia south through the Pacific Northwest and Rocky Mountains into New Mexico and Arizona.

**Life Cycle**—The fir engraver has a 1-year life cycle except in cooler portions of its range where complete development takes 2 years. In Colorado, flight may occur any time from early spring until early fall. The fir engraver is monogamous, and females initiate attack on host trees. Apparently,

**Figure 307.** Fir engraver egg and larval galleries. *Photo: Intermountain Region, USDA Forest Service, Bugwood.org.*

this species does not utilize aggregating pheromones during its attack, and the dynamics of attack appear to be associated with primary host volatiles alone. The fir engraver transports a brown-staining fungus that is important for successful development of the brood larvae.

**Damage**—The fir engraver infests boles, large branches, slash, and windthrown trees larger than 4 inches (10 cm) in diameter. Trees infected with root disease or defoliated by Douglas-fir tussock moth or western spruce budworm are especially subject to attack. Trees may be killed outright (fig. 308) or attacked repeatedly for multiple years, leading to patches of dead bark. The beetles also may attack the tops of trees, causing top-kill.

**Management**—Maintaining good tree vigor in ornamental settings is important for preventing losses to fir engraver. Also, care during construction and soil disturbing activities should be practiced to limit root and lower stem damage. Natural controls include predators and parasitoids. Direct control methods are considered impractical due to the

**Figure 308.** White fir mortality caused by fir engraver on the Rio Grande National Forest. *Photo: William M. Ciesla, Forest Health Management International, Bugwood.org.*

tendency of beetles to be common in portions of living trees that are out of view from the ground. Insecticides labeled for bark beetles will provide protection from fir engraver attack.

**Reference: 190**

# Limber Pine Engraver
## Attacks branches and boles

**Name and Description**—*Ips woodi* Thatcher [Coleoptera: Curculionidae: Scolytinae]

The limber pine engraver is a common bark beetle in the central Rocky Mountains. The beetle is found across the range of limber pine, including in the states of Arizona, California, Colorado, Idaho, Montana, New Mexico, Utah, and Wyoming and in the province of Alberta, Canada. The beetle is approximately 1/8-1/5 inch (3.5-4.7 mm) long (fig. 309). The larva is a C-shaped, legless grub.

**Hosts**—Host trees for *I. woodi* include limber pine (fig. 310) and southwestern white pine. It seems likely that *I. woodi* also will infest Rocky Mountain bristlecone pine, which often occurs with limber pine at high elevations in the central Rocky Mountains, although this species is not listed as a host tree.

**Figure 309.** *Ips* spp. beetle adult. *Image: Whitney Cranshaw and David Leatherman, Colorado State University.*

**Life Cycle**—There is no available information concerning the biology of *I. woodi*. Beetles have been observed attacking white pine blister rust-infected limber pine in spring, and brood beetles have been observed under the bark in late summer.

**Damage**—*Ips woodi* infests large limbs and boles of weakened or drought-stressed limber pines and fallen trees. The beetle is common in limber pines

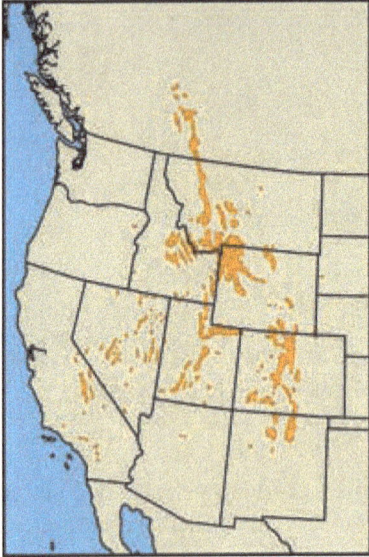

Figure 310. Range of limber pine in western North America. *Image: Anna Schoettle, USDA Forest Service.*

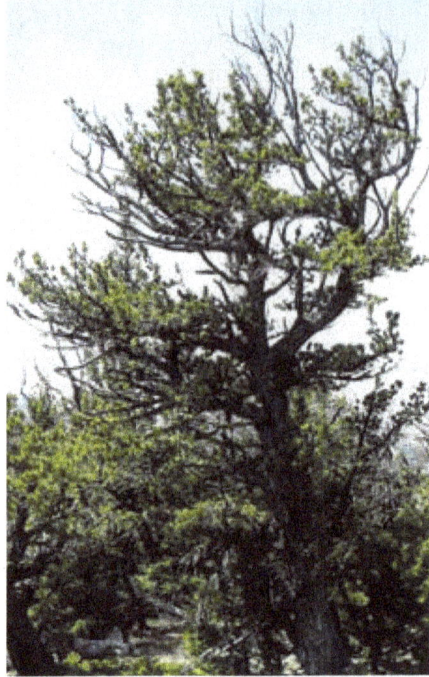

Figure 311. Limber pine tree in Colorado. *Photo: Anna Schoettle, USDA Forest Service.*

severely infected with white pine blister rust. The beetle also can be found in the upper crowns of limber pines attacked by the mountain pine beetle, *Dendroctonus ponderosae*. The galleries are typical *Ips* spp. galleries—longitudinal in orientation, resembling a narrow tuning fork.

**Management**—Although this bark beetle has not received much attention by entomologists, it is of some importance with regard to managing limber pines (fig. 311) for potential genetic resistance to white pine blister rust. Several limber pines have been identified with significant levels of resistance to this exotic fungal pathogen. Maintaining these trees in a healthy, cone-producing condition will be important in characterizing the source of this resistance.

**Reference: 190**

# Lodgepole Pine Beetle

## An uncommon lodgepole pine bark beetle

**Name and Description**—*Dendroctonus murrayanae* Hopkins [Coleoptera: Curculionidae: Scolytinae]

The lodgepole pine beetle (fig. 312) is a rarely encountered species in the genus. This bark beetle is not thought to be an aggressive species, although it is sometimes associated with tree mortality. The lodgepole pine beetle resembles the more common spruce beetle, *D. rufipennis*, based on external features and coloration of the adult and the general shape of the egg galleries. The beetles are most commonly encountered in the lower 2 ft (61 cm) of the bole of lodgepole pine trees and may be associated with trees attacked by the mountain pine beetle, *D. ponderosae*. The adults are approximately 1/4 inch (5.0-7.3 mm) long and are dark brown with reddish brown wing covers. The larvae are white, legless grubs.

**Figure 312.** Lodgepole pine beetle adult. *Photo: Dan Jensen, University of Alberta.*

**Hosts**—Hosts for the lodgepole pine beetle include lodgepole pine, jack pine, and eastern white pine. In the central Rocky Mountains, the beetle has been collected on the Pike National Forest in Jefferson County, Colorado, and on the Medicine Bow, Bighorn, and Shoshone National Forests in Wyoming.

**Figure 313.** Egg galleries of the lodgepole pine beetle. *Photo: Brian Howell, USDA Forest Service.*

**Life Cycle**—The biology of this species is not well-documented. The life cycle is thought to be approximately one generation per year. The adults disperse in June or early July through early September, with initial attacks observed by mid-July. The egg galleries (fig. 313) are vertical and approximately 5 inches (13 cm) long, and the eggs are deposited in groups of 20-50 in broad niches on either side of the egg gallery. The larvae feed together collectively, beginning perpendicular to the egg gallery and then turning up or down. Larvae may finish feeding in separate galleries and prepare a chamber for pupation, or they may pupate within the frass of the main larval feeding area.

**Damage**—The beetle has been found infesting stumps, windfalls (where they prefer the underside of the stem), and the lower 2 ft (61 cm) of weakened trees larger than 8 inches (20 cm) DBH (fig. 314).

**Management**—Although the lodgepole pine beetle has been associated with epidemics of other bark beetles, its relative significance in these events is not fully understood. In most situations, this species attacks and kills trees that are severely damaged or are dying from other causes. In managed stands, the beetle is found infesting the stumps and larger roots of cut trees.

The lodgepole pine beetle has never been associated with significant economic losses. As a result, management options for this bark beetle have never been developed.

Significant controversy surrounds this beetle because of its similarity to the spruce beetle and because of reports of spruce beetle killing lodgepole pine during the Flat Tops spruce beetle epidemic in Colorado in the late 1940s and early 1950s. Although the lodgepole pine mortality was attributed to the spruce beetle, specimens that were collected during the epidemic and placed in museums all proved to be the lodgepole pine beetle. In a recent spruce beetle epidemic in the Medicine Bow Mountains of

**Figure 314.** Lodgepole pine trees attacked at the base by the lodgepole pine beetle. *Photo: Brian Howell, USDA Forest Service.*

Wyoming, spruce beetles were observed attacking lodgepole pine trees where this species was mixed with Engelmann spruce. Surprisingly, only trees that were also attacked by the lodgepole pine beetle died. In the absence of lodgepole pine beetle, spruce beetle-attacked lodgepole pine trees survived and spruce beetles died inside their galleries, exited their galleries and died at the base of the tree, or abandoned their galleries.

**References: 51, 190**

# Mountain Pine Beetle

## Aggressive bark beetle of western pines

**Name and Description**—*Dendroctonus ponderosae* Hopkins [Coleoptera: Curculionidae: Scolytinae]

The mountain pine beetle is the most aggressive, persistent, and destructive bark beetle in the western United States and Canada. Adult beetles are dark brown to black, cylindrically-shaped, and 1/4 inch (4-7.5 mm) long (fig. 315). Larvae are small, white grubs with tan head capsules.

**Hosts**—Most native and introduced species of pines are hosts for mountain pine beetle. In the Rocky Mountain Region, ponderosa, lodgepole, whitebark, limber, and bristlecone pines are all attacked. During large outbreaks, Engelmann and blue spruce have been attacked and successfully colonized.

**Life Cycle**—Mountain pine beetles overwinter mostly as larvae beneath (or within) the inner bark of host trees. Occasionally, pupae and callow adults may also overwinter. In most lodgepole and ponderosa pine stands, larvae pupate at the ends of their feeding galleries in late spring. Adults emerge and attack from about early July through August, depending on elevation and temperature (fig. 316). Egg galleries are more or less straight and vertical and may be up to 30 inches (76 cm) long. Eggs are laid along each side of the gallery in individual niches. Both niches and egg galleries are tightly packed with frass. Eggs hatch and larvae feed until freezing temperatures cause dormancy. Extremely cold temperatures (prolonged periods below -30 °F) can

Figure 315. Adult mountain pine beetle. *Photo: Erich G. Vallery, USDA Forest Service, Bugwood. org.*

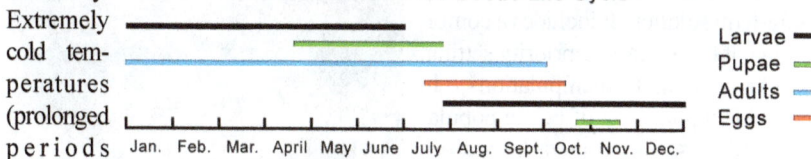

Figure 316. Mountain pine beetle life cycle (from Johnson 1982).

cause significant mortality and can lead to an outbreak reduction or end.

**Damage**—This is one of the few bark beetles that usually make obvious pitch tubes on bark surface at the attack site (fig. 317). Pitch tubes are masses of red, amorphous resin mixed with bark and wood borings. Boring dust is evident in bark crevices and around the base of infested trees. Under bark, look for straight, vertical egg galleries with a crook or J shape at the beginning, which can extend upward 30 inches (76 cm) or more (fig. 318). Galleries are packed tightly with boring dust. Larvae (grubs) are present during fall and winter. Most pupate in late spring, and adults emerge from the

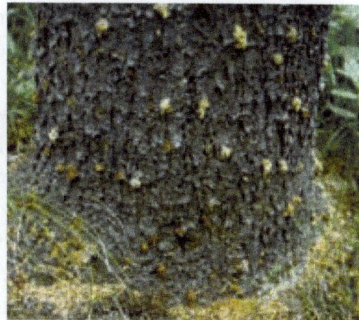

Figure 317. Pitch tubes and boring dust at site of mountain pine beetle attack. *Photo: Kenneth E. Gibson, USDA Forest Service, Bugwood.org.*

bark in midsummer to attack new trees. Infested trees fade within a year from yellow-green to red-brown. Hosts may have their bark removed by woodpeckers searching for larvae.

**Management**—Outbreaks usually develop in mature to over-mature forests. Large reserves of these forests pose a constant hazard in areas climatically favorable for the mountain pine beetle. In addition, management plans for reserved areas such as parks and wildernesses should consider the need for protection against destructive outbreaks. Management should focus on forests and not on the beetle by altering stand conditions that favor buildup of beetle populations. There are two basic approaches to reducing losses from mountain pine beetle in pine forests: (1) long-term (preventive) forest management, and (2) direct control.

The strategy of preventive management is to keep beetle populations below injurious levels by limiting the beetles' food supply through forestry practices designed to maintain or increase tree/stand resistance. Preventive management addresses the basic cause of epidemics—stand susceptibility—and is considered the most satisfactory long-term solution. It includes a combination of hazard rating, priority setting, and silvicultural manipulations. In contrast, suppression of beetle populations, (that is, killing them by various methods of direct control) treats only the problem of too many beetles.

Methods of direct control include felling and burning, debarking, or solarizing infested trees. Effects of direct control are only temporary, so the treatment must be implemented yearly as

**Figure 318.** Mountain pine beetle gallery has a crook at the end. Larvae are in feeding galleries. *Photo: Ladd Livingston, Idaho Department of Lands, Bugwood.org.*

long as beetle infestations exist. When properly used, direct control might be effective both in reducing the rate of the spread and the intensification of infestations; but it should be considered only a holding action until susceptible stands can be altered silviculturally. Individual trees can also be protected from fatal attacks through the use of chemicals applied prior to tree infestation (preventive sprays). This treatment must be repeated as long as beetle infestations exist in the area. Preventive sprays have been used in combination with treatments such as sanitation (removing beetle-infested trees) and thinning to increase overall stand health while protecting individual high-value trees. Use of preventive spray treatments in Rocky Mountain Region recreation areas should be supported by a site-specific vegetation management plan.

**References: 3, 50, 93**

# Pine *Ips* Species (Engraver Beetles)

## Attracted to green slash

**Name and Description**—*Ips* spp. [Coleoptera: Curculionidae: Scolytinae]

Adult beetles are cylindrical, dark red-brown to black, and typically $1/8$-$3/16$ inch (3-5 mm) long. Some species may be as long as $1/4$ inch (6 mm). They have a dish-shaped depression on the end of their abdomens with spines along each side (fig. 319). Larvae are typical C-shaped and are indistinguishable from other bark beetle species larvae.

**Hosts**—All pine species are attacked.

**Life Cycle**—Normally, there are two generations of the beetle each year (fig. 320). In dry years, three or even four generations may occur. Winter is passed primarily in the adult stage beneath the duff on the forest floor or within infested material. Adults become active early in the spring, infesting fresh slash or winter-damaged trees. Initial flights vary with weather but probably occur most often in late April to early May. This brood develops into adults after 40-55 days, and they attack slash and standing trees by August. Attacks are initiated by male beetles, which construct nuptial chambers beneath the bark. Each one then attracts several females, which, after mating, construct egg galleries radiating from the nuptial chamber (fig. 321).

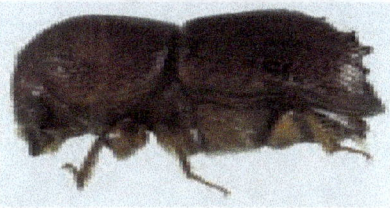

**Figure 319.** Adult *Ips pini* beetle with depression and spines on end of elytra. *Photo: Natasha Wright, Florida Department of Agriculture and Consumer Services, Bugwood.org.*

Egg galleries are kept free of boring dust and frass, unlike those of many other bark beetles. Beetles prefer fresh debris from logging, construction activity, or natural events, but living trees may be attacked during outbreaks.

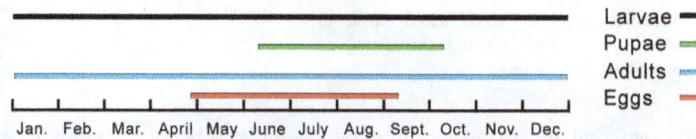

**Figure 320.** Life cycle of pine engraver beetles (from Johnson 1982).

**Damage**—In standing trees, fading tops of large trees or whole crowns in small trees can be indicators of *Ips* spp. infestation (fig. 322). Other external evidence consists of accumulations of boring dust in bark crevices and at the base of the tree (fig. 323). Occasionally, pitch tubes can be found on the trunk. Characteristic egg galleries may be found under the bark, slightly engraving

**Figure 321.** Egg galleries of *Ips* sp. radiating from nuptial chamber. *Photo: Jerald E. Dewey, USDA Forest Service, Bugwood.org.*

**Figure 322.** Top-kill from *Ips pini* attack. *Photo: William M. Ciesla, Forest Health Management International, Bugwood.org.*

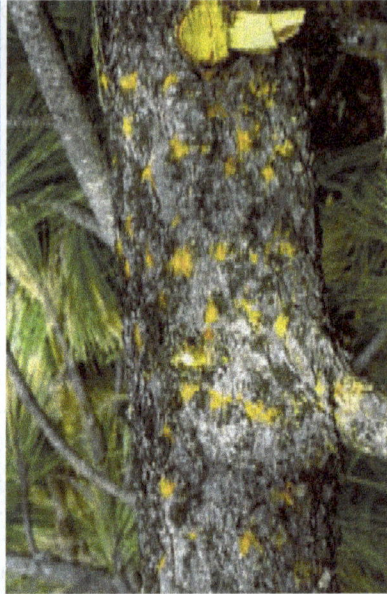

**Figure 323.** Boring dust on bark of pine attacked by pine engraver beetles. *Photo: Kenneth E. Gibson, USDA Forest Service, Bugwood.org.*

the sapwood, hence the common name, engraver beetle. In slash, look for boring dust and galleries.

**Management**—Most pine engraver problems are associated with disturbances such as windthrow and ice breakage, drought in spring and early summer, thinning, logging, fires, road construction, housing development, or tops of trees being weakened or killed by other agents. Pine engraver beetles overwinter in the adult stage and normally infest green slash only in the spring. Logging slash created from December through June can be especially hazardous because it provides large amounts of breeding material. Slash should not be created during this time period unless it can be treated prior to beetle emergence. During years of extremely low spring soil moisture, overwintering beetles have been known to attack and kill living trees. Silvicultural strategies are effective, particularly thinning stands to maintain tree resistance to attack. Thinned, vigorous stands of ponderosa pine are less attractive to pine engraver beetles. During drought years, stand vigor is even more important. Stands in which basal area has been reduced to 80-100 square ft per acre (18-23 square m per hectare) have been found to be less susceptible to beetle attack. Recently thinned stands may temporarily be more attractive because of the presence of fresh slash or logging damage to residual trees.

**References: 50, 93, 103**

# Pinyon Ips

## Causes death of mature pinyon pines

**Name and Description**—*Ips confusus* (LeConte) [Coleoptera: Curculionidae: Scolytinae]

The pinyon ips (fig. 324) is a small, brown, cylindrical bark beetle with spines on the distal portion of the abdomen, which is typical for this genus of bark beetles. Pinyon ips has five such spines. The adult's length ranges between 1/8 and 1/4 inch (3 and 6 mm).

**Host**—The pinyon ips is recognized as a primary cause of mortality for mature pinyon pines. Within Colorado, it attacks two-leaf pinyon (*Pinus edulis*), but other pinyon pine species serve as hosts to this insect in other areas.

**Figure 324.** Pinyon ips adult (*Ips confusus*). *Photo: William M. Ciesla, Forest Health Management International.*

**Life Cycle**—Adult beetles attack potential hosts in the warmer months of the year, typically from March until the end of October. There are two and one-half to three generations per year, although overlapping broods are quite common. The adult male initiates attacks in the bole and larger branches of the host and constructs a central nuptial chamber. Males release pheromones that attract two to four responding females who then mate with the male and construct individual egg galleries radiating away from the central nuptial chamber. The resulting gallery pattern frequently appears to have a Y shape to it (fig. 325). The females lay numerous eggs in the galleries, and the resulting larvae feed beneath the bark until they are ready to pupate. The mature larvae create individual pupal chambers, from which they emerge through the bark to seek out new hosts. During the winter months, large numbers of adult beetles seek overwintering sites in the lower boles of host trees. While no brood is produced from these "feeding galleries," (fig. 326) large numbers of overwintering beetles can effectively girdle and kill the host tree.

**Figure 325.** Typical Y-shaped gallery of pinyon ips. *Photo: Tom Eager, USDA Forest Service.*

**Damage**—Trees that have been attacked are fairly conspicuous. The initial phases

**Figure 326.** Overwintering or "feeding galleries" of pinyon ips. *Photo: Tom Eager, USDA Forest Service.*

of attack are notable for the large amounts of resin or "pitch" that readily flow from the attack site. This pitch flow constitutes a major component of the host tree's defense system, and the phenomenon of mass attack that is driven by pheromone release serves to attract enough beetles to exhaust this defense. Periods of low moisture availability mean that less resin is produced by the tree, thus decreasing the tree's defenses.

Like other bark beetles, pinyon ips can operate in either an endemic or epidemic fashion. Most of the time, populations of this insect are sparse, and the insects persist in pinyon stands by attacking damaged or stressed host trees. Mechanical damage, fire injury, drought, and stress created by other damaging agents (particularly black stain root disease) often increase host tree susceptibility. Human activity that results in tree damage to any portion of the trees, including roots and branches, can create habitat suitable for the beetles. In addition to the stress created by damage to host trees, mechanical wounding of trees releases volatile compounds found in tree resin that are particularly attractive to *Ips* spp. beetles. Small outbreaks of the beetle are often initiated by thinning, road or structural construction, or other similar activities.

When large numbers of suitably stressed host trees become available, pinyon ips populations can increase rapidly and kill large numbers of pinyon trees. During periods of outbreak conditions, beetle populations can increase in susceptible stands and then spread to adjacent, unsusceptible stands, killing large numbers of host trees. Vast acreages of pinyon trees can be affected at times; a noteworthy outbreak from 2002 to 2004 killed millions of pinyon pines over an area that included portions of six southwestern states.

**Management**—Over the long run, the most economical and efficient means of protection is maintenance of pinyon trees and stands in as thrifty a condition as possible. Reduced stocking and sanitation of damaged or diseased trees will reduce the chance that pinyon ips can build up in susceptible hosts and then emerge to attack additional trees. High-value trees in landscaping and recreational settings can be protected by using a chemical protective spray. Care must be taken to ensure complete coverage of the tree, as insufficient treatments will result in the death of the tree. During periods of drought, supplemental watering may also provide a degree of protection to stressed trees.

**References: 42, 46, 50, 123**

# Pinyon Twig Beetles

## Often overlooked but important branch and tree killer

**Name and Description**—*Pityophthorus* spp. [Coleoptera: Curculionidae: Scolytinae]

A number of different bark beetles will attack the branches and twigs of pinyon pine trees, but it is the members of the genus *Pityophthorus* that are most commonly referred to as pinyon twig beetles (fig. 327). Despite their small size, these scolytids behave in a manner similar to larger bark beetles that attack the boles and large branches of host trees.

**Figure 327.** *Pityophthorus* sp. *Photo: E. Richard Hoebeke, Cornell University, Bugwood.org.*

*Pityophthorus* is the largest genus of bark beetles in North America with over 120 different species. The very small size of these insects (less than $1/10$ inch [1.5-2 mm] long), combined with the large degree of diversity within the genus caused the authors of ref. 50 to advise their readers that "Identification is for experts and students of taxonomy." However, the similarity of behavior within this group allows the practitioner to treat these beetles as if they were all the same.

**Host**—Pinyon

**Life Cycle**—Twig beetles become active in the early spring, and, depending on local temperature regimes, this can be as early as mid-March. The adult beetles attack the branches at varying distances from the branch tip. If large numbers of beetles are present, they will utilize sections of the branch further away from the tips. The beetles will bore through the thin bark of the branches, creating small galleries in the phloem. Several females will mate with a single male, but the confined spaces of the branch tips result in a non-descript gallery pattern. Pheromones are used by twig beetles to coordinate attacks and recruit large numbers of beetles to specific hosts, but the exact constituents of the pheromone system are generally unknown. Attack sites are notable by the presence of tiny pitch tubes (fig. 328) in the generally pitchy branches of pinyon pine. Tiny eggs are laid within the branches that hatch quickly. Beetle larvae feed on the phloem within the branch, finally pupating within the bark. The emerging adult beetle will feed for a short time within the branch before emerging to attack new host sites. Generally, there are several generations over the course of the warmer months, but the length and number of generations are highly dependent on local weather conditions. In general, there are between two and four generations per year.

**Figure 328.** Pitch tubes on pinyon pine from attack by *Pityophthorus* spp. *Photo: Tom Eager, USDA Forest Service.*

The quick generation time of twig beetles is matched by rapid changes in the host's appearance. As noted above, the first indication of twig beetle activity is the appearance of tiny pitch tubes at each attack site. As time passes and the beetles consume more of the host tissues, the needles of an affected branch will start to dry out and turn to a golden yellow. As feeding continues, the branch tip will die, and individual needles will start to drop off of the branches. About the time the new adult beetles are ready to leave the infested material, the affected branches will consist of a light husk of bark, with the phloem having been completely consumed and a wooden pith remaining. Tearing away the very thin bark will reveal some very fine powder (frass) that has been left behind by the feeding beetles. Though twig beetle activity is sometimes confused with the attacks of the pinyon ips (*Ips confusus*), with some practice, it is easy to distinguish the two, even from a distance. Trees attacked and killed by pinyon ips have numerous pitch tubes and leave boring dust on the bole of the tree. The coloration of the foliage of a tree killed by pinyon ips quickly fades from green to light brown to a distinctive reddish brown. In the case of pinyon twig beetles, the fact that these beetles typically do not attack portions of the tree with thick bark means that only the smaller twigs and branches will be affected, leaving a portion of the foliage unaffected. The distal tips of the branches will be attacked, and the crown of the tree will have a frosted appearance with the overall color being more yellow to tan rather than red (fig. 329). Final diagnosis can be determined by examining the branch tips for the previously mentioned signs of beetle activity. During epidemics, all of the branches on a tree may be attacked, ultimately killing the entire tree.

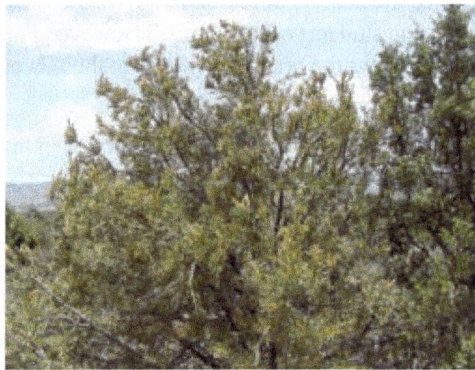

**Figure 329.** Pinyon tree under attack by twig beetles. Note the "frosted" appearance, with outer twigs affected and inner branches still green. *Photo: Tom Eager, USDA Forest Service.*

**Damage**—In general, pinyon twig beetles do not cause much of a problem in pinyon stands, but they can achieve pest status under certain conditions. During periods of drought, their numbers can become large as they are quick to take advantage of drought-stricken hosts. The close relationship between moisture availability and the host tree's ability to produce pitch means that the ability of pinyon trees to defend themselves declines under drought conditions. Twig beetles can also cause problems in the wake of human activity. Cutting, pruning, or disrupting the root system of pinyon trees can weaken them to the point that they become more susceptible to the twig beetle. In addition, the volatile odors that are created by host wounding can attract and concentrate twig beetles that then attack and damage the residual trees.

**Management**—Management techniques that maintain pinyon trees and stands in a healthy condition are probably the best defense against twig beetles. Proper techniques should be exercised if tree cutting or disturbances are unavoidable within a pinyon stand. These techniques include limiting disturbance to the colder months of the year when the beetles are inactive, removing slash that would otherwise attract beetles, and monitoring affected stands for signs of beetle attack for a period following such activities. Supplemental watering during drought periods helps the trees avoid stress. However, if beetle numbers appear to be increasing in a general area, this additional care may not be enough to prevent attack. In such cases, preventive sprays may be used to protect individual trees. Such sprays are also suitable for protecting against pinyon ips.

**Reference: 50**

# Red Turpentine Beetle

## Large pitch tubes low on tree trunks

**Name and Description**—*Dendroctonus valens* LeConte [Coleoptera: Curculionidae: Scolytinae]

The red turpentine beetle is the largest (1/4-3/8 inch [6-10 mm] long) and most widely distributed bark beetle in North America. It has a distinctively red-brown color (fig. 330) and is a common pest of forest, shade, and park trees 8 inches (20 cm) or larger in diameter.

**Figure 330.** Red turpentine beetle adult. *Photo: Ladd Livingston, Idaho Department of Lands, Bugwood.org.*

**Hosts**—Can infest any pine species and is infrequently found in other conifers

**Life Cycle**—Red turpentine beetle peak flight and attack activity usually occurs in the spring to early summer (fig. 331). Beetles emerge from recently cut stumps and dying trees and attack trees, exposed roots, or freshly cut stumps. In summer, eggs hatch in 1-3 weeks. A unique feature of red turpentine beetle is that the small larvae feed gregariously, whereas most other bark beetle larvae maintain separate feeding tunnels. Red turpentine beetle larvae tunnel away from the adult gallery. As they grow, they feed more extensively and make an irregularly margined, fan-shaped gallery (fig. 332). The beetles overwinter as larvae, new adults, or parent adults.

Red Turpentine Beetle Life Cycle

Larvae ▬ Pupae ▬ Adults ▭ Eggs ▬

Jan. Feb. Mar. April May June July Aug. Sept. Oct. Nov. Dec.

**Figure 331.** Life cycle of the red turpentine beetle (from Johnson 1982).

**Figure 332.** Red turpentine beetle larvae and feeding gallery. *Photo: Ladd Livingston, Idaho Department of Lands, Bugwood.org.*

**Figure 333.** Large pitch tubes showing attack sites of red turpentine beetle. Attacks are usually confined to the lower 6 ft (1.8 m) of the bole. *Photo: Kenneth E. Gibson, USDA Forest Service, Bugwood. org.*

**Damage**—Red turpentine beetle attacks generally start near ground level and rarely occur above 8 ft (2.4 m) (fig. 333). Attacks are often accompanied by the presence of light pink to reddish brown pitch tubes around the base of the tree and/or white granular material on the ground. On pines, red turpentine beetle pitch tubes may be as large as 2 inches (5 cm) in diameter, much larger than the pitch tubes of other pine infesting bark beetles. The large pitch tubes, galleries, and beetle size distinguish red turpentine beetle from other bark beetles. Trees that have been scorched by fire or stressed by drought are frequently attacked by red turpentine beetles. Attack by red turpentine beetles often predisposes trees to attack by other bark beetles.

**Management**—The most effective way to prevent red turpentine beetle attacks is to maintain tree vigor and avoid practices that attract beetles. Red turpentine beetle attacks frequently occur on pines that have been damaged by fire, mechanical wounding, or root disease. Damage to stands or individual trees should be minimized through improved logging, construction, and management practices. Fresh stumps, slow-dying trees, fire scorched trees, exposed roots of live trees, and trees with compacted soil around them should be treated or removed. Certain pesticide formulations containing carbaryl, chlorpyrifos, or permethrin have been proven effective at preventing bark beetle attacks when applied to the bark of a tree. Pesticide applied to the lower 6-8 ft (2-2.5 m) of the tree trunk can be used to prevent red turpentine beetle attacks, but realize that other species of bark beetles may pose a threat to the tree.

**References: 50, 93, 160**

# Roundheaded Pine Beetle

## Elongated, L-shaped galleries beneath bark

**Name and Description**—*Dendroctonus adjunctus* Blandford [Coleoptera: Curculionidae: Scolytinae]

The roundheaded pine beetle is a bark beetle that kills pine trees throughout its range. This insect's common name should not be confused with the roundheaded borer, which is the common name for the insect family Cerambycidae. The roundheaded pine beetle averages about 1/5 inch (5 mm) long and is a shiny, dark brown. By comparison, it is slightly larger than the western pine beetle (*D. brevicomis*) and slightly more slender than the mountain pine beetle (*D. ponderosae*).

Identification of the roundheaded pine beetle is most easily accomplished by recognizing its unique gallery pattern (fig. 334). From the entrance site, the gallery will proceed horizontally (i.e., across the grain, either right or left) for a short distance of 1-2 inches (2.5-5 cm). The gallery will then wind vertically for an average distance of 12 inches (30 cm), forming an elongated L shape. Eggs are laid in alternating niches along the egg gallery. Once the larvae hatch out, they will mine vertically across the grain until they reach the third instar, then they will turn out into the outer bark where they will continue their development.

**Figure 334.** Roundheaded pine beetle galleries. Note the narrower, more sinuous nature of this gallery compared to mountain pine beetle gallery to the right. *Photo: USDA Forest Service Archive, USDA Forest Service, Bugwood.org.*

**Hosts**—Within Colorado, it primarily attacks ponderosa pine (*Pinus ponderosa*) but has also been collected from limber pine (*P. flexilis*).

**Life Cycle**—The beetle is primarily active in the southwestern United States, but its range extends south through central Mexico. Because southern Colorado represents the northern limits of its distribution, several aspects of its life cycle differ from the life cycles in other portions of its range. There is typically one generation per year, although in some locales the life cycle may take more than 1 year. Roundheaded pine beetles are unique in the United States in that they almost always overwinter in the egg stage. This means that they also attack quite late in the year, frequently in late October and November. However, this schedule is different in the southern reaches of the beetle's range, where there are typically two generations per year. Egg hatch occurs in mid to late March. The larvae feed in the phloem for three instars before turning out into the bark. Pupation occurs

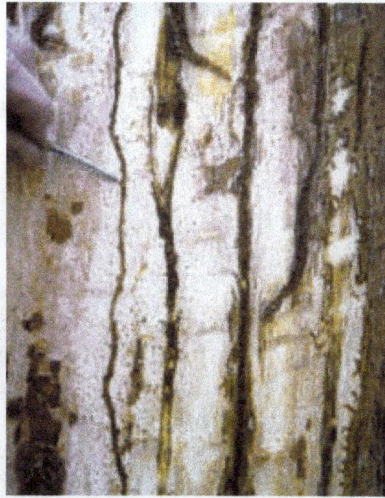

in midsummer, and the adults may remain within the host tree for 2-3 months before they emerge to attack new hosts.

**Damage**—Outbreaks of this beetle tend to be sporadic and short-lived, and they will infest trees of a wide range of size classes. They are most often active on ridge tops or other sites with dry, shallow soils. They are frequently found in "mixed broods" in trees that are simultaneously under attack by several species of bark beetles. See the Western Pine Beetles entry for further discussion of mixed broods.

**Management**—Because outbreaks are often short-lived, management is seldom warranted. In areas where management is desired, thinning stands, eliminating infested trees, and applying a preventive spray to non-infested trees can help protect trees.

**Reference: 50**

# Spruce Beetle

## Windthrown trees can set up outbreaks

**Name and Description**—*Dendroctonus rufipennis* (Kirby) [Coleoptera: Curculionidae: Scolytinae]

Adult spruce beetles are dark brown to black, with reddish brown to black wing covers (fig. 335). Beetles are approximately 1/4 inch (6 mm) long. The rear margins of their wing covers are evenly rounded. Eggs are approximately 1/16 inch (1.5 mm) long and look like tiny pearls. Larvae are legless grubs and are approximately 1/4 inch (6 mm) long at full development. Pupae are creamy white, show some body features similar to adults, and are found in individual chambers at the end of larval galleries.

**Figure 335.** Spruce beetle adult. *Photo: Steve Valley, Oregon Department of Agriculture.*

**Hosts**—Engelmann and blue spruce, *Picea engelmannii* and *P. pungens*. During very large outbreaks, this beetle has also attacked lodgepole pine, though such occurrences are not common.

**Life Cycle**—A 2-year life cycle is most common, but 1- and 3-year generations have occurred. Adults emerge from May through July, depending on local factors that influence beetle development such as temperature, aspect, and elevation. The period of attack may last as long as 5-6 weeks. With a 2-year life cycle, brood spend their first winter as larvae and their second as adults.

Female beetles bore through the bark of trees (standing or fresh, cut or fallen) and deposit eggs on either side of constructed egg galleries. Egg galleries vary from a few to 12 inches (30 cm). Galleries are packed with frass. Larvae emerge from eggs and feed in phloem.

In the 2-year life cycle, the first winter is spent in the larval stage. Larvae develop into pupae in summer (approximately 1 year after initial attack). The second winter is spent as adult beetles. Some of these beetles exit and colonize the base of trees, where snow insulates them from extremely cold temperatures. Beetles emerge and colonize new hosts in spring/summer, 2 years after initial attack.

**Damage**—The spruce beetle, *Dendroctonus rufipennis*, is the most significant natural mortality agent of mature spruce. Outbreaks have occurred in spruce forests from Alaska to Arizona. Following a 1939 windthrow event, a very large spruce beetle outbreak that impacted thousands of acres and spanned more than a decade occurred on the White River and Grand Mesa National Forests and adjoining lands. Outbreaks cause extensive tree mortality (fig. 336) and can alter stand structure and composition. Average tree diameter, tree height, and stand density are all reduced following large outbreaks. As a result of the 1940s outbreak, many stands once dominated by mature Engelmann spruce are now dominated by subalpine fir.

The earliest sign of infestation is the presence of fine, bark-colored boring dust in bark crevices and around the base of standing trees. Pitch tubes may or may not be evident. Spruce beetles prefer down spruce to standing trees. On down (windthrown and cut) trees, spruce beetles commonly colonize the lower, well-shaded surfaces and may colonize the entire length of the trunk, typically up to an 8-inch (20-cm) top. In standing trees, beetle activity is most common in the lower 30 ft (9 m) or so of the trunk. Strip attacks (attacks that impact only a portion of the tree's circumference) may be common, especially when beetle populations are small. Trees may live despite a strip attack, but such trees are often colonized again and killed in subsequent years.

Tree crowns typically remain green for up to a year after attack. By the second year, needles have faded and soon fall from the tree. The aerial "signature" of spruce beetle-infested spruce is not as striking or long-lasting as that of pine

**Figure 336.** Spruce beetle activity in Engelmann spruce. *Photo: A. Steven Munson, USDA Forest Service, Bugwood.org.*

beetles in pines. Therefore, aerial detection of spruce beetle is extremely difficult. In the winter, infested trees are often easily identified by the abundance of bark flakes on the snow, which is evidence of feeding activity by woodpeckers.

Forest stands most susceptible to attack are located along drainage bottoms, have an average DBH of 16 inches (40 cm) or more, have a basal area of over 150 square ft per acre (34 square m per hectare) and have a canopy comprised of more than 65% spruce.

**Management**—Goals should be based on existing and desired forest conditions. For endemic spruce beetle populations:
- From stand exam data, determine spruce beetle stand hazard ratings. Prioritize stand treatments to reduce susceptibility among moderate to high hazard stands.
- During stand treatments, minimize stump heights, maximize utilization of green spruce, cut green spruce cull logs into short lengths, and lop and scatter slash to maximize exposure to sunlight.
- Remove windthrown spruce before it is colonized by spruce beetles, or remove such trees after beetle colonization and before brood beetles develop and exit. If beetle response to recent windthrow is not known, conduct systematic sampling following beetle flight periods.

For growing spruce beetle populations:
- Registered insecticides can be applied to high-value trees to prevent beetle infestation.
- Identify and fell trap trees (large-diameter green spruce) into well-shaded areas. Trap trees cut in the fall before the beetle flight typically serve as optimal spruce beetle habitat. Do not remove limbs. Trap trees are often deployed in advance of larger-scale harvest operations and in combination with sanitation efforts to absorb beetles that might otherwise colonize green, standing trees. Remove infested trap trees in a timely manner.
- Remove spruce beetle infested trees. This can be accomplished in combination with beetle attack susceptibility reduction such as stand basal area reduction. These activities should be designed in order to minimize future windthrow potential.
- Following windthrow events, remove windthrown spruce after beetle colonization and before brood beetles develop and exit.
- Beetle-killed spruce may stand for decades, and many trees will remain viable for products such as house logs. Dead spruce can be salvaged to meet management objectives. Salvage activities alone do not influence spruce beetle populations.

For epidemic spruce beetle populations:
- Once spruce beetle populations reach epidemic proportions and impact large landscapes, it is not possible to stop such an occurrence with management activities. But certain management activities may still be successful at limited scales.

**References: 85, 147, 148, 149**

# Spruce Engraver Beetles

## Colonizes upper surface of windthrown and weakened trees

**Name and Description—***Ips pilifrons* (Swaine), *I. borealis* (Swaine), *I. hunteri* (Swaine), and *I. tridens* (Mannerheim) [Coleoptera: Curculionidae: Scolytinae]

All of the four most common adult spruce engraver beetles are four-spined and have a distinct declivity at the rear of their abdomen with varying numbers of spines along the margins (fig. 337). Adult beetles are shiny, reddish brown to black, and up to approximately 1/6 inch (4 mm) long, depending on the species.

*Ips pilifrons* has been reported as the most abundant species of spruce engraver beetles in Colorado.

**Hosts—**Engelmann and blue and white spruces, depending on the *Ips* species

**Figure 337.** Adult Ips beetle. Note the declivity at the rear of the abdomen with spines along each side. These are two characteristics that distinguish members of this genus from those of the genus *Dendroctonus*. Photo: Darren Blackford, USDA Forest Service, Bugwood.org.

**Life Cycle—**Life cycles vary by species, from one to multiple generations per year. Most overwinter beneath the bark or in forest litter as adults. In spring or summer, male beetles fly and initiate attacks on new hosts. They construct a nuptial chamber beneath the bark and attract multiple females. Mated females construct egg galleries and place their eggs along them. Egg galleries radiate away from the nuptial chamber, commonly forming a Y-shaped pattern (exception: *I. hunteri* typically has only two egg galleries, which often radiate away from the nuptial chamber in opposite directions). Adult beetles keep the galleries free of frass/boring dust, a characteristic that distinguishes them from galleries of *Dendroctonus* species, which are tightly packed with frass. Bark beetles in the genus *Ips* are referred to as engravers as their gallery patterns are lightly or sometimes deeply "engraved" into the face of the tree's sapwood (fig. 338). Larvae feed away from the egg galleries.

**Damage—**These bark beetles are not known as aggressive tree killers. Their preferred habitat is downed trees and broken branches and tops. However, they do cause some tree mortality

**Figure 338.** Ips galleries beneath the bark of spruce. Egg galleries are free of frass. These beetles are called engravers because their galleries are "engraved" into the face of the host's sapwood. Photo: Whitney Cranshaw, Colorado State University, Bugwood.org.

among stressed trees but, more commonly, only kill the tops of such trees. Outbreaks in standing, healthy trees are uncommon and short-lived.

Spruce engraver beetles are usually found colonizing the upper surfaces of windthrown trees and broken tops. They are much more tolerant of sun-exposed surfaces than are spruce beetles. Fresh piles of boring dust are key indicators of recent infestation. Pitch tubes may be found on green, standing trees, but they are not always present. If external signs of infestation are present, removing a small section of bark will likely reveal the Y-shaped, frass-free galleries in the inner bark; the same pattern engraved on the face of the sapwood; and possibly various life stages (larvae, pupae, or adults) of the beetles. Once beetles have developed and flown, the outer surface of the bark is peppered with many small, round exit holes.

Engraver beetles are often found in downed spruce in combination with other bark beetles, including spruce beetle, *D. rufipennis*. Spruce engraver beetles often occupy habitat beneath the bark of fallen spruce that might otherwise be used by the more aggressive spruce beetle. This can be significant, as the availability of suitable breeding habitat is the major factor contributing to substantial bark beetle population increase.

**Management**—Management of spruce engraver beetles in spruce is typically not necessary in general forest settings, although prudent and timely slash management directed at spruce beetle prevention might also serve to reduce spruce engraver beetle populations. Spruce engraver beetle activity can be a periodic problem in high-use recreation areas where spruce are stressed from soil compaction, wounded by human activity, etc. In such settings, timely removal of infested trees (sanitation) is usually sufficient to prevent/avoid other substantial tree mortality. In urban settings, preventive sprays can be used to protect ornamentals from ips attack (e.g., *Ips hunteri* [blue spruce engraver] often colonizes and kills the tops of blue spruce).

**References: 50, 176**

# Twig Beetles

## Attack small diameters and can kill trees

**Name and Description**—*Pityophthorus* spp., *P. confertus* Swaine, *P. confinis* LeConte, *P. pseudotsugae* Swaine, *Pityogenes* spp., *P. knechteli* (Swaine), *P. carinulatus* (LeConte) [Coleoptera: Curculionidae: Scolytinae]

Twig beetles are a large group of difficult to identify tiny bark beetles in several genera. Adult beetles are light brown to almost black and less than 1/10 inch (<2.5 mm) long. Twig beetles in the genus *Pityogenes* have posterior spines, and those in the genus *Pityophthorus* have a smooth posterior. Most go unnoticed, feeding in shaded branches, weakened trees, newly felled small trees, or broken limbs. When trees are weakened by drought or disease, some species of twig beetles can kill trees by attacking the bole and/or by killing all the large branches.

Tiny holes, small piles of orange-colored boring dust, and, occasionally, pitch tubes mark the beetles' entrances into the tree (fig. 339). Star-shaped galleries score the sapwood and inner bark (fig. 340).

**Hosts**—Hosts vary with twig beetle species. Pines and Douglas-fir are more affected than spruce and true firs.

**Life Cycle**—Life cycles vary by species and have not been well-studied. Amman and others (ref. 2) reported that *P. confertus* has only two larval instars and can complete development in eight weeks. Koch (ref. 105) reported one to two generations for twig beetles in lodgepole pine, and Furniss and Carolin (ref. 50) reported two or more generations per year, varying with species and locality. All twig beetles are polygymous, and female egg galleries radiate out from a central nuptial chamber just below the bark. Eggs are deposited in individual niches, and larvae mine away from the egg gallery in irregular patterns. Pupation occurs in the cambium region, and young adults emerge from individual exit holes.

**Damage**—Twig beetles are generally innocuous on healthy trees but can cause top-kill and tree mortality of drought-stressed or diseased trees. Populations of twig beetles can build in slash from activities like spacing and pruning and then can move into and kill small trees. Similarly, large populations of twig beetles can develop in the tops and branches of trees attacked by other bark beetles. The twig beetle, *P. knechteli*, has caused notable mortality in young lodgepole pines in Colorado in association with mountain pine beetle epidemics (fig. 341). Twig beetles may also create habitat for endemic mountain pine beetle populations by attacking the upper portions of small-diameter suppressed or diseased trees. This allows mountain pine beetle to attack the basal foot or two of the tree trunk.

Figure 339. Twig beetle frass at entrance holes in lodgepole pine; twig beetle activity is associated with a mountain pine beetle outbreak. *Photo: Robert Cain, USDA Forest Service.*

Figure 340. Twig beetle gallery on lodgepole pine. *Photo: L. Machauchlan, British Columbia Ministry of Forests, Kamloops Forest Region.*

**Figure 341.** Twig beetles contribute to mortality detected in younger lodgepole pine stands adjacent to mature stands killed by mountain pine beetle. *Photo: Brian Howell, USDA Forest Service.*

**Management**—Management techniques that maintain trees and stands in a healthy condition are probably the best defense against twig beetles. Strategies suggested for pinyon twig beetle may also be effective for other twig beetles.

**References: 2, 50, 105**

# Western Balsam Bark Beetle

## Star-shaped egg galleries

**Name and Description**—*Dryocoetes confusus* (Swaine) [Coleoptera: Curculionidae: Scolytinae]

Adult western balsam bark beetles are shiny, dark-colored, and approximately 1/6 inch (3.4-4.3 mm) long. The rear edges of the wing covers are abruptly rounded and do not have spines. Both male and female beetles have patches of "hairs" on the fronts of their heads. Females have a dense and distinctive patch of these "hairs" (fig. 342).

**Hosts**—Subalpine fir, *Abies lasiocarpa*, other true firs, and very rarely, Engelmann spruce and lodgepole pine

**Life Cycle**—The biology of this beetle is not well known, but a 2-year life cycle has been documented in this Region. The first winter is spent beneath the bark as larvae and the second as near fully developed adults. Approximately midsummer, male beetles initiate attacks on trees and construct nuptial chambers beneath the bark. They attract several female beetles. After mating, female beetles construct egg galleries that radiate from the nuptial chamber, forming a stellate gallery pattern (fig. 343). Larvae feed away from the egg galleries. Females are capable of producing eggs during the initial summer of attack and in the following spring. The resulting gallery pattern and brood makeup is often confusing, hence the name, *Dryocoetes confusus*. Two flight periods have been observed in

Figure 342. Western balsam bark beetle adults; female on the left, male on the right. Note the dense patch of "hairs" at front of female's head. *Photo: Rocky Mountain Region, USDA Forest Service, Bugwood.org.*

Utah: a large flight in early to midsummer and a smaller flight in the fall.

**Damage**—The western balsam bark beetle is part of a group of organisms that contribute to large amounts of subalpine fir mortality in the West. Western balsam bark beetles frequent trees weakened by root disease, drought, wind breakage, and other damaging factors. As beetle populations grow, they become more able to take advantage of large areas of susceptible host trees. However, unlike mountain pine beetle and spruce beetle, resulting tree mortality is not usually uniform across the impacted landscape, although cumulative mortality over years can result in extensive landscape mortality (fig. 344). Large concentrations of subalpine fir mortality are particularly troublesome in high-elevation developments such as ski areas.

External evidence of beetle attack is often difficult to find. If beetle attacks are fresh, small amounts of boring dust may be visible along the bark and in proximity to entrance holes. Small amounts of sap may stream for about an inch below the entrance hole, but this symptom is not always present. If a segment of bark can be removed, identification of beetle brood, the stellate gallery pattern, and slight etching of the sapwood will confirm the

Figure 343. Stellate (star-shaped) gallery pattern of western balsam bark beetle evident on the face of subalpine fir sapwood. *Photo: Dave Leatherman, Gillette Entomology Club, Colorado State University.*

**Figure 344.** Western balsam bark beetle contributed to widespread subalpine fir mortality. *Photo: Mike Blakeman, Rio Grande National Forest, USDA Forest Service.*

presence of western balsam bark beetle. Beetle-killed firs retain their red needles for 3 or more years, creating a "signature" that is evident from both the ground and the air. This needle retention characteristic can complicate aerial detection of most recent infestations.

Endemic populations are evident among weakened trees and recently broken branches/tops. Little is known about factors contributing to outbreaks, but McMillin and others (ref. 120) documented that windthrown subalpine fir is readily colonized by western balsam bark beetles. The authors (along with Gibson and others, ref. 55) concluded that because western balsam bark beetle responds to windthrow, addressing such material may reduce the possibility of a future epidemic. Beetle activity is often common among trees with Armillaria root disease. This beetle and root disease, along with with some lesser known organisms, account for large amounts of mortality among mature subalpine fir stands. Western balsam bark beetle activity has been described as chronic in some locations with abundant numbers of host trees.

**Management**—Management directed specifically at subalpine fir forest susceptibility and western balsam bark beetle/Armillaria root disease activity has been primarily limited to treatments that favor spruce. Dead and dying subalpine fir have been harvested and sites have been prepared to promote residual spruce health and to favor spruce regeneration. Timely salvage of subalpine fir windthrow may help to avoid western balsam bark beetle outbreaks in adjoining forests. Aggregate and anti-aggregate pheromones are known for these beetles, but they are not yet available for operational management purposes.

**References: 50, 52, 55, 102, 120**

# Western Pine Beetle

## Meandering egg galleries

**Name and Description**—*Dendroctonus brevicomis* LeConte [Coleoptera: Curculionidae: Scolytinae]

The western pine beetle attacks and kills mature ponderosa pine throughout much of its range. It is one of the smaller members of the genus *Dendroctonus* (only about 1/8 inch [3-5 mm] long), and is a fairly non-descript bark beetle; it is dark brown and cylindrical in shape (fig. 345).

**Figure 345.** Western pine beetle adult (*Dendroctonus brevicomis*). *Photo: Erich Vallery, USDA Forest Service-SRS-4552, Bugwood.org.*

**Host**—Ponderosa pine; however, this beetle is never found east of the Continental Divide, and, within the Rocky Mountain Region, it is active only in a narrow band in western Colorado.

**Live Cycle**—There are several generations per year of western pine beetle. This number varies with location, depending upon elevation and latitude. The number of generations is tied directly to the length of the "growing season," so in some portions of their distribution, they will have two generations per year, whereas in other areas, there may be as many as four generations per year.

Western pine beetles are virtually inactive during the winter months but as soon as the weather becomes warmer, they will increase their activity. The first sign of western pine beetle attack is the production of pitch tubes on ponderosa pines. The egg gallery that is constructed by western pine beetles is described as being "serpentine;" that is, it is a sinuous, winding gallery that can even cross itself at times (fig. 346). The beetles mate, and eggs are deposited along the margins of the central gallery. *Dendroctonus* beetles pack these galleries with frass (boring dust and beetle excrement) once the eggs have been deposited. These eggs hatch out, and the tiny larvae feed briefly within the inner bark on the phloem layer. However, after this brief period, the larvae turn out into the bark of the host tree and continue to tunnel within the outer bark. Pupation occurs within the outer bark of the host, and the next generation of adult beetles will emerge to renew the cycle. The combination of long, serpentine

**Figure 346.** Western pine beetle galleries. *Photo: Ladd Livingston, Idaho Department of Lands, Bugwood.org.*

galleries and the absence of larval feeding beneath the bark of the host tree are key characteristics of western pine beetle galleries that allow positive identification even after the beetles have left the host.

**Mixed Broods**—One of the more confusing aspects of western pine beetles is the occurrence of mixed broods in a single tree under attack. This term refers to the presence of several different species of bark beetles attacking the same host simultaneously or within the same season. While in some cases, the various species will concentrate their activities in separate portions of the tree (e.g., *Ips* spp. in the top, *Dendroctonus adjunctus* in the crown, and *D. brevicomis* in the bole), in many cases, the galleries of different species can be found immediately adjacent to each other. This phenomenon is poorly studied but has been noted for some time. Furniss and Carolin (ref. 50) list a number of species that will co-attack a ponderosa pine. Within Colorado, the western pine beetle's most common associates include the roundheaded pine beetle (*D. adjunctus*); the red turpentine beetle (*D. valens*); the pine engraver (*Ips pini*); and, occasionally, various borers (e.g., Cerambycidae and Buprestidae). The occurrence of mixed broods increases the difficulty of identification in the field, and it also implies that some aspects of bark beetle biology are still not well understood.

**Damage**—Most of the time, the populations of these native insects are at low levels, and the beetles attack stressed, damaged, or weakened ponderosa pines. Trees damaged mechanically or by lightning or fire are often targets of attack, as are diseased host trees. Western pine beetle populations often increase dramatically during periods of drought, and over-stocked stands are also subject to western pine beetle attack. These beetles kill host trees in a wide range of size classes, from 6-inch (15-cm), pole-sized ponderosa pines to very large "yellow-bellies" with diameters over 3 ft (0.9 m). Populations of western pine beetles can increase when large numbers of susceptible hosts are present, and the large number of beetles can spread to other stands with generally low susceptibility. These outbreaks can last for several years and can affect forests over widespread areas.

**Management**—As with most bark beetles, the most economical and efficient means of management is to maintain trees and stands in a healthy condition. Stocking reduction and creation of diverse stand conditions reduce overall susceptibility to western pine beetles. During times of drought, vigilance and prompt sanitation of infested hosts can reduce the overall impacts of western pine beetles. Programs that seek to remove infested trees require that managers have a good understanding of western pine beetle biology. They should be familiar with the occurrence of mixed broods and also be able to recognize the various stages of beetle development so that treatments can be prioritized.

Supplemental watering of high-value trees and stands has shown some efficacy in reducing western pine beetle attack. In addition, the use of chemical sprays in landscape or recreational settings has proven to be highly effective at preserving individual high-value trees.

**References: 37, 46, 50**

# Introduction to Defoliating Insects

## Chewed, mined, and missing foliage

Insects that consume leaves or needles are classified as defoliators and are indicated by chewed, mined, or missing foliage. Damage from defoliating insects varies considerably with tree and insect species, feeding intensity, and the time of year that feeding occurs (table 17, figs. 347-349). Although the effects of most defoliating insects are negligible, some may cause significant damage to individual trees. A few defoliators are important forest pests that can kill trees across large landscapes during outbreaks. Most of the important defoliating insects in conifers are the larval stages of moths, butterflies, or sawflies. In hardwoods, beetle larvae and leaf-cutting bees are also damaging agents. Occasional defoliation is noted from other insects that tend to be more general feeders on rangeland, such as grasshoppers or adult scarab beetles.

**General Features**

General features of defoliating insects:

- Defoliation can kill trees directly or can predispose trees to other insects or pathogens that can kill the trees.
- Generally, high levels of defoliation will occur for several consecutive years before trees are killed.
- Defoliation that occurs later in the summer is typically less stressful to trees than defoliation that occurs early in the growing season.
- Deciduous hardwood trees are more tolerant of defoliation because they can refoliate after early season defoliation, whereas coniferous tree cannot.

**Figure 347.** Douglas-fir tussock moths feed on foliage of Douglas-fir, true firs, and spruce. *Photo: Ladd Livingston, Idaho Department of Lands; Bugwood.org.*

**Figure 348.** Western spruce budworm is another important defoliator of spruce, Douglas-fir, and true firs. *Photo: Scott Tunnock, USDA Forest Service, Bugwood.org.*

**Figure 349.** Western tent caterpillar is an important defoliator of aspen in the Rocky Mountain Region. *Photo: Southwestern Region, USDA Forest Service.*

**Table 17.** Common defoliating insects in the Rocky Mountain Region and their active feeding seasons.

| Defoliator | Host | Feeding occurs |
|---|---|---|
| Aspen leafminer (*Phyllocnistis populiella*) | Aspen | Late spring to late summer |
| Bagworm (*Thyridopteryx ephemeraeformis*) | Junipers, arborvitaes, and various hardwoods | Late spring to late summer |
| Boxelder leafminer[a] (*Caloptilia negundella*) | Boxelder | Late spring to early fall; two generations per year |
| Boxelder leafroller[a] (*Archips negunanus*) | Boxelder; sometimes alder and honeysuckle | Late spring to summer |
| Cottonwood leaf beetle (*Chrysomela scripta*) | Cottonwood | Late spring to early fall |
| Douglas-fir tussock moth (fig. 347) (*Orgyia pseudotsugata*) | Douglas-fir, true firs, and spruce | Late spring (bud burst) to midsummer |
| Eastern tent caterpillar[a] (*Malacosoma americanum*) | Chokecherry, fruit trees, and other hardwood trees and shrubs | Spring to midsummer |
| Elm leaf beetle (*Xanthogaleruca luteola*) | Elm | Late spring to early fall |
| Elm leafminer (*Kaliofenusa ulmi*) | American elm | Late spring-summer |
| Fall webworm (*Hyphantria cunea*) | Hardwoods; many species | Mid to late summer |
| Forest tent caterpillar (*Malacosoma disstria*) | Aspen and other hardwoods | Late spring (bud burst) to early summer |
| Large aspen tortrix (*Choristoneura conflictana*) | Aspen | Early spring to early summer |
| Leafcutter bee[a] | Ash, redbud, lilac, rose, and others | Early summer |
| Needleminers (*Coleotechnites* and others) | Ponderosa pine, pinyon pine, white fir, and blue spruce | Generally, late summer to late spring; some variation with species |
| Oak Leafroller (*Archips semiferana*) | Gambel oak | Early spring (bud burst) to early summer |
| Pandora Moth (*Coloradia pandora*) | Lodgepole and ponderosa pines | Mid to late summer to midsummer the next year |
| Pine butterfly (*Neophasia menapia*) | Ponderosa pine | Late spring (bud burst) to midsummer |
| Pine sawflies (*Neodiprion* spp. *Zadiprion* spp.) | Ponderosa and pinyon pines | All seasons, depending on species |
| Sonoran tent caterpillar[a] (*Malacosoma tigris*) | Oaks | Spring to midsummer |
| Tiger moth (*Lophocampa* spp.) | Ponderosa pine, pinyon pine, Douglas-fir, white fir, and junipers | Fall to early spring |
| Uglynest caterpillar[a] (*Archips cerasivorana*) | Chokecherry and others | Spring to late summer |
| Western pine budworm (*Choristoneura lambertiana ponderosana*) | Ponderosa pine | Spring to early or midsummer |
| Western pine tussock moth (*Dasychira grisefacta*) | Ponderosa pine | Early spring to early summer |
| Western spruce budworm (fig. 348) (*Choristoneura occidentalis*) | Douglas-fir, true firs, spruce | Spring (bud burst) to early summer |
| Western tent caterpillar (fig. 349) (*Malacosoma californicum*) | Aspen and other hardwoods | Late spring (bud burst) to early summer |

[a] Not discussed in this guide.

196

- Repeated years of defoliation become increasingly stressful to both deciduous and coniferous trees.
- A significant amount of defoliation can slow and even stop tree growth.
- Defoliation occurring in drought years or following a late freeze can be more stressful to trees.
- Some defoliators feed while protected in silken tents; in webbed, clipped foliage; or in folded foliage, and some mine while protected in the leaf tissue. Others feed in groups or may be covered with irritating or toxic hairs.
- Some species of defoliators feed in all seasons, including winter.
- Many defoliating insects have a preference for the current or previous year's foliage, and their feeding locations and life cycles are timed accordingly.

# Aspen Leafminer

## Serpentine mines impart silvery hue

**Name and Description**—*Phyllocnistis populiella* Chambers [Lepidoptera: Gracillariidae]

Adult aspen leafminers are tiny moths with a wingspan of about 1/4 inch (6 mm); narrow, lance-shaped, white wings mottled with brown and black markings; and relatively long, thread-like antennae. Moths have an unusual resting posture of standing tall on their front legs with their bodies slanted downward with the tips of their hair-fringed wings resting on the surface behind. Larvae are a pale, yellowish white color; very flat; and 1/10-1/4 inch (3-6 mm) long when mature. They develop internally in serpentine tunnels they create within aspen leaves

**Figure 350.** Characteristic serpentine leaf mine with central line made by larva of *Phyllocnistis populiella* on quaking aspen. *Photo: William M. Ciesla, Forest Health Management International, Bugwood.org.*

(fig. 350). Dark-colored pupae are found within the larval mine. Aspen leafminer is a transcontinental species that apparently can be found everywhere aspen occurs across the western United States.

**Host**—Aspen; similar mines are found on cottonwood, poplar, and willow but are probably caused by other species

**Life Cycle**—Aspen leafminer has one generation per year. In early spring, more or less concurrent with aspen leaf flushing, moths emerge from hibernation and feed for several weeks on nectar produced by glands near the base of young aspen leaves. After mating, single eggs are deposited on the edge of newly opened aspen leaves, and the female then folds the leaf edge over to form a protective shelter for her eggs until they hatch. When populations are low, one or two eggs are laid per leaf, but more will be deposited per leaf during outbreaks. Hatching larvae cut through the bottom of the egg directly into the leaf and begin feeding. They spend their immature life inside this leaf, creating a meandering or serpentine mine that has a center line composed of excrement. Larvae feed with sickle-shaped jaws on the tissue between the two leaf surfaces without breaking the leaf cuticle. The last larval stage does not feed. Pupation occurs inside the mine within a silken cell the larva constructs. New adults emerge in late summer before or during leaf senescence and are active for several weeks. In the fall, adult moths locate protected sites to overwinter. Other species of moths and flies also mine aspen leaves but do not create the characteristic mines and pupate within them.

**Damage**—Mining of leaf tissue causes the leaves to dry out and turn brown, which may lead to premature leaf drop, especially in severe infestations. Large populations render a silvery hue to the appearance of aspen stands viewed from a distance (fig. 351). Periodic epidemics occur in Wyoming, and sustained outbreaks that cover up to half a million acres in 1 year have occurred several times in Alaska. During outbreaks, moths seeking overwintering shelter can become a minor nuisance by entering homes. Damage due to aspen leafminer feeding is described as being primarily cosmetic. However, heavily mined leaves have been shown to lose much of their photosynthetic capacity, which may reduce aspen tree growth. Sustained aspen leafminer outbreaks are suspected to cause branch dieback and top-kill, but this has not yet been proven. Similar leaf mining damage, in the shape of irregular blotches, can be caused by the larvae of the poplar blackmine beetle, *Zeugophora scutellaris* Suffrian.

**Management**—As with many leaf mining insects, natural enemies, including diseases, parasites, and predators, are the main causes for collapsing outbreak populations of aspen leafminer. Cannibalism can also occur in dense populations with many larval mines per leaf. It is also suspected that climate is influential in determining aspen leafminer populations.

Aspen can tolerate significant defoliation. The fact that outbreaks collapse on their own, combined with a lack of evidence indicating economic damage, means that there is little need for active management of aspen leafminer populations. In addition, there are no effective control measures for large-scale outbreaks.

**Figure 351.** Aspen showing the silvery hue indicative of a dense population of aspen leaf miner. *Photo: Intermountain Region, Ogden Archive, USDA Forest Service, Bugwood.org.*

Keeping trees in good health helps reduce the impacts of insect pests in general. Contact insecticides are not effective against leafmining insects during their most damaging stage because they are protected inside the leaf. Some insecticides that are carried systemically through the tree can be effective in controlling leaf mining insects.

**References: 35, 50, 88, 109**

# Bagworm

## Carries shelter made of clipped foliage and twigs

**Name and Description**—*Thyridopteryx ephemeraeformis* (Haworth) [Lepidoptera: Psychidae]

Male bagworm moths are sooty black with a densely hairy body, clear wings, and a wingspread of about 1 inch (25 mm) (fig. 352). Female adults are soft-bodied and grub-like. They are naked except for a circle of woolly posterior hairs and are yellowish-white in color with no functional legs, eyes, or antennae (fig. 353). Larvae, pupae, and adult females live inside a bag constructed of silk, twigs, and leaves (figs. 354-357). Mature larvae are dark brown and 3/4-1 inch (18-25 mm) long. The head and thoracic plates are yellowish and spotted with black (fig. 356). Bags of mature larvae are 1 1/4-2 inches (3-5 cm) long and vary in appearance due to the use of host material in bag construction (figs. 356-357). Eggs are small, white, and found inside female bags. Bagworm is a native defoliator that occurs throughout the eastern half of the United States, is commonly encountered in Kansas and Nebraska, and has been occasionally reported in South Dakota.

**Figure 352.** Bagworm adult male. *Photo: Pennsylvania Department of Conservation and Natural Resources, Forestry Archive, Bugwood.org.*

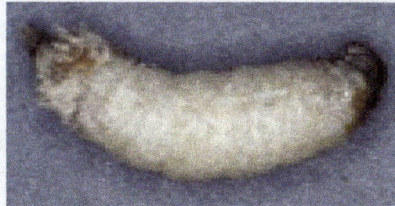

**Figure 353.** Bagworm adult female removed from her bag. *Photo: Lacy L. Hyche, Auburn University, Bugwood.org.*

**Figure 354.** Young bagworm larva with bag on Colorado blue spruce. *Photo: Rayanne Lehman, Pennsylvania Department of Agriculture, Bugwood.org.*

**Hosts**—Bagworm feeds on a wide variety of trees and shrubs but is primarily a pest on arborvitae and eastern redcedar. Other coniferous and broadleaf hosts include juniper, pines, spruces, apple, basswood, boxelder, elms, black locust, maple, oaks, persimmon, poplars, and willows.

Figure 355. Mature bagworm larva in its silk-lined bag that was cut open. *Photo: Lacy L. Hyche, Auburn University, Bugwood.org.*

Figure 356. Bagworm larva crawling out of its bag. *Photo: Connecticut Agricultural Experiment Station, Bugwood. org.*

Figure 357. Several larval bags that show white, silk anchors around the branch. *Photo: Lacy L. Hyche, Auburn University, Bugwood. org.*

**Life Cycle**—Bagworm has one generation per year across almost its entire range. In September or October, adult males emerge and begin their mating flight, seeking the wingless females. Females remain within their bag, so mating takes place through the bag opening. The female's bag also contains her empty pupal case, into which eggs are laid. Adults die, and eggs overwinter within their mother's bag that is attached to a twig with silk. Eggs hatch in May or June, yielding 1/25 inch (1 mm) long glossy black or brownish larvae that almost immediately begin to spin their own bag. Young larvae, buoyed by long strands of silk, can be dispersed short distances by wind. The bag begins as a garland of pellets on a long silken strand around the larva's body behind the legs, to which it adds twigs and leaf material, eventually becoming large enough to enclose the entire immature insect (fig. 354). Larvae move about carrying their tough, silken bag with them. An opening is maintained at the top of the bag, through which the head and several segments protrude when the larva is feeding, moving, or enlarging the bag (fig. 356). There is a smaller opening on the bottom of the bag through which excrement is passed. Larvae mature from August through September, firmly anchoring their bag prior to pupation (fig. 357). Despite having wingless females, bagworms are widely distributed due to wind-aided larval dispersal, unintended movement by humans, an ability to go long periods without food, a wide host range, and the production of many eggs by a single female.

**Damage**—Bagworm is seldom a forest pest of economic concern, but it is commonly a problem on trees and shrubs in urban settings, parks, and shelterbelts. Arborvitae and juniper can be killed by complete defoliation. Less severe defoliation results in reduced growth and weakens host plants, increasing susceptibility to opportunistic damaging agents. Mortality and gouting of twigs has been attributed to the girdling effect of the tight silken bands with which larvae attach the bags (fig. 357). Periodically, bagworm becomes exceedingly abundant, defoliating most any plant it can contact. Although young larvae are inconspicuous and have little impact, the seemingly sudden defoliation of tree tops and presence of unsightly, though highly visible, bags of large larvae firmly attached to twigs and leaves causes considerable consternation. Bags may remain attached and intact long after the insects that constructed them have died.

**Management**—Bagworm populations vary widely in size from year to year. Disease, low winter temperatures, bird predation, and, especially, parasitic wasps affect population size and collapse the short-lived epidemics.

Because they are so conspicuous, overwintering bags and the eggs they contain can be picked from small trees and shrubs and then destroyed. Discarding, but not completely destroying removed bags may result in eggs surviving, hatching, and dispersing larvae that reinfest treated trees. Labeled, registered insecticides are available to use as an alternative means of control against small bagworm larvae in spring or early summer. However, by about early July when larvae are more than ¹/₂ inch (13 mm) long, it is nearly impossible to kill them using insecticide. It is often at this time or later when bagworm infestations and associated defoliation become apparent.

**References: 32, 174, 189**

# Cottonwood Leaf Beetle

## Skeletonized leaves

**Name and Description**—*Chrysomela scripta* Fabricius [Coleoptera: Chrysomelidae]

Mature larvae of the cottonwood leaf beetle are about ¹/₂ inch (13 mm) long, black, and have two white spots on each side (fig. 358). Adults are about ¹/₄ inch (6 mm) long and have yellow wing covers with black stripes and black heads (fig. 359).

**Host**—Cottonwood, willow, and alder

**Figure 358.** Cottonwood leaf beetle larvae. *Photo: Whitney Cranshaw, Colorado State Forest Service, Bugwood.org.*

Figure 359. Cottonwood leaf beetle adult. *Photo: Whitney Cranshaw, Colorado State University, Bugwood.org.*

Figure 360. Windowpane feeding by cottonwood leaf beetle larvae. *Photo: Lacy L. Hyche, Auburn University, Bugwood.org.*

Figure 361. Cottonwood leaf beetle damage. *Photo: James Solomon, USDA Forest Service, Bugwood.org.*

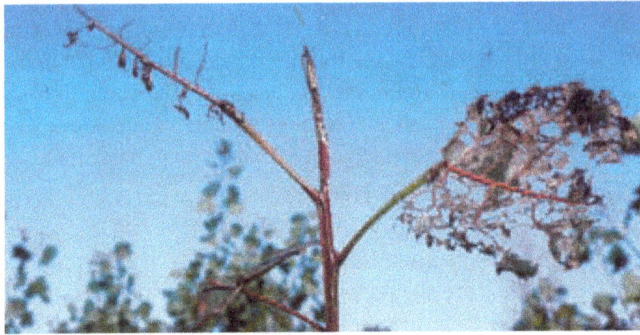

**Life Cycle**—There can be three to four generations per year, depending on location. All life stages can be found throughout the growing season. Bright yellow eggs are laid in groups on the undersides of leaves. Early instar larvae hatch and feed gregariously, skeletonizing leaves and creating a windowpane effect (fig. 360). Older larvae and adults consume the entire leaf except for major veins and petioles (fig. 361); adults also feed on the succulent new shoots. Adults overwinter in duff around the base of host trees or under loose bark on the tree.

**Damage**—Young trees are generally more affected than older trees. Heavily defoliated trees are weakened. Defoliation also reduces growth rate and can lead to deformed trees when young leaders are attacked. Young trees can be killed by repeated defoliation events.

**Management**—Management is not typically warranted. However, large populations can be a nuisance, especially in landscape situations. In heavy infestations, chemical sprays can be used to kill both adults and larvae.

**References: 124, 174**

# Douglas-Fir Tussock Moth

## Defoliates Douglas-fir, true firs, and spruce

**Name and Description**—*Orgyia pseudotsugata* (McDunnough) [Lepidoptera: Lymantriidae]

USDA Forest Service RMRS-GTR-241. 2010.

The Douglas-fir tussock moth is a common defoliator of Douglas-fir and true firs. Adult male moths are a non-descript, gray-brown moth with feathery antennae and a wingspread of 1-1 $1/4$ inches (25-32 mm) (fig. 362). The female is flightless and notably different from the male in that it has rudimentary wings and a large abdomen, usually about $3/4$ inch (19 mm) long. Young larvae possess fine hairs; older larvae have two tufts behind the head, one posterior tuft, and four dense tussocks located along the back (fig. 363). Larvae grow up to 1 $1/4$ inches (32 mm). Eggs are laid in a mass on top of the cocoon from which the female moth emerged (fig. 364).

**Figure 362.** Douglas-fir tussock moth adults: male (left) and female (right). *Photo: Rocky Mountain Region, USDA Forest Service.*

**Hosts**—In forested settings, the Douglas-fir tussock moth prefers Douglas-fir. It is also occasionally found on true firs or spruce. Forest infestations can be intense, and other species of conifers surrounding Douglas-fir trees are often also defoliated. In urban settings, blue spruce is attacked. Urban infestations are often confined to individual trees, and the same trees may be attacked year after year, which can cause considerable damage or mortality.

**Figure 363.** Douglas-fir tussock moth larva. *Photo: Rocky Mountain Region, USDA Forest Service.*

**Life Cycle**—Douglas-fir tussock moth has a 1-year life cycle and overwinters as eggs. Egg hatch coincides with bud burst. Larvae pass through four to six

**Figure 364.** Tussock moth female, egg mass, and cocoon (from left to right). *Photo: William M. Ciesla, Forest Health International, Bugwood.org.*

molts. Pupation occurs any time from late July to the end of August inside a thin cocoon of silken webbing mixed with larval hairs. Adults appear from late July into November, depending on the location. The female moth emits a sex pheromone that attracts males.

**Damage**—The first sign of attack appears in late spring as young larvae feed on current year's foliage, causing it to shrivel and turn reddish brown (fig. 365). As larvae mature, they feed on older needles. Defoliation occurs first at tops of trees and outer branches and then, as the season progresses, on lower crowns and inner branches of the host tree. During a severe defoliation event, trees will appear as skeletons once the damaged needles have fallen off, and cocoons and egg masses will be visible year-round in the lower tree canopy. Damage from severe defoliation can lead to tree death or predispose trees to subsequent bark beetle attack. Douglas-fir tussock moth can be one of the most damaging of western defoliators.

**Management**—Natural controls, including predators, parasitoids, and a nuclear polyhedrosis virus (NPV), keep the tussock moth populations low most of the time. The natural controls, especially the NPV, also act to bring populations back under control during an outbreak. Douglas-fir tussock moth populations seem to follow a cyclical outbreak pattern, with outbreaks occurring every 8-12 years and lasting for 2-4 years. If applied control is desired, there are registered insecticides that might be used to reduce outbreak populations. The NPV has been made into a biocontrol (under the name TM-Biocontrol) and has been used in areas where rare Lepidoptera co-occur with the tussock moth. Availaiblity of this producted is limited. The microbial pesticide *Bacillus thuringiensis* var. *kurstaki* (B.t.k.) is not hazardous to most beneficial insects, birds, small mammals, and aquatic systems. However, B.t.k. results against the tussock moth have not been consistent. Other contact chemical insecticides are also available for tussock moth management.

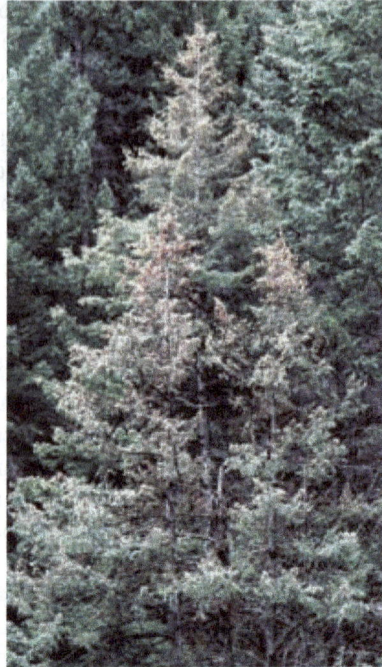

**Figure 365.** Tussock moth damage on Douglas-fir tree. *Photo: William M. Ciesla, Forest Health International, Bugwood.org.*

**Tussockosis**—Hairs on the tussock moth larvae can cause an allergic reaction in humans. The most common reaction is a skin irritation. Rashes, watery eyes, and sneezing are common symptoms. Avoid handling the larvae, and wash after exposure.

**References: 50, 185**

# Elm Leaf Beetle

## Chewed leaves look scorched

**Name and Description**—*Xanthogaleruca luteola* (Müller) (= *Pyrrhalta luteola* [Müller]) [Coleoptera: Chrysomelidae]

The wing covers of overwintering adult elm leaf beetles are dark olive-green (fig. 366). The wing covers change in color during spring to yellow or a dull yellowish green with a longitudinal, dark stripe along the edges and a dark spot at the base of each wing cover (fig. 367). Adults are about 1/4 inch (6 mm) long with three black spots present on the thoracic shield behind their heads. Eggs are bright yellow to yellowish orange, droplet-shaped, and laid in clumps or irregular rows. Small larvae are dark and hairy. Mature larvae are ¹/₂ inch (13 mm) long and dull yellow with two dark, longitudinal stripes along their side and black-colored heads, legs, and tubercles (fig. 368). Pupae are bright orange-yellow with a few black bristles and are about ¹/₅ inch (5 mm) long. Elm leaf beetle is a European species established in North America for over 150 years and now occurs throughout most of the United States and eastern Canada.

Figure 366. Overwintering elm leaf beetles in reproductive diapause. *Photo: Whitney Cranshaw, Colorado State University, Bugwood.org.*

**Hosts**—Elm leaf beetle feeds upon all elms, particularly Siberian and English, as well as American, rock, slippery, and other native, Asian, and European species of elm.

**Life Cycle**—Elm leaf beetle has from one to three generations per year, depending upon weather and location, with two per year being most common in the Great Plains. About the time elm buds begin to swell in late April and early May, overwintering adults emerge from protected sites and move to elm trees to feed and mate. After a few weeks, beetles will begin to lay eggs, typically in masses of 5-25 eggs, each attached to veins on the undersides of lower leaves, totaling 400-800 eggs per female. Larvae hatch in 10-14 days, feed for about three weeks, and then crawl down the trunk in search of pupation sites. Most pupate at the base of the tree, but some do so in bark crevices. Adults emerge in 10-15 days. Most will mate and reproduce, beginning a second generation, but some adults may seek a sheltered location

Figure 367. Adult elm leaf beetles displaying coloration typical of the growing season. *Photo: Clemson University, USDA Cooperative Extension Slide Series, Bugwood.org.*

Figure 368. Mature elm leaf beetle larva showing skeletonizing feeding pattern. *Photo: William M. Ciesla, Forest Health Management International, Bugwood.org.*

and enter diapause to overwinter. The proportion of the elm leaf beetle population becoming dormant after one generation is weather-dependent, and larger second generations are associated with warmer weather. A partial or complete third generation may occur rarely. At the end of summer, all adult beetles move to dry, sheltered areas and go dormant until spring.

**Figure 369.** Larval leaf skeletonizing dries and browns leaves, shown on Siberian elm. *Photo: Whitney Cranshaw, Colorado State University, Bugwood.org.*

**Damage**—Elm leaf beetle is a serious pest of shade and ornamental elms but is of little significance in the forest. Larvae feed on the undersides of leaves between the major veins, seldom penetrating the upper surface, in a behavior called skeletonizing (fig. 368). Damaged leaf areas dry out and turn brown, giving severely affected leaves and branches a scorched appearance (fig. 369). Leaf feeding by adults results in a "shot hole" effect (fig. 370). Heavily injured leaves may be shed by the tree prematurely. Because young larvae require tender leaf tissue, the first generation is usually the most damaging. Elm leaf beetle feeding can greatly affect the appearance of elms, rendering them unsightly, and can reduce growth and weaken them, particularly when large beetle populations persist for several years. A secondary impact occurs when overwintering elm leaf beetles enter homes in swarms, particularly during warm winter episodes, which can be so severe that it can rival the box elder bug as the primary nuisance insect invader of homes in the Great Plains.

**Figure 370.** Adult elm leaf beetle feeding produces a "shot hole" appearance, shown on American elm. *Photo: Pennsylvania Department of Conservation and Natural Resources, Forestry Archive, Bugwood.org.*

**Management**—Important natural controls include extreme winter temperatures and late spring frosts, as well as a native parasitic wasp and native insect predators, including earwigs, stink bugs, and a ladybird beetle. Attempts to introduce natural enemies from Europe have met with some success in California, but there are no reports of this being attempted in the Great Plains.

Destroying egg masses on leaves by hand can reduce localized, small populations, but care must be exercised to avoid damaging the ladybird beetle eggs that are somewhat similar in appearance.

There are several labeled, registered chemical and biological insecticides that readily control elm leaf beetle populations when sprayed on the foliage. Effective applications against the first generation generally suppress populations sufficiently such that a second application is not necessary. In addition, soil systemic application of appropriate insecticide can also provide effective control.

Another strategy is to apply a contact insecticide in about a foot-wide band around the trunk of infested trees just as larvae begin moving down to pupate. If this strategy is employed area-wide, defoliation by the second generation and nuisance populations of overwintering elm leaf beetles can be reduced. Migration from untreated trees in the area will accomplish rapid reinfestation.

Elm species, hybrids, and horticultural varieties vary in their resistance to elm leaf beetle feeding. In ornamental settings, planting the most beetle-resistant and climate-appropriate elms will minimize elm leaf beetle problems.

**References: 35, 50, 174**

# Fall Webworm

## Wraps branches in silk in late summer

**Name and Description**—*Hyphantria cunea* (Drury) [Lepidoptera: Arctiidae]

Adult fall webworms are white with some orange markings on the body and legs. The wings often have some black spots and an expanse of about 1 1/4 inches (3.2 cm) (fig. 371). Eggs are light green or yellow, globular, and laid in flat masses consisting of several hundred eggs. Larvae are hairy with distinct, paired, dark spots on each segment of the back and are 1 1/4-1 1/2 inches (3.2-3.8 cm) long at maturity. Larval color is variable, with two races recognized (figs. 372-373). The black-headed, or northern race, has a black head with a pale yellowish or greenish body that has a dark stripe on the back and long white hairs rising from red or black tubercles (humps). Larvae of the

Figure 371. Fall webworm adult. *Photo: Gerald J. Lenhard, Bugwood.org.*

Figure 372. Two forms of the red-headed race of the fall webworm larva. *Photo: Lacy L. Hyche, Auburn University, Bugwood.org.*

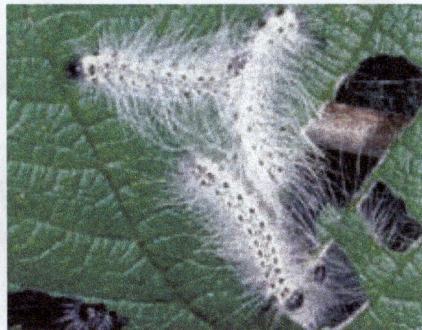

Figure 373. Examples of the black-headed race of fall webworm larvae. Note that both races can be present together and each is variable in coloration. *Photo: Lacy L. Hyche, Auburn University, Bugwood.org.*

red-headed, or southern race, are yellowish tan with red or orange-colored heads and brownish hair that arises from reddish brown tubercles. The distribution of fall webworm in North America is transcontinental, extending from southern Canada to northern Mexico.

**Hosts**—Over 100 hardwood tree species, including alder, ash, choke-cherry, cottonwood, elm, maple, various fruit trees, walnut, and willow

**Life Cycle**—Fall webworm generally has one generation per year in the Rocky Mountain Region. Adults appear from late June to early July, fly at night, and lay eggs on the underside of leaves. After hatching, the young larvae feed gregariously, at first on the epidermis but not the veins of both leaf surfaces, producing a skeletonized pattern. As the larvae grow, they begin to feed on entire leaves, spinning silk wherever they go. Eventually, they enshroud leaves and then whole branches in a loosely spun tent of silk, within which larval development and feeding occurs, so that leaf fragments, cast skins, and droppings become incorporated. The larvae, when disturbed, often twitch and wave their bodies synchronously in what is thought to be predator avoidance. Feeding continues until mid-September, when full-grown larvae wander from their host plant in search of protected pupation sites. The insect overwinters in the pupal stage within a light silken cocoon in the soil, leaf litter, on the sides of buildings, or on tree trunks.

**Damage**—This insect causes minor defoliation in most forested situations. Large infestations can envelop and defoliate entire trees, which seems a relatively frequent occurrence on narrowleaf cottonwood along streams in Colorado. Substantial defoliation may weaken tree defenses to other opportunistic agents. As the most common tent-making defoliator in much of North America, fall webworm can cause loss of visual quality in ornamental plantings due to the unsightly tents (figs. 374-375). Wandering hordes of hairy, full-grown fall webworm larvae can be a substantial, if brief, nuisance.

Figure 374. Silken tents that contain fall webworm colonies can envelop whole branches or entire trees. *Photo: Ronald F. Billings, Texas Forest Service, Bugwood.org.*

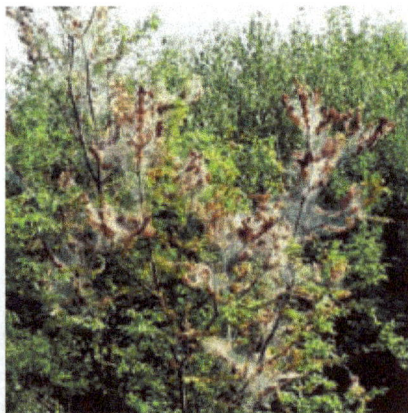

Figure 375. Silken tents that contain fall webworm colonies on chokecherry. *Photo: Whitney Cranshaw, Colorado State University, Bugwood.org.*

**Management**—Generally, population control need not be attempted, as biological impact from fall webworm defoliation is inconsequential. Deciduous tree species can recover from substantial defoliation, particularly if impacted late in the season when most growth has already occurred. Typically, impacts to plant health cannot be observed until defoliation involves more than 20% of the foliage. Fall webworm populations are usually not maintained at high levels for consecutive years, although local outbreaks can be relatively common.

Fall webworm is food for many natural enemy species whose actions reduce population size and impact. Conservation of natural enemies includes restraint from using contact insecticides or judicious, well-timed applications when such beneficial organisms are not present.

Stressed or recently planted, smaller trees can be substantially weakened by intense webworm defoliation, potentially rendering them more vulnerable to opportunistic pathogens and/or insects. Aesthetic and nuisance considerations may prompt control attempts as well.

Fall webworm larvae and colonies are most susceptible to direct control when they are small. Hand picking larvae or destroying their tents can significantly reduce a local population. Destroying tents or larvae by burning should not be done, as it can easily cause more damage to the tree than defoliation. Several insecticides are registered for fall webworm control and, to be most effective, must be applied before tents and the larvae they contain become large. Some registered insecticides have activity against Lepidoptera only, sparing the natural enemies. Limitations of these products include: the material must be ingested to kill webworms, may be washed away by rain, and will only kill larvae that are very small. Pesticide coverage on tall trees can be especially challenging to achieve. Area-wide control in a forest setting is not warranted.

**References: 32, 35, 50**

# Forest Tent Caterpillar

## Aspen defoliator makes no tent

**Name and Description**—*Malacosoma disstria* Hübner [Lepidoptera: Lasiocampidae]

Adult forest tent caterpillars are stout-bodied and light yellow to yellow-brown and have a wingspread of 1-1 1/2 inches (2.5-3.8 cm) (fig. 376). The forewings have two darker oblique lines near the middle. Eggs are cemented together and coated with a frothy, glue-like substance that hardens and turns a glossy dark brown (fig. 377). Newly hatched larvae are nearly uniformly black, about 1/8 inch (3 mm) long, and conspicuously hairy. With each successive molt, the characteristic markings of mature larvae become more evident—pale blue lines along the sides of a brownish body with a mid-dorsal row of keyhole- or footprint-shaped, whitish spots on a black background (fig. 378). When full-grown, caterpillars are about 2 inches (50 mm) long. Cocoons are composed of dense, yellowish silk with a powdery material dispersed between the strands.

**Hosts**—Hosts include many hardwood tree species and preference varies by location. In the western United States, aspen is the preferred host, although alder, basswood, birch, cherry, cottonwood, elm, oak, poplar, various fruit trees, and willow are acceptable. If trees are defoliated, atypical hosts such as understory shrubs and the leaves of fruit and vegetable plantings may be fed upon by hungry, wandering larvae.

**Life Cycle**—All *Malacosoma* spp. have one generation per year and similar life cycles. Young larvae hatch in spring when leaves are beginning to unfold and are gregarious. Colonies stay together and move about in a line, following silk trails that are laid down by the leaders. Unlike other North American species of *Malacosoma*, the

Figure 376. Adult forest tent caterpillar. *Photo: Whitney Cranshaw, Colorado State University, Bugwood.org.*

larvae do not construct tents. Instead, they spin silken mats on tree trunks or branches where they congregate when not feeding (fig. 379). The larvae feed for 5-6 weeks, becoming solitary in habit during the last stage. If large populations result in complete tree defoliation, the last two larval stages will wander in search of food. When full-grown, larvae disperse and spin dense silk cocoons, within

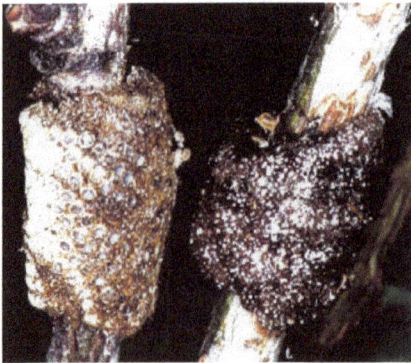

Figure 377. Egg masses of forest tent caterpillar, showing an old, hatched mass on the left and a new mass on the right. *Photo: James B. Hanson, USDA Forest Service, Bugwood.org.*

Figure 378. Mature larva of forest tent caterpillar displays the characteristic "keyhole" pattern down the back and distinctive blue markings. *Photo: Whitney Cranshaw, Colorado State University, Bugwood.org.*

Figure 379. When not feeding, forest tent caterpillars congregate on a silken mat to rest and to molt. *Photo: James Solomon, USDA Forest Service, Bugwood.org.*

which they pupate. Cocoons can be found in folded leaves, bark crevices, or other sheltered sites. The moths, which emerge about 10 days later in late July, are most active at night. Strong winds can carry the moths over long distances, and great numbers are attracted to lights. The eggs are laid mostly on upper-crown branches in masses of 100-350, which completely encircle small-diameter twigs. Inside the eggs, embryos develop into larvae that overwinter and hatch in the spring.

**Damage**—Forest tent caterpillar is the most widely distributed and destructive tent caterpillar in North America. In many areas, short-duration outbreaks result in intense defoliation, particularly to aspen. Forest tent caterpillar outbreaks have been known to occur across vast areas of the Lake States and prairie provinces of Canada, but less so in Colorado and Wyoming. Defoliation can result in reduced tree growth, top-kill, tree mortality, and impairment of aesthetic values. During years when forest tent caterpillars are especially abundant, they strip host plants of foliage and then wander across open ground in search of additional food sources, becoming a hazard for automotive traffic, trains, and pedestrians. A major natural enemy of forest tent caterpillar, the "friendly fly" *Sarcophaga aldrichi* Parker [Diptera: Sarcophagidae], becomes extremely abundant and contributes to the termination of forest tent caterpillar outbreaks in aspen. Although they do not bite, clouds of this friendly fly alighting on people and regurgitating on them and their clothing can be exceedingly annoying. During forest tent caterpillar outbreaks, the loss of shade, the nuisance occasioned by wandering caterpillar hordes, and incessant harassment from the friendly fly can negatively impact recreational use of an area.

**Management**—Forest tent caterpillar outbreaks are usually of short duration, lasting 3-5 years. Many factors regulate population size and contribute to outbreak collapse, including potent viral and fungal diseases, parasitic flies and wasps, predatory insects and birds, starvation, and, occasionally, unseasonable cold weather immediately prior to, during, and right after egg hatch.

On small, ornamental trees and shrubs, the egg masses can be removed by hand and destroyed between July and the following spring. After hatching, young colonies of larvae can be removed from branch tips or squashed while they are resting in dark clusters on the main stem, especially in the evening or on cool days. In June, when the caterpillars are migrating, a sticky or slippery collar can be placed around the base of trees as a barrier, affording trees some protection.

Several insecticides are registered for forest tent caterpillar control, all of which are most effective if applied when the larvae are small. Some registered insecticides have activity against Lepidoptera only, which spares important insect natural enemies because they are immune. Limitations of such products include: some defoliation occurs after the insecticide is applied because the material must be ingested to take effect; the product is easily washed away by rain; and it will only kill larvae that are very small. Area-wide control in a forest setting using insecticide has been and will likely continue to be employed in locations with a significant outbreak history in order to protect values and activities at risk such as growth increment on high-value trees and recreation.

**References: 11, 50**

# Large Aspen Tortrix

## Larvae in rolled leaves

**Name and Description**—*Choristoneura conflictana* (Walker) [Lepidoptera: Tortricidae]

The adult large aspen tortrix is a moth with a wingspread of 1-1 1/4 inches (2.5-3.0 cm) (fig. 380). The forewing is grayish with basal, middle, and outer brownish patches. Oval, pale green eggs are laid in rows that overlap like fish scales, forming large, flat masses (fig. 381). Young larvae have yellowish or pale green bodies and black heads. With each molt, larvae darken in color. Mature larvae are about 3/4 inch (15-21 mm) long and are grey-green to nearly black with dark spots on each segment and a dark shield and head (figs. 382-383). Pupae are light green when first formed but soon change to a reddish brown to black color (fig. 384). Large aspen tortrix is a transcontinental, principally boreal species, occurring from Alaska to New Mexico, distributed along with its principal host, aspen.

Figure 380. Adult large aspen tortrix. *Photo: Steven Katovich, USDA Forest Service, Bugwood.org.*

**Hosts**—Aspen. During epidemics when aspen is completely defoliated, larvae will feed on other associated broad-leaved trees such as alder, birch, cottonwood, chokecherry, and willow and may consume the foliage of understory plants when starving.

Figure 381. Unhatched egg mass of large aspen tortrix. *Photo: Steven Katovich, USDA Forest Service, Bugwood.org.*

Figure 382. Mature, light green larva of large aspen tortrix. *Photo: William M. Ciesla, Forest Health Management International, Bugwood.org.*

Figure 383. Dark green larva of large aspen tortrix. *Photo: K.B. Jamieson, Canadian Forest Service, Bugwood.org.*

**Life Cycle**—There is one generation per year. Eggs are laid in mid-June through July, mainly on the upper surfaces of leaves but almost anywhere if no aspen foliage is available. Newly hatched larvae feed on leaf tissue but not veins, skeletonizing the leaf. Young larvae are often gregarious and can be found feeding within previously rolled leaves in old pupation sites. Dispersal to hibernation sites such as bark crevices and other protected places begins in early August and is completed by September. Larvae molt once and then overwinter within hibernaculae, which are white silk casings that each larva spins around itself. On warm days in early spring, the small larvae emerge from hibernaculae and move up the stems of aspen. The larvae mine into the buds, feeding on the young tissue. At this stage, secondary hibernaculae are often constructed at the base of the buds. In later stages, the larvae roll (fig. 384), fold (fig. 385), or web leaves into shelters, within which they feed and eventually pupate. When disturbed, larvae writhe vigorously and often drop, suspended by a silk thread. During epidemics, larvae frequently wander in search of food, resulting in extensive webbing that covers grass and understory plants. Pupation occurs from mid-June to mid-July, depending upon location and weather, and lasts about 10 days. Adults are active for about a 2-week period from late June through July. Females are sluggish and stay in the area of emergence, whereas males are more energetic and erratic in flight.

**Damage**—Large aspen tortrix becomes a problem only where aspen is a major component of the forest stand. Entire aspen stands and vast aspen forests have been completely defoliated during outbreaks. Larval feeding in early spring sometimes causes complete defoliation before the buds have expanded.

Healthy, defoliated aspen usually grow new leaves by midsummer, but they are often smaller and fewer than normal, resulting in thinned crowns. Copious amounts of silk webbing in the understory can be annoying to people walking through heavily defoliated aspen stands. Growth loss is the major impact in the early stages of an outbreak. Repeated defoliation over a period of years can cause branch death that is most noticeable in trees growing on marginal sites. Weakened aspen are more susceptible to opportunistic agents such as plant diseases and insects. Despite growth loss, aspen usually withstand large aspen tortrix outbreaks with little tree mortality.

**Figure 384.** Leaf roll held together by silk. *Photo: Steven Katovich, USDA Forest Service, Bugwood.org.*

**Figure 385.** Pupa of large aspen tortrix protruding from leaf shelter. *Photo: William M. Ciesla, Forest Health Management International, Bugwood.org.*

The impact of an outbreak depends upon many factors, primarily the frequency and severity of defoliation. Other considerations include wildlife use and recreational and aesthetic values of the aspen stand. Defoliation alone can reduce values, especially in high-use areas.

There are numerous similar looking caterpillar species on aspen that roll, tie, or fold leaves, many of which are also in the family Tortricidae. All are of minor consequence. Increases in large aspen tortrix populations are sometimes associated with increases in the populations of other aspen feeding insect species, such as tent caterpillars or leafminers.

**Management**—Outbreaks are characterized by the build-up of large populations that persist for 2 or 3 years and then suddenly collapse without human intervention. Though natural enemies, including insects, diseases, and birds, feed upon large aspen tortrix, the most significant contributors to outbreak collapse are thought to be starvation and other debilitating effects of foliage depletion, as well as unseasonable cold weather early in spring that either kills young larvae or reduces their food supply by harming leaf tissue. Aspen can tolerate significant defoliation. Because of the short duration of outbreaks of large aspen tortrix and associated insect defoliators, direct control is not generally recommended. In high-use areas like campgrounds, applying registered chemical or biological insecticides may be appropriate in order to reduce defoliation, masses of webbing in the understory, or aesthetic damage.

**References: 13, 35, 50**

# Needleminers

## Hollowed out needles

**Name and Description**—Pinyon needleminer—*Coleotechnites edulicola* Hodges and Stevens [Lepidoptera: Gelechiidae]

Ponderosa pine needleminer—*C. ponderosae* Hodges and Stevens

White fir needleminer—*Epinotia meritana* (Heinrich) [Lepidoptera: Tortricidae]

Spruce needleminers—*Endothenia albolineana* (Kearfott) [Lepidoptera: Tortricidae] and *Coleotechnites piceaella* (Kearfott)

Needleminers are tiny moth larvae that feed inside conifer needles. They are locally common on pinyon and ponderosa pine and are less commonly found on white fir and spruce. Adult moths are rarely noticed, but the hollowed out needles with tiny larvae or frass are distinctive (fig. 386). Larvae are particularly visible in ponderosa pine needles when held into the sunlight (fig. 387). Exit holes are also diagnostic on older hollowed out needles (fig. 388).

**Hosts**—Pinyon pine, ponderosa pine, white fir, and spruce are hosts to different species of needleminers in the Rocky Mountain Region. Lodgepole pine needleminers occur in California and Oregon but have not been reported in the Rocky Mountain Region.

**Figure 386.** Ponderosa pine needleminer. *Photo: Scott Tunnock, USDA Forest Service, Bugwood.org.*

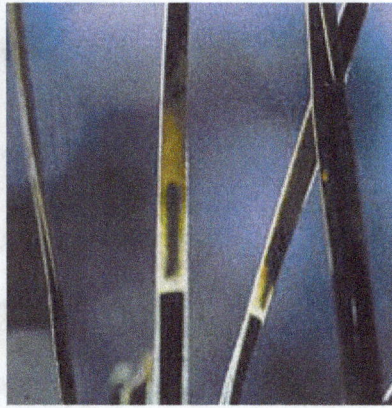

**Figure 387.** Ponderosa pine needleminer. *Photo: Southwestern Region, USDA Forest Service.*

**Figure 388.** Pinyon needleminer-damaged needles. *Photo: Robert Cain, USDA Forest Service.*

**Life Cycle**—Pinyon needleminers lay eggs from early June through mid-July. Larvae emerge soon after eggs are laid and bore into uninfested needles where they feed until fall. They overwinter inside the needles as dormant larvae. Feeding resumes in the spring, and larvae grow to about 3/8 inch (5 mm) long. Pupation occurs in late May. Ponderosa pine needleminers lay eggs in late summer inside previously mined needles. The newly hatched larva bores into the tip of a green needle and mines slowly through the winter, developing more rapidly as the weather warms and pupating in midsummer. White fir needleminers have a similar life cycle but will mine about six needles before pupating in June or early July. White fir needleminers hatch in August and September and spend the winter within a needle. The two species of spruce needleminer that occur in the Rocky Mountain Region also overwinter as larvae. *E. albolineana* spends the winter protected in mats of webbing and dead needles attached to twigs (fig. 389). *Coleotechnites piceaella* spends the winter inside of mined needles. Both species resume feeding in early spring and complete their development by late spring. Eggs are laid in May and June.

**Figure 389.** Webbed needles from spruce needleminer. *Photo: Whitney Cranshaw, Colorado State University, Bugwood.org.*

**Damage**—Damage first becomes evident as foliage fades to yellow and brown. Early needle drop, reduced growth, and tree mortality can all result from needleminer infestations. The severity of the infestation can vary significantly from tree to tree, suggesting that individual trees have some resistance to these pests. Foliage thinning by needleminers can reduce tree vigor and increase susceptibility to bark beetle attack (fig. 390).

**Figure 390.** White fir needleminer damage. *Photo: Intermountain Region, Ogden Archive, USDA Forest Service, Bugwood.org.*

**Management**—Trees usually recover from needleminer damage without suffering serious injury. Properly timed insecticide applications have effectively controlled populations on landscape trees. *Endothenia albolineana* overwintering nests can be removed from infested spruce with a strong stream of water from a garden hose in early spring before bud flush.

**References: 22, 35, 50**

# Oak Leafroller

## Defoliates red and white oaks

**Name and Description**—*Archips semiferana* (Walker) [Lepidoptera: Tortricidae]

The oak leafroller is a common defoliator of oaks. In the central Rocky Mountains, the oak leafroller defoliates gambel oak. Moths are a light olive brown and have a diagonal rusty band on the forewing (fig. 391). The wingspan is approximately 7/8 inch (22 mm) long. Mature larvae are approximately 4/5 inch (21 mm) long and are greenish yellow (figs. 392-393). Eggs are laid in a mass containing 15-125 eggs and are covered with tan scales from the female's abdomen.

**Hosts**—Gambel oak is the primary host in the Rocky Mountain Region, although ornamental plantings of various red and white oaks may be subject to infestation by the oak leafroller. Occasionally, the insect may be found on maples.

**Figure 391.** Oak leafroller adult. *Photo: John Himmelman, http://booksandnature.homestead.com/moth133.html.*

**Life Cycle**—The oak leafroller egg masses are deposited in the canopy of trees in July and August. A majority of the egg masses are deposited on and overwinter in the middle half of the bole and branches of the canopy, and more occur on the southern side of the tree. Larvae emerge from eggs in spring at

**Figure 392.** Oak leafroller larva. *Image: Texas Agriculture Extension Service.*

about the time that buds are bursting. Larvae feed within the protective folds of young leaf clusters and later roll or fold one or more entire leaves or leaf parts (fig. 394). The larvae pass through five instars. Pupation takes place in the leaf roll in June to early July, and the adults appear in July.

**Damage**—The oak leafroller defoliates oaks and is of particular significance in the mid-continent oak woodlands, where red, scarlet, northern pin, white, and chestnut oaks have been defoliated. Outbreaks of this insect are infrequent in the Midwest and East

**Figure 393.** Oak leafroller larvae and damaged leaf. *Photo: Northern Research Station, USDA Forest Service.*

and may last from 3-5 years. In the central Rocky Mountains, outbreaks of the oak leafroller are sporadic. Defoliation may lead to weakened trees that succumb to other factors such as late frosts or wood borers. First-instar larvae produce webbing within folds of the young leaf cluster; later instars will web together leaves or roll leaves to provide for shelter and food.

**Management**—There is little information on the natural control agents of the oak leafroller, but natural enemies are thought to be important in preventing damaging populations. Mature larvae exhibit behavior typical of the members of its moth family, Tortricidae, when disturbed. They retreat rapidly, exit the shelter, and drop on a silken thread. To manage a local problem, horticultural oil is considered effective in controlling the overwintering egg stage of this insect. Applications should target the twigs where the egg masses are deposited.

**Reference: 186**

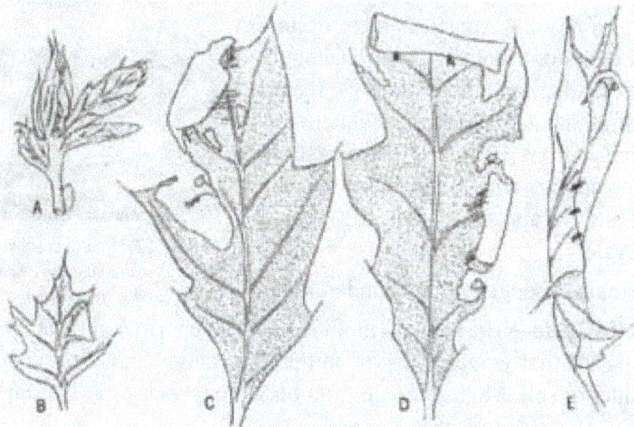

**Figure 394.** Rolled oak leaves (from Wilson 1972).

# Pandora Moth

## Rare outbreaks leave pines tufted

**Name and Description**—*Coloradia pandora* (Blake) [Lepidoptera: Saturniidae]

The pandora moth is one of the largest forest insects in North America. Adults have a thick, heavy body 1-1 $1/2$ inches (2.5-3.8 cm) long and a wingspread of 3-5 inches (7-13 cm). Forewings are brownish grey, and hindwings are a light pinkish grey, each marked with a dark spot near the center above a transverse wavy line (figs. 395-396). Eggs are globular, about $1/8$ inch (3 mm) in diameter, and are bluish green at first, later turning bluish grey. Eggs are laid in clusters of 2-50 each. Newly hatched larvae have shiny black heads and dark brown to black, spiny bodies about $1/4$ inch (6 mm) long. Intermediate stage caterpillars have two narrow white lines marking their upper surface and an overall brown color. Mature larvae have an orange-brown head with a pale yellow-brown collar and are about 2-3 inches (60-80 mm) long (figs. 397-398). Their bodies start out dark brown and then change to greenish yellow, are marked with transverse yellow bands and a longitudinal white stripe and bear a few branched spines on each segment. These spines can produce a stinging sensation upon contact. The pupae are stout, dark reddish or purplish brown (fig. 399) and naked in the soil beneath pines in a cell formed by the larvae.

The similar looking Black Hills pandora moth, *C. doris* Barnes, occurs in Colorado, Montana, South Dakota, and Wyoming, although little is recorded of its habits beyond a documented outbreak from 1938 to 1939 near Osage, Wyoming, and in the northern Black Hills. It apparently has a 1-year life cycle.

**Figure 395.** Female adult pandora moth with thread-like, narrow antennae. *Photo: Intermountain Region, Ogden Archive, USDA Forest Service, Bugwood.org.*

**Figure 396.** Male adult pandora moth with feathery antennae. *Photo: Terry Spivey, USDA Forest Service, Bugwood.org.*

**Figure 397.** Mature, healthy larva of pandora moth. *Photo: Rocky Mountain Region Archive, USDA Forest Service, Bugwood.org.*

**Hosts**—Lodgepole and ponderosa pines

**Life Cycle**—The pandora moth requires 2 years to complete its life cycle, spending the first winter as larvae in the tree canopy and the second as pupae in the litter or soil. Moths emerge from buried pupal cases in late June or July. During

**Figure 399.** Pupae of pandora moth with smaller pupae of parasitic flies, visible at lower right. *Photo: Southwestern Region, USDA Forest Service.*

**Figure 398.** Mature, healthy larva of pandora moth (on fingers) and diseased larva with viral wilt (hanging). *Photo: Donald Owen, California Department of Forestry and Fire Protection, Bugwood.org.*

epidemics, thousands of these moths can be seen flying in the forest canopy. Males seek the females, who generally do not fly until after mating. Females lay an average of 80 eggs that are distributed in several clusters, usually on the bark and needles of pines. Eggs are occasionally deposited on ground litter or brush. Eggs hatch in August following approximately 40-50 days of incubation. Young larvae crawl up the trees and feed in clusters on needles of the terminal shoots on the outer branches. In moving from one feeding site to another, larvae travel together in single file. In autumn, larvae disperse and become solitary. Larvae overwinter primarily in the second stage at the base of the needles, usually amidst some strands of silk, and may feed as weather permits. With the onset of warmer weather in spring, feeding resumes and larvae grow rapidly. Masses of greenish or brownish droppings and cast skins can be found under infested overstory trees due to the enormous quantity of needles consumed by caterpillars. Needles of all age classes are consumed, but the buds are not damaged. When fully grown in July or early August, larvae crawl down tree trunks and enter loose material to pupate, preferring pumice or decomposed granite soils. Pupation typically lasts for about a year, but, in some areas at least, a substantial portion of the generation remains in the soil for 2 years and some individuals remain for up to 5 years.

**Damage**—Outbreaks, some covering vast acreage, have occurred only on areas with soils loose enough for larvae to bury themselves prior to pupation. This insect is rarely seen except during outbreaks, which occur at 20-30-year intervals and may last 6-8 years. The most recent outbreak in the Rocky Mountains occurred from about 1959 to 1966 in Colorado and Wyoming. During outbreaks, growth loss and mortality can be significant. Pandora moth has a 2-year life cycle, with feeding and moth flight occurring in alternate years, so that most of the defoliation occurs every other year. Larvae feed primarily on older foliage, leaving trees with a tufted appearance. Because terminal buds are not damaged, even severely defoliated trees usually recover completely. Heavy defoliation can predispose trees to attack by bark beetles, and repeated defoliation alone can kill pines, especially those stressed by other agents such as heavy dwarf mistletoe infection or drought. Moth flights to nearby sources of light may be the first indication of an increasing population of pandora moth.

**Management**—The pandora moth has many natural enemies, including insects (fig. 400), diseases, rodents that dig up and eat pupae, and birds that consume all life stages. High soil temperatures during the pupation period can cause substantial mortality. Perhaps most significant in causing the collapse of outbreaks is a viral wilt disease that devastates dense populations.

Insecticides are not generally needed because natural controls suppress outbreaks before significant mortality occurs. An appropriate, registered chemical insecticide would likely reduce larval populations, considering how exposed the feeding larvae are. Several prescribed burning treatments have been attempted to kill pupae and larvae with variable success. Pandora moth is one of many insects used as human food, although sporadic abundance limits its use as a staple.

**References: 24, 35, 50**

**Figure 400.** Mature larva of pandora moth, a victim of multi-brooded parasitic wasp, showing wasp cocoons. *Photo: Donald Owen, California Department of Forestry and Fire Protection, Bugwood.org.*

# Pine Butterfly

## Larvae feed on older needles

**Name and Description**— *Neophasia menapia* (Felder and Felder) [Lepidoptera: Pieridae]

This insect can be identified easily during any season. Look for single rows of emerald green eggs from September to June on needles, colonies of immature pale green larvae with black heads, or individual full-grown larvae about 1 inch (2.5 cm) long with two white, lateral stripes and green heads (fig. 401) from June to August. Pupae, usually found during August, are also green with white stripes and are attached to needles, branches, or stems. Adults are white butterflies with black wing markings and are seen flying around tree canopies from August through September (fig. 402).

**Hosts**—Ponderosa pine

**Figure 402.** Pine butterfly adult. *Photo: Jerald E. Dewey, USDA Forest Service, Bugwood.org.*

**Figure 403.** Pine butterfly eggs. *Photo: Intermountain Region, Ogden Archive, USDA Forest Service, Bugwood.org.*

**Figure 401.** Pine butterfly larva. *Photo: Ladd Livingston, Idaho Department of Lands, Bugwood.org.*

**Life Cycle**—Adults fly and lay their green eggs from August through October (fig. 403). The eggs overwinter, and the larvae hatch the following June about the time new foliage appears. The small larvae feed in clusters with their black heads oriented toward the tip of the needle. As the larvae mature, they feed singly. They pupate in late July for approximately 15-20 days (fig. 404).

**Damage**—Repeated defoliation can kill ponderosa pine, reduce radial growth up to 70%, or weaken trees enough to attract bark beetles. Under ordinary conditions, larvae feed only on older needles and are not a big problem, but during epidemics, they also consume new needles. In normal years, pine butterfly feeding causes the twigs to be defoliated with the exception of the current year's growth. Tree crowns then appear thin, and only tufts of new needles remain (fig. 405).

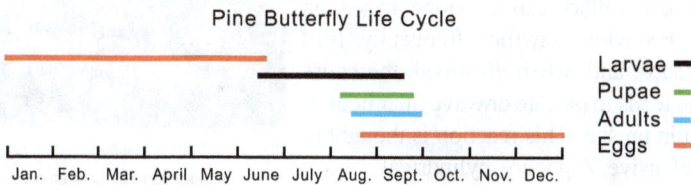

Pine Butterfly Life Cycle

**Figure 404.** Pine butterfly life cycle (from Johnson 1982).

**Management**—Management is generally not warranted. Natural control agents often keep populations in check and end epidemics. No correlation has been found between crown classes or diameters and resultant mortality from defoliation. Tree vigor prior to defoliation is important in a tree's recovery from defoliation.

**References: 28, 50, 93**

**Figure 405.** Pine butterfly defoliation. *Photo: Intermountain Region, Ogden Archive, USDA Forest Service, Bugwood.org.*

# Pine Sawflies

## Feed on older needles, leaving distinctive stubs

**Name and Description**—*Neodiprion* spp. and *Zadiprion* spp. [Hymenoptera: Diprionidae]

Adult pine sawflies are thick-waisted wasps about 1/2 inch (10-12 mm) long (fig. 406). Females are yellowish brown, and the smaller males are mostly black, bearing feathery antennae. Pine sawfly larval appearance varies by species and by larval stage, but most are green or yellowish green in color with black, tan, or orange head capsules (figs. 407-408). Mature larvae are about 3/8 inch (10 mm) long. Both caterpillars (order Lepidoptera) and sawfly larvae have three pairs of true legs, but sawfly larvae have six or more pairs of fleshy prolegs behind their legs along the abdomen, while caterpillars have two to five pairs of prolegs. The head and jaws of sawfly larvae point downward, perpendicular to the long axis of the body, while caterpillar heads and jaws point forward, in line with the body length. Although not completely accurate, a quick test to distinguish sawfly larvae from caterpillars can be done based on their behavior—sawflies frequently feed in groups, and, when disturbed, the entire group tends to pulsate or wave their bodies, often in unison. This reaction is thought to be defensive. A papery, cylindrical cocoon covers sawfly pupae, which form in the soil.

**Figure 406.** Adult sawfly female (*Neodiprion* spp.) laying an egg in a ponderosa pine needle. *Photo: Michael R. Wagner, Northern Arizona University.*

**Figure 408.** Late stage larvae of *Zadiprion rohweri* feeding on pinyon pine. *Photo: Michael R. Wagner, Northern Arizona University.*

**Figure 409.** Rows of *Neodiprion autumnalis* eggs inserted into ponderosa pine needles. *Photo: Mark Harrell, Nebraska Forest Service.*

**Figure 407.** Late stage larvae of *Neodiprion gillettei* feeding on ponderosa pine. *Photo: Southwestern Region, USDA Forest Service.*

Eggs are laid in slits cut in the edge of living pine needles (fig. 409). There are six pine-feeding *Neodiprion* spp. in the region and two pine-feeding *Zadiprion* spp. Differentiating among *Neodiprion* spp., especially among the five that feed on ponderosa pine, can be difficult given the considerable resemblance and variation among closely related species and recent taxonomic changes.

**Hosts**—Pines, particularly ponderosa pine and pinyon, but rarely on other pines and Douglas-fir

**Life Cycle**—*Neodiprion* spp. pine sawflies have a 1-year life cycle that is similar among species but differs in timing of life stages. They overwinter either as eggs inserted into pine needles or as full-grown larvae within cocoons in the ground litter or soil. The species that overwinter as eggs begin feeding upon hatching early in spring, while species that overwinter as larvae pupate in spring, emerge and mate as adults, and lay eggs in pine needles. Consequently, young *Neodiprion* spp. sawfly larvae are found in either spring-summer or summer-fall, feeding gregariously on older foliage. Young larvae consume only the outer needle tissue and leave the central ribs intact. The central ribs later turn yellow-brown and break off. Older larvae feed singly and consume most of the needle, often leaving a distinctive stub. Species of spring-summer feeders generally consume only old needles, completing their feeding before pine shoots have elongated, and enter the soil to pupate. Adults emerge late in summer and lay eggs in live needles to overwinter. Species of summer-fall feeders consume both old and new pine needles, complete feeding in late summer, and enter the litter or soil, forming cocoons in which they overwinter.

*Zadiprion townsendi* (Cockerell), the bull pine sawfly, is the only sawfly species whose larvae feed on ponderosa pine in winter, weather permitting. The larvae are wormlike, dark green to black with gold flecking, and feed gregariously (fig. 410). Larvae are about half-grown by the onset of winter and are mature by May or June. The mature larvae burrow into the soil and spin a cocoon. Some pupate, emerge, mate, and lay eggs in pine needles, concentrating on the upper areas of trees that hatch that

Figure 410. Late stage larvae of *Zadiprion townsendii* feeding on ponderosa pine. *Photo: Michael R. Wagner, Northern Arizona University.*

same summer. Others delay pupation and subsequent events until late summer or the following season. The entire life cycle takes 1-2 years to complete.

*Zadiprion rohweri* (Middleton) (fig. 408) reportedly has two generations per year on pinyon.

**Damage**—Infested trees have sparse foliage and thin crowns. Consumption of older needles gives pines a tufted appearance that is created by current year needles growing on the tips of branches stripped of older foliage (fig. 411). Consumption of all old and new needles in one season has a greater impact on pine health. When dense larval populations descend together to enter the soil and spin cocoons, they can cause considerable consternation. Different species have different preferences for the size of host attacked and location on the host where they feed. Pine sawflies typically attack open-grown trees or areas where pine is growing at a low density, often on poorer sites with dry, shallow soils. In some cases, prolonged drought is accompanied by pine sawfly outbreaks. The same trees are frequently defoliated year after year while others nearby may remain largely unaffected. In general, defoliation causes slower growth. Repeated defoliation can result in top-kill. Although relatively rare, pine mortality may occur, particularly if bark beetles attack trees weakened by defoliation.

**Management**— Typically, predators, parasites, viral disease, and foliage depletion manage to reduce pine sawfly outbreak populations within 2-3 years. Consequently, active management is often not necessary. When only a few colonies of larvae

Figure 411. Feeding on old needles by *Neodiprion autumnalis* in the Nebraska Panhandle resulted in a tufted appearance on defoliated ponderosa pine. *Photo: Todd Nordeen, Nebraska Game and Parks Commission.*

are present on small pines, they can be picked off by hand or washed off with a high-pressure hose. Registered insecticides can be used effectively when larvae are present across an area in large numbers and some control is appropriate. Insecticides should be applied as soon as possible after egg hatch for best control on high-value pines. Note that biological insecticides that only target Lepidoptera are ineffective against sawflies. Because the last larval stages consume the most foliage, control is often sought after most of the annual defoliation has occurred and larvae are too large to be killed efficiently by insecticide.

**References: 35, 50**

# Tiger Moth

## Tree-top tents appear early in conifers

**Name and Description**—*Lophocampa ingens* (H. Edwards) (= *Halisidota ingens* H. Edwards) [Lepidoptera: Arctiidae]

Adult moths have a wingspan of about 3 inches (8 cm); dark, reddish brown forewings that bear large, white oval splotches (fig. 412); and white hindwings. Mature caterpillars are about 1 1/2 inches (3.8 cm) long and reddish brown to black in color with tufts of black and yellow hairs on their backs (fig. 413). These hairs have been known to irritate the skin of some people on contact, causing a rash.

A subspecies of the similar looking silver spotted tiger moth *L. argentata subalpina* (French) feeds primarily on juniper and occasionally on pinyon.

**Hosts**—All native pines are hosts. Less commonly, Douglas-fir and white fir are fed upon during outbreaks.

**Life Cycle**—The tiger moth has one generation per year. Adult moths emerge and lay eggs in July and August. During September and October, the small, dark-colored, hairy caterpillars hatch from eggs and begin feeding gregariously on needles in the upper branches of host pines, producing silk webbing. Larvae overwinter in the webs in groups and will feed during warmer periods of winter

**Figure 412.** Tiger moth adult. *Photo: Rocky Mountain Region Archive, USDA Forest Service.*

**Figure 413.** Mature larva of the tiger moth, *Lophocampa ingens*. *Photo: Whitney Cranshaw, Colorado State University, Bugwood.org.*

**Figure 414.** Tent with larvae of tiger moth in the top of ponderosa pine. *Photo: William M. Ciesla, Forest Health Management International, Bugwood.org.*

**Figure 415.** Tent with larvae of tiger moth in the top of ponderosa pine. Note the dense, silk webbing and consumption of older needles. *Photo: Southwestern Region, USDA Forest Service.*

and early spring. A life cycle that includes larval feeding during winter and early spring is quite rare among temperate climate forest defoliator species. In April and May, larvae resume feeding continuously on older needles and expand their webs, binding needles with dense white silk (figs. 414-415). At this time, the larvae and their tent-like webs become very conspicuous. Tiger moth larvae reach maturity very early in the season and are among the largest caterpillars one can find on conifers in spring. Pupation occurs in June.

**Damage**—Larvae feed on older foliage, sometimes defoliating trees by early spring. They make large silk tents concentrated on the upper branches and tree tops, often enclosing the terminal growth. Tiger moth defoliation tends to be more common on younger pines, but generally occurs in natural, rather than landscape, settings. Although the webs and larvae can be very noticeable in late winter and spring, this insect causes minor defoliation. Permanent tree injury rarely results from feeding, which is usually limited to the upper foliage. Outbreaks are somewhat infrequent and usually collapse within 1 or 2 years.

**Management**—Populations of tiger moth usually remain at low levels due to the action of predators, parasites, diseases, and cold winters. Because outbreaks are brief, defoliation is rare in landscape settings, and plant injury is unlikely, management efforts against tiger moth are rarely warranted.

Due to the potential for skin irritation and the preference for upper foliage, hand picking probably will not be a successful method of population reduction. Pruning to remove tents and applying insecticide treatments are unnecessary because new shoots will elongate in late spring and early summer and will grow to obscure defoliated portions of the uninjured tree.

**References: 35, 50**

# Western Pine Budworm

## Messy, webbed, clipped foliage

**Name and Description**—*Choristoneura lambertiana* (Busck) [Lepidoptera: Tortricidae]

The appearance and life cycle of this pine-feeding species complex is very similar to that of the Douglas-fir, spruce, and true fir-feeding western spruce budworm, *C. occidentalis*, which is generally larger and rarely found feeding on pines. Forewings of adult western pine budworm are about $1/2$ inch (11-15 mm) long and yellowish brown, red, cream, and silver (fig. 416). Eggs are oval, light green, and laid in 25-50-egg clusters, overlapping shingle-like along pine needles. Young larvae are yellow-brown with a dark head capsule. Mature larvae are about $3/4$ inch (17-19 mm) long and have a dark shield and head capsule and a smooth, brown-green body with paired white spots on each segment (fig. 417). Characteristic of many Tortricidae, larvae wriggle vigorously when disturbed and will drop, suspended on a silken thread. Pupae are yellow-brown with darker brown abdominal segments (fig. 418). Distributed across western North America, the western pine budworm species complex includes three described subspecies and an as yet unassigned additional form.

**Hosts**—Limber, lodgepole, and ponderosa pines; rarely whitebark pine; white fir if mixed in stands of lodgepole pine. In other regions,

Figure 416. Adult western pine budworm. *Photo: Bernard J. Raimo, USDA Forest Service, Bugwood. org.*

Figure 417. Mature larva of western pine budworm. *Photo: Scott Tunnock, USDA Forest Service, Bugwood.org.*

Figure 418. Pupa of western pine budworm. *Photo: Intermountain Region, Ogden Archive, USDA Forest Service, Bugwood.org.*

sugar pine is a host, hence the common name, sugar-pine tortrix.

**Life Cycle**—There is one generation per year. Eggs hatch in late summer, and the tiny larvae do not feed but migrate to sheltered overwintering sites in branch and stem crevices. The larvae molt once and then hibernate through winter within silken shelters they spin around themselves. Larvae emerge in spring when shoot development is nearly complete, mining needle sheaths, old needles, and developing cones. As they grow, larvae feed on new, elongating needles. However, in some cases, larvae may continue to feed primarily on male and female cones. One to five larvae per shoot will web new needles together and feed in this protective enclosure, which becomes littered with droppings and brown, partially chewed needles (fig. 419). Larvae pupate in these loose silk and needle enclosures and emerge as adults in July and August. Eggs are deposited on the concave side of older needles, preferentially in the upper crown, hatching 2 or 3 weeks later.

**Figure 419.** Feeding damage on new growth. *Photo: Intermountain Region, Ogden Archive, USDA Forest Service, Bugwood.org.*

**Damage**—Occasional outbreaks lasting 1-3 years on limber, lodgepole, and ponderosa pines have been documented. Impacts include the loss of current year needles, destruction of shoots and developing cones, growth loss, and top-kill, but little or no tree mortality. Successive years of defoliation from 1973 to 1975 in Colorado resulted in branch deformity, although severe damage was localized and of short duration. Pines weakened by western pine budworm defoliation may be more susceptible to mountain pine beetle and other bark beetles.

**Management**—Western pine budworm populations are subject to regulation by a large complex of insect and avian natural enemies as well as the effects of unseasonable temperature extremes and, during sustained outbreaks, starvation due to depletion of new growth.

Because populations are naturally regulated and impacts tend to be short-term and localized, active management is generally not attempted. No silvicultural methods have been developed against western pine budworm. When direct control is deemed necessary to reduce aesthetic or economic injury, application of appropriate, registered insecticide should be made in spring when larvae emerge from hibernation and begin feeding. This is usually just after buds begin to expand in May or June.

**References: 35, 50, 88**

# Western Pine Tussock Moth

## Eruptive defoliator on poor sites

**Name and Description**—*Dasychira grisefacta* (Dyar) (= *Paraorgyia grisefacta* [Dyar]) [Lepidoptera: Lymantriidae]

Adults are grayish brown moths with light and dark bands across the forewings and a wingspread of 1 1/4-1 1/2 inches (3.2-3.8 cm) (fig. 420). Eggs are bluish white, relatively large at about 1/13 inch (2 mm) in diameter, and laid in lines

**Figure 420.** Western pine tussock moth adults. *Photo: Whitney Cranshaw, Colorado State University, Bugwood.org.*

or clusters (fig. 421). Mature caterpillars are rusty brown with a dark head, four grey hair brushes on top, and numerous tufts of white and black hairs along the body. They are 2-2 3/4 inches (5-7 cm) long and have distinctive bundles of black hairs projecting in front of and behind the body (fig. 422). Silken cocoons have larval hairs incorporated into them (fig. 423) and envelop black pupae. *Dasychira grisefacta* occurs from the Dakotas and southern Alberta west to the coast and south to Arizona. The closely related, conifer-feeding species *D. pinicola* (Dyar) and *D. plagiata* (Walker) are found farther to the east.

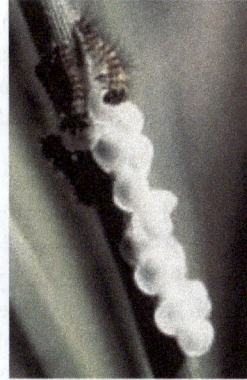

**Figure 421.** Eggs with newly hatched larvae of the western pine tussock moth. *Photo: Bill Schaupp, USDA Forest Service.*

**Figure 422.** Mature larva of the western pine tussock moth. *Photo: William M. Ciesla, Forest Health Management International, Bugwood.org.*

**Figure 423.** Characteristic *Dasychira* spp. cocoon with incorporated larval hairs. *Photo: Herbert A. "Joe" Pase III, Texas Forest Service, Bugwood.org.*

**Hosts**—Primarily ponderosa pine, although other known or reported hosts include pinyon, Douglas-fir, white and Engelmann spruces, and additional conifers elsewhere

**Life Cycle**—The western pine tussock moth has a 1-year life cycle, with adults present in late July to mid-August. Eggs are laid in small clusters, usually on pine needles, and hatch in 1-2 weeks. Young caterpillars feed externally on the top layers of the center of pine needles for a short period. After feeding, they seek sheltered locations, generally on the tree stem under bark scales, where they spend the winter. As soon as temperatures warm in the spring, the larvae resume feeding along needles and on staminate cones, growing large enough to consume entire needles by early summer. Most of the defoliation occurs then because older caterpillars consume the most foliage. The mature larvae spin a whitish cocoon with incorporated body hairs. Within the cocoon they undergo metamorphosis into a black pupa and emerge as a moth. The moths exit the pupal skin, mate, and females lay conspicuous, large eggs on pine foliage and elsewhere by late summer.

**Damage**—Because feeding begins before shoot expansion, old foliage is consumed first. However, eventually, pine foliage of all age classes will be eaten. Extensive feeding gives a pine tree a distinct reddish brown hue due to retention of discolored needles that were incompletely consumed and/or killed (fig. 424). Growth loss often accompanies defoliation. Because the pine tussock moth will eat all age classes of pine needles, it can completely strip a tree of foliage. This could kill a coniferous tree in 1 year. Partially defoliated trees might die if weakened by drought, previous defoliation, root disease, or other stresses. Defoliation-stressed trees may be at greater risk of attack from bark beetle. Two recent outbreaks in South Dakota, Nebraska, and Wyoming have resulted in some patches of ponderosa pine mortality, often associated with bark beetle attack on older, larger-diameter defoliated trees, although most of the defoliated ponderosa pine survived (fig. 425). These two episodes occurred on ridges and escarpments of lands of low or marginal quality for pine growth, whereas within the Black Hills, better quality pine sites were not defoliated. One consequence of heavy defoliation is

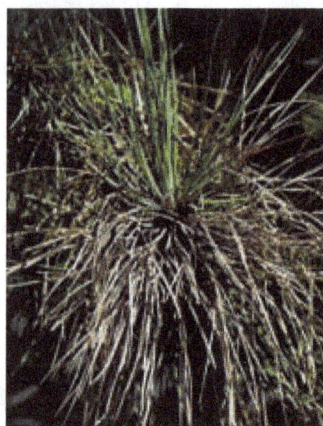

**Figure 424.** Extensive feeding by young western pine tussock moth larvae down the center of older ponderosa pine needles results in needle mortality and a reddish appearance. *Photo: William M. Ciesla, Forest Health Management International, Bugwood.org.*

the litter fall of significant quantities of caterpillar excreta, and cast skins rich in readily available nutrients may play a role in host tree recovery and compensatory growth once an outbreak subsides.

**Management**—A viral disease appears to be important in natural control of dense populations, with infected larvae dying in a characteristic deflated pose clinging to, yet limply dangling from, branches and twigs. Outbreak populations typically show a gradual decline over a several-year period, with documented outbreaks lasting no more than a few years. In many instances, no direct management action is needed.

Other than those promoting overall stand and tree health and vigor, specific silvicultural management tactics are not available. Depending upon resources and values at risk, pine tussock moths can be con-

Figure 425. Consecutive years of severe defoliation on ponderosa pine can give trees a tufted appearance. These pines survived almost complete defoliation for several consecutive years. *Photo: Bill Schaupp, USDA Forest Service.*

trolled with insecticide applications. However, because populations can increase rapidly and late-stage larvae do the most feeding, interest in direct control action often peaks each year after most defoliation has already occurred. At such times, when a mature larval population has nearly finished feeding, spraying insecticides will not result in population suppression or resource protection. It is best to make the determination of purpose and need and to prepare for spraying before spring. Then monitor population levels as temperatures warm, spraying only when potential benefits outweigh costs and risks, and apply material as early in the growing season as possible. Most of the time, this showy insect is rarely encountered.

**Reference: 50**

# Western Spruce Budworm

## Imparts reddish brown cast to forest

**Name and Description**—*Choristoneura occidentalis* Freeman [Lepidoptera: Tortricidae]

The Western spruce budworm is a widely distributed defoliator. The nondescript, brownish moths are 1/2 inch (13 mm) long and have a wingspan of 7/8-1 1/8 inches (22-28 mm). Both sexes are similar in appearance and are able to fly (unlike the Douglas-fir tussock moth) (fig. 426). Young larvae are yellow-green with brown heads and distinctive white spots. Larvae change colors as they develop

**Figure 426.** Western spruce budworm adult. *Photo: Rocky Mountain Region, USDA Forest Service.*

**Figure 427.** Western spruce budworm larva. *Photo: Scott Tunnock, USDA Forest Service, Bugwood.org.*

through six stages (fig. 427). Mature larvae are 1-1 $^1$/4 inches (2.5-3.2 cm) long, have tan to light-colored heads and olive-brown or reddish brown bodies with large, ivory-colored areas. Eggs are oval, light green, about $^1$/25 inch (1 mm) long, and overlap like shingles on needles.

**Hosts**—Commonly found on Douglas-fir, Engelmann spruce, blue spruce, white fir, and subalpine fir; occasionally found on pines when mixed with other hosts

**Life Cycle**—Western spruce budworm has a 1-year life cycle. Moths usually emerge from late July to early August. Eggs are laid on the underside of conifer needles and hatch in about 10 days. Young larvae seek a sheltered place and overwinter in a silken casing called a hibernaculum. From early May to late June, larvae begin to feed within closed buds, 1-year-old needles, and new foliage. New foliage is preferred, followed by older needles. Larvae mature in 30-40 days and pupate in early July. Pupation usually lasts 10 days and is followed by adult moth emergence.

**Damage**—Young larvae will feed by mining needles or newly swelling buds. Larvae feed mainly on new foliage but will feed on old foliage if all the new foliage has been destroyed. Cones and seeds can also be destroyed. Larvae and pupae can be seen in silken nests of webbed, chewed needles (fig. 428). Defoliation occurs at tops of trees and outer branches. During a defoliation event, entire stands will have a brown appearance from the needle damage. Understory trees are most severely impacted (fig. 429). Repeated severe defoliation (4-5 years) can decrease growth, cause tree mortality, or render weakened trees more susceptible to other damaging agents such as bark beetles.

**Management**—Western spruce budworm can be managed through silvicultural methods. Because larvae disperse on silken threads and often impact the

**Figure 428.** Early instar webbing and feeding. *Photo: Ladd Livingston, Idaho Department of Lands, Bugwood.org.*

understory more intensively, removing the understory (thinning from below), lowering stand density, and maintaining tree species diversity can reduce the budworm populations. If more short-term, direct control is desired, then insecticides can be applied. The microbial insecticide *Bacillus thuringiensis* (B.t.) is available and is not hazardous to most beneficial insects, birds, small mammals, and aquatic systems. Other contact chemical insecticides are also available for budworm management.

**References: 45, 50**

**Figure 429.** Western spruce budworm damage. *Photo: Rocky Mountain Region, USDA Forest Service.*

# Western Tent Caterpillar

## Silken tents in early spring

**Name and Description**—*Malacosoma californicum* (Packard) [Lepidoptera: Lasiocampidae]

Adults are heavy bodied with a wingspan of 1-2 inches (2.5-5 cm) (fig. 430). The color of the moths of both sexes varies from dark red-brown to yellow, tan, or gray. Forewings have a pair of lines that can be either darker or lighter than the wing color. Egg masses are covered with a hardened, glossy material that varies in color from dark brown to pale gray (fig. 431). When first hatched,

**Figure 430.** Adults of western tent caterpillar: female (left) and a male (right). *Photo: Jerald E. Dewey, USDA Forest Service, Bugwood.org.*

larvae are about 1/8 inch (3 mm) long and are dark brown to black in color with whitish hairs (fig. 432). Mature larvae are about 2 inches (4.5-5.1 cm) long and are highly variable in color and markings. Most larvae have a pale blue head capsule and body speckled with black markings (fig. 433). They have a mid-dorsal stripe formed by a blue-white to pale blue dash on each body segment. The mid-dorsal stripe is edged with two bands that

Figure 431. Female western tent caterpillar adult with an egg mass. *Photo: Jerald E. Dewey, USDA Forest Service, Bugwood.org.*

may be black or yellowish orange, banded with black. The body is covered with orange or orange-brown hairs with white tips. When full-grown, caterpillars are about 2 inches (50 mm) long. Cocoons are of white silk dusted with a white to yellow powder. This insect is the most variable of the North American species of *Malacosoma*, in which six subspecies and several unclassified forms have been recognized.

Figure 432. Hatchling larvae of western tent caterpillar. *Photo: Whitney Cranshaw, Colorado State University, Bugwood.org.*

**Hosts**—Hosts include a wide range of tree and shrub species. In the central and southern Rocky Mountains, quaking aspen is the preferred host. Other reported tree hosts include alder, cottonwood, crabapple, fruit trees, oak, poplar, and willow. Chokecherry is a host throughout its range, especially on the western slopes of Colorado. Other shrub hosts include bitterbrush, *Ceanothus* spp., mountain mahogany, nine-bark, serviceberry, sumac, wild currant, and wild rose.

**Life Cycle**—Western tent caterpillar has one generation per year. White silken tents in the branches of host plants provide evidence of western tent caterpillar presence in spring. These are generally visible shortly after bud burst. Larvae live and feed as a colony, enlarging the tent as they grow (figs. 432, 434-435). If disturbed, larvae wave their bodies and twitch. When not feeding, larvae remain in the tent, which they use as shelter, protection, and a molting site. Late-stage larvae become solitary feeders and no longer use the tent. The larvae mature in 30-42 days, depending on weather conditions. Pupation

Figure 433. Mature larvae of western tent caterpillar. *Photo: William M. Ciesla, Forest Health Management International, Bugwood.org.*

Figure 434. Tents indicate western tent caterpillar infestation, viewed from within an aspen stand. *Photo: William M. Ciesla, Forest Health Management International, Bugwood.org.*

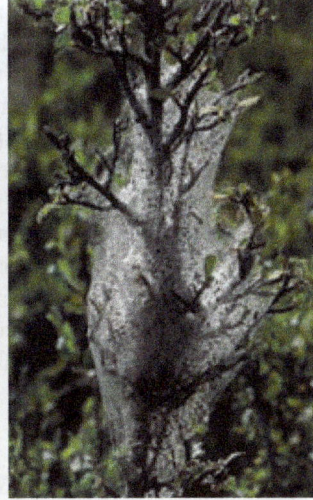

Figure 435. Tent of western tent caterpillar on a mountain mahogany stem. *Photo: William M. Ciesla, Forest Health Management International, Bugwood.org.*

occurs within silken cocoons spun on branches in the remaining leaves of host trees, on non-hosts, or in leaf litter. The duration of the pupal stage is from 12-18 days. Adult emergence usually takes place in late July to early August. Both males and females are strong fliers and, during outbreaks, swarms of moths are common. Eggs are laid in flat, clasping masses on the twigs and branches of host trees. Each female lays a single egg mass. Live branches less than 3/4 inch (19 mm) in diameter are the preferred egg laying sites. Larvae hatch 3-4 weeks later but remain inside the egg until the following spring.

**Damage**—Successive years of defoliation of aspen and other hosts can cause reduced diameter growth, branch dieback and top-kill, reduced production on fruit trees, and, in rare cases, mortality. In Colorado, entire aspen stands across large areas have been defoliated. During outbreaks, western tent caterpillar may be only one component in a complex of factors causing such widespread aspen defoliation, along with other insect defoliators and leaf diseases. Chokecherry fruit is an important food source for some indigenous cultures, and intense defoliation reduces fruit set. Chokecherry is also used as a food source by many wildlife species. Defoliation of range shrubs can reduce browse. Defoliated trees and tents are unsightly, and large hordes of wandering larvae are a nuisance in residential and recreation areas.

**Management**—Because western tent caterpillar is more of a nuisance than a damaging pest, infestations normally are allowed to run their course without intervention. A large complex of natural enemy species feeds upon western tent caterpillar and, together with unfavorable weather and foliage depletion, regulates population size. High populations usually collapse within a year or two due to natural controls, although outbreaks in the Southwest have lasted longer and caused significant aspen mortality.

When populations are low, the egg masses on small plants can be removed by hand and destroyed between July and the following spring. Hand picking can inflict significant larval mortality. Destruction of occupied tents by hand or with a brush leads to colony collapse. Pruning to remove egg masses, tents, or groups of larvae must be done sparingly to avoid injuring the plant. Using fire to accomplish this is not recommended as it damages the plant more than the defoliation.

Outbreaks in fruit trees, forests designated for timber or fiber production, heavily used recreation sites, or around homes in urban-wildland interface areas may require direct control. Several chemical and biological insecticides are registered for western tent caterpillar control.

References: 27, 35, 50

# Introduction to Wood Borers

## Bore holes and tunnels into the wood

In general, wood-boring insects (figs. 436-440) feed on a wide variety of stressed, dying, or dead trees. Exceptions include some hardwood borers that attack apparently healthy, live trees and non-native wood borers for which natural enemies or host resistance are missing or absent. Adult wood borers lay eggs in or on a suitable tree, and the eggs hatch into larvae that mine under the bark. Some species will tunnel into the wood of the tree. Once larvae have completed development, new

**Figure 436.** Adult striped ambrosia beetle. *Photo: Maja Jurc, University of Ljubljana, Slovenia, Bugwood.org.*

adults will chew an exit hole through the bark and emerge. Wood borers serve an important ecological role in facilitating tree decomposition by the holes and channels they produce in the woody tissue, phloem, and bark and by the decay fungi they bring with them.

**Figure 437.** Adult cottonwood borer. *Photo: Charles T. Bryson, USDA Agricultural Research Service, Bugwood.org.*

**Figure 438.** Adult green flatheaded pine borer. *Photo: Sheryl Costello, USDA Forest Service.*

**General Features**—General features of wood borers:

- Most only attack trees under stress or in decline or those that are dying or dead.
- They develop under the bark and the life cycle is usually 1-3 years; adults occur outside the bark; larvae are found in stems, branches, or major roots.
- Larvae create meandering galleries that are packed with coarse to fine frass and that often obscure galleries made by other insects.
- Cerambicidae form an oval-shaped exit hole, Buprestidae form a D-shaped exit hole, and Siricidae form a round exit hole.
- Some species are very large (wood wasps look like very large wasps) and some have long antennae (longhorned beetles can have antennae longer than their body).
- Holes in the wood caused by wood borers can be a weak point in the tree, leaving the tree prone to wind damage and can reduce the timber value.

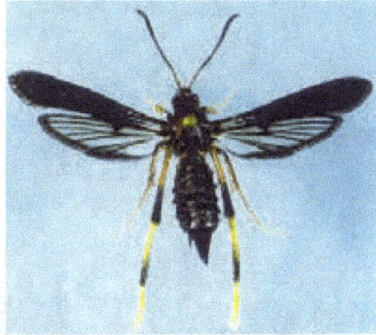

**Figure 439.** Adult clear-wing moth. *Photo: James Solomon, USDA Forest Service, Bugwood.org.*

**Figure 440.** Adult wood wasp, pigeon tremex. *Photo: Kenneth R. Law, USDA APHIS PPQ, Bugwood.org.*

**Conifer Borers**—In the Rocky Mountain Region, wood borers found in conifers only attack recently felled trees, dead or dying trees, trees severely stressed by drought, or trees attacked by bark beetles. For this reason, they are not considered a primary cause of tree mortality. Wood borers are often very active after large disturbances such as a fire. Where wood borer populations are high, larvae can be heard chewing under the bark. Some species only mine in the outer layers of the tree, and others mine into the sapwood and heartwood of a tree. The types of hole that wood borers create distinguishes this group from other insects. Different insect families create differently shaped exit holes (see "General Features" above). Most wood borers in conifers are either beetles or wasps (table 18).

**Hardwood Borers**—Wood borers found in hardwoods in this region usually attack trees that are experiencing some form of stress, with the exception of clear-wing moths (Sesiidae), which attack apparently healthy trees. Borers found in hardwoods are problematic because they can be one of the factors leading to tree decline or death. Wood borers may attack hardwoods year after year, and attacks may eventually cause branch death or create a weak point in the tree that is prone to wind breakage. Some hardwood borers like the cottonwood borer develop in roots and can cause severe damage to young trees. Wood borers found in hardwoods can be beetles, moths, or a single wasp species (table 18).

**References: 29, 35, 50**

Table 18. Common wood-boring insects in the Rocky Mountain Region.

| Insect | Family | Host | Signs |
|---|---|---|---|
| Agrilus quercicola | Buprestidae | Oak species, specifically Gambel oak | Wide, meandering galleries under the bark; D-shaped holes in the wood |
| Alder borer[a] (Saperda obliqua) | Cerambycidae | Alder and birch | Oval-shaped holes in the wood |
| Ambrosia beetles (fig. 436) | Scolytidae, Platypodidae | Most conifers | Fine, white sawdust at base of dead/dying trees |
| Bronze poplar borer (Agrilus liragus) | Buprestidae | Aspen | Meandering or zig-zag galleries under bark; D-shaped exit holes |
| Carpenterworm (Prionoxystus robiniae) | Cossidae | Certain hardwoods, depending on the region | Sap spots and fine frass mixed with sap |
| Cottonwood borer (fig. 437) (Plectrodera scalator) | Cerambycidae | Cottonwood, poplars, and occasionally willows | Found in roots; small seedlings can snap at root collar |
| Elm borer (Saperda tridentata) | Cerambycidae | American elm and other elms | Meandering galleries; oval-shaped exit hole |
| Flatheaded borers (fig. 438) | Buprestidae | Most conifers and hardwoods | Meandering galleries; D-shaped exit hole |
| Juniper borers | Cerambycidae, Buprestidae | Juniper | Meandering galleries; oval-shaped exit hole |
| Lilac (Ash) borer (fig. 439) (Podosesia syringae) | Sesiidae | Ash and lilac | Rough, gouging wounds under bark of large branches and trunk; old pupal skin attached to the tree |
| Poplar borer (Saperda calcarata) | Cerambycidae | Aspen; occasionally cottonwood and poplar | Oozing wounds on bark, often associated with some frass; chronic infestations |
| Roundheaded borers | Cerambycidae | Most conifers and hardwoods | Meandering galleries; oval-shaped exit hole |
| Pitch moths (discussed in Bud, Shoot, Branch, and Terminal Insects section) | Pyralidae | Pinyon, ponderosa, and other pines, occasionally | Large gouges that ooze light pinkish pitch on trunk and large branches |
| Wood wasps (fig. 440) | Siricidae | Most conifers and some hardwoods | Meandering galleries; round exit hole |

[a] Not discussed in this guide.

# *Agrilus quercicola*

## Attacks boles and branches of stressed oaks

**Name and Description**—*Agrilus quercicola* Fisher [Coleoptera: Buprestidae]

*Agrilus quercicola*, a flatheaded wood borer, attacks stressed oaks. Adults are small, approximately 1/3 inch (8 mm) long, and have small antennae and the characteristic oval body shape. The head, thorax, and abdomen coloration is consistently metallic and varies in shades of green and copper, with the thorax usually more copper-colored and the abdomen usually greener. The elytra (wing covers) are consistently black (fig. 441). Larvae are white, legless grubs with the appearance of a flattened head.

**Figure 441.** Adults of *Agrilus quercicola*. *Photo: Andrea Sever, Colorado State University.*

**Hosts**—Oak species; specifically associated with native stands of Gambel oak, *Quercus gambelii*

**Life Cycle**—There is one generation annually. Adults begin to emerge in late May from D-shaped exit holes (fig. 442). Peak flight is mid-June, and flight continues into August. Eggs are laid in the outer layers of the bark. Larvae develop under the bark in the cambium layer and then tunnel into the sapwood and heartwood (fig. 443). *Agrilus quercicola* overwinters as larvae under the bark.

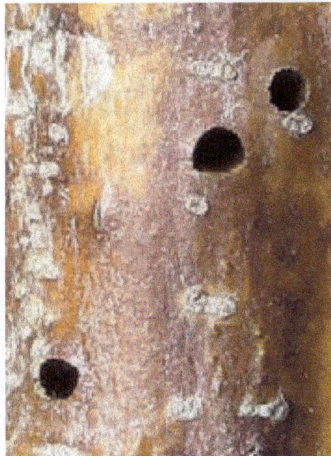

**Figure 442.** D-shaped exit holes of *Agrilus quercicola* and round-shaped exit hole of a parasitoid (upper right). *Photo: Andrea Sever, Colorado State University.*

**Figure 443.** *Agrilus quercicola* larval galleries. *Photo: Andrea Sever, Colorado State University.*

**Figure 444.** Damage to gambel oak in Douglas County, Colorado, due to *Agrilus quercicola*. *Photo: Andrea Sever, Colorado State University.*

**Damage**—In 2003 and 2004, there were numerous reports of a borer attacking and damaging nursery and ornamental oaks along the Colorado Front Range and Durango-Cortez areas. It was determined that these attacks were caused by an outbreak of *A. quercicola*, which was also attacking and damaging surrounding native Gambel oak (fig. 444). A build-up of borer populations in Gambel oak was due to years of drought conditions combined with late frost, causing stress and dieback of Gambel oaks. The most obvious sign of an *A. quercicola* attack was the wide, meandering galleries under the bark. Holes that penetrate into the wood were also present. Emerging adults left D-shaped, cleanly cut exit holes.

**Management**—This native insect rarely achieves pest status. Populations of *A. quercicola* naturally decline with normal rainfall that lessens stress on oak trees.
**Reference: 153**

# Ambrosia Beetles

## White frass on boles

**Name and Description**—*Trypodendron* sp., *Gnathotrichus* sp., *Xyleborus* sp. [Curculionidae: Scolytinae]

*Platypus* sp. [Curculionidae: Platypodidae]

Ambrosia beetles attack stressed, dying, or dead trees. There are several species that belong to two different bark beetle families (figs. 445-446). Adult ambrosia beetles are generally small, reddish brown to nearly

**Figure 445.** Adult striped ambrosia beetle, *Trypodendron lineatum*. *Photo: Maja Jurc, University of Ljubljana, Slovenia, Bugwood.org.*

black, cylindrical beetles from about 1/8-3/16 inch (3-5 mm) long. Larvae are small, white, legless grubs similar to bark beetle larvae.

**Hosts**—Most western conifers

**Figure 446.** Adult Wilson's white-headed ambrosia beetle, *Platypus wilsoni*. *Photo: Pest and Diseases Image Library, Australia, Bugwood.org.*

**Life Cycle**—There may be one or multiple generations per year. Ambrosia beetle adults attack spring through fall, depending on the species. Ambrosia beetles develop through four stages typical to bark beetles: egg, larva, pupa, and adult. For some species, all stages overwinter in the wood, and other species overwinter as adults in the duff and litter on the forest floor. Most species have a fairly wide host range.

**Damage**—Ambrosia beetles attack weakened, dying, and recently cut or killed trees. They can attack freshly cut lumber and lumber in decks before it is dried, and they can cause pinhole defects and dark staining in the outer wood. Galleries are formed in the sapwood or heartwood and damage the wood. Because ambrosia beetles tunnel into the wood, they are considered wood borers rather than bark beetles in this guide. Adults introduce ambrosia fungi that stain the wood, and lower its value. Ambrosia beetles feed on the fungus rather than the wood. The most obvious sign of an ambrosia beetle attack is the fine, white boring dust that accumulates at the base of the tree and in the bark crevices (fig. 447). Adults bore straight into the tree, creating perfectly round, small-diameter holes. If the bark is removed, the entrance points of adult ambrosia beetles and galleries are distinctive and are often surrounded by a dark brown

**Figure 447.** Fine, white boring dust produced by ambrosia beetles. *Photo: Ronald F. Billings, Texas Forest Service, Bugwood.org.*

**Figure 448.** Entrance holes of adult ambrosia beetles. *Photo: W.H. Bennett, USDA Forest Service, Bugwood.org.*

or black fungal stain (fig. 448). Damage caused by ambrosia beetles can vary greatly among locales. In some areas, aggressive control programs are required to reduce economic damage to wood products.

**Management**—Management is focused on preventing attacks because ambrosia beetles attack weakened or recently dead trees. Removing and processing wood quickly is the best way to prevent damage. Management can also be done through proper handling of wood products. Proper handling methods include: milling or debarking susceptible logs prior to the attack period, storing logs in an area safe from attack, and creating mill conditions unfavorable to beetle development.

**Reference: 50**

# Bronze Poplar Borer

## Attacks stressed aspen

**Name and Description**—*Agrilus liragus* Barter and Brown [Coleoptera: Buprestidae]

The bronze poplar borer is a member of a family commonly known as flatheaded or metallic boring beetles. The adult beetle is blackish with a faint metallic green lustre and shiny, bronzed underside, 1/4-1/2 inch (7-13 mm) long, narrow, and slightly flattened. Eggs are cream-colored, oval, and about 1/20 by 1/33 inch (1.2 by 0.8 mm). The larva is much longer than the adult, narrow and flat, about 1 1/2 by 1/10 inch (30-40 by 2-3.5 mm), and its the prothorax is slightly wider than the rest of the body (figs. 449-450). The rear segment has a pair of dark, anal spines.

**Host**—The bronze poplar borer typically attacks weakened aspen. Trees that are partially girdled, such as those heavily gnawed by elk or carved by campers, are commonly attacked. Overmature or injured trees, those damaged by poplar borer (*Saperda calcarata*) or cankers, and young trees released from the dominance of other trees are also susceptible to attack.

**Life Cycle**—The life cycle typically takes 2 years, though borers in severely weakened trees in southerly locations usually complete a generation in 1 year. Adults emerge from June to August. They feed on

**Figure 450.** Larva of the bronze poplar borer. *Photo: Jim Worrall, USDA Forest Service.*

**Figure 449.** Larva of the bronze poplar borer, slightly displaced from its gallery on the underside of live bark. *Photo: Mike Ostry, USDA Forest Service, Bugwood.org.*

USDA Forest Service RMRS-GTR-241. 2010.

Figure 451. External appearance of *Agrilus* sp. attack on a live aspen. *Photo: Jim Worrall, USDA Forest Service.*

Figure 452. Fresh gallery in live bark with frass. *Photo: Jim Worrall, USDA Forest Service.*

aspen foliage for about one week before laying eggs in bark crevices in groups of five to eight. Eggs hatch in about two weeks. Young larvae bore through bark to the cambium. Feeding is mostly along the cambium, though larvae occasionally move into the phloem and cortex. Generally, larvae bore into the wood to molt and then come back to the cambium. A completed gallery typically shows four such departures, suggesting five instars. Pupation occurs in outer sapwood or thick bark in spring.

**Damage**—Larval galleries are $1/50$-$1/8$ inch (0.6-3.0 mm) wide and can be over 39 inches (1 m long), but they are never straight. In severely weakened trees, galleries usually meander without any distinct pattern. In vigorous hosts, a zig-zag or sinuate gallery is the rule (figs. 451-454). Galleries weave back and forth

Figure 454. Typical zig-zag gallery. A gallery generally extends down from the origin and is packed with frass. Galleries tend to be more random in severely weakened trees. *Photo: James Solomon, USDA Forest Service, Bugwood.org.*

Figure 453. When the cambium produces callus in a gallery in live bark, a raised pattern of wood is left behind after the bark deteriorates. *Photo: Jim Worrall, USDA Forest Service.*

across the grain with successive loops closer together in the most vigorous hosts. Adult emergence holes in the outer bark are distinctly D-shaped. Multiple attacks may contribute to mortality or result in girdling and direct mortality.

**Management**—Maintenance of high tree vigor reduces the likelihood of attack, as does prevention of mechanical injuries and diseases. Natural control includes a variety of egg and larval parasites. Woodpeckers are particularly important. According to one study, up to 40% of larvae, pupae, and young adults are taken by woodpeckers.

**References: 99, 161**

# Carpenter Ants

## Wood dust and honeycombed wood

**Name and Description**—*Camponotus* spp. [Hymenoptera: Formicidae]

Carpenter ants tunnel into the wood of stumps, logs, dead standing trees, the dead interior of living trees, and wooden portions of buildings. They do not eat wood; rather, they excavate cavities in decayed wood for nests to raise their young. Adult carpenter ants are this Region's largest ants at $1/4$-$1/2$ inch (6-13 mm) long and they can be black or black and red ants (fig. 455). Adults have elbowed antennae and a constricted waist. Larvae are small, white, legless grubs (fig. 456).

**Figure 455.** Carpenter ant adult, *Camponotus* sp. Photo: Edward H. Holsten, USDA Forest Service, Bugwood.org.

**Hosts**—Dead and decaying wood of all tree species

**Life Cycle**—The queen carpenter ant works alone in founding the colony and lays eggs that develop into the first worker ants in about 30-40 days. Worker ants (infertile females) carry food (insects, honeydew, etc.) into the colony for the developing larvae. The queen continues to lay eggs, and colonies can become quite large. Completely developed and fertile queen females are produced spring to midsummer and can been seen flying in swarms. Winged

**Figure 456.** Carpenter ant larvae, *Camponotus* sp. Photo: Whitney Cranshaw, Colorado State University, Bugwood.org.

males are also produced at this time and they mate with the queens. Queens then disperse and establish a new colony. Queens shed their wings shortly after finding a suitable site for beginning a colony.

**Damage**—Carpenter ant damage to forest trees is usually very minor. In some cases, carpenter ant excavations can be so extensive that they can cause a loss of

**Figure 457.** Sawdust-like borings from carpenter ants.
*Photo: Dave Powell, USDA Forest Service, Bugwood.org.*

**Figure 458.** Carpenter ant damage.
*Photo: Edward R. Werner, USDA Forest Service, Bugwood.org.*

structural integrity in the tree and lead to wind breakage. Carpenter ants sometimes damage young conifers by gnawing around the root collars. The most obvious sign of a carpenter ant colony is the large amount of sawdust-like borings piled on the ground beneath entrance holes (fig. 457). The galleries are also distinctive with their vertical, honeycomb appearance and smooth walls that are free of boring dust (fig. 458). These might be confused with termite activity, but termites do not typically occur at elevations higher than the pinyon-juniper woodlands in the forests of the Rocky Mountain Region, and fecal pellets from termites are distinctively different from carpenter ant boring dust.

**Management**—Carpenter ants play important, beneficial roles in that they contribute to decomposition of woody debris by excavating decayed wood, prey on small insects, and serve as food sources for wildlife species. Management in the forest is seldom needed, but removing and processing wood quickly is the best way to prevent damage. Because carpenter ant excavations lead to a loss in structural integrity, trees should be treated as falling hazards. In rare cases, carpenter ants will infest wooden structures, and control in these situations can be difficult.

**References: 35, 50**

# Carpenterworm

## Sapstaining on bark of hardwoods and frass at base of tree

**Name and Description**—*Prionoxystus robiniae* (Peck) [Lepidoptera: Cossidae]

The carpenterworm is widely distributed in the United States and southern Canada where it breeds in a variety of hardwoods. In Colorado, the insect is most common in shelterbelt settings in the eastern portion of the state. The adult is a large, stout-bodied moth with mottled forewings (fig. 459). The female is of lighter coloration than the male and has a wingspan of about 3 inches (7.6 cm). Full-grown larvae are greenish white with a shiny, dark brown head, approximately 2-3 inches (5.0-7.6 cm) long, and they have very few, simple hairs on their surface (fig. 460).

**Hosts**—Hosts include oak, elm, maple, willow, cottonwood, black locust, box-elder, sycamore, and ash, but the insect exhibits regional host preferences. In the Rocky Mountain Region, poplars are the major hosts in the mountain areas and green ash and elms are the major hosts in the Great Plains.

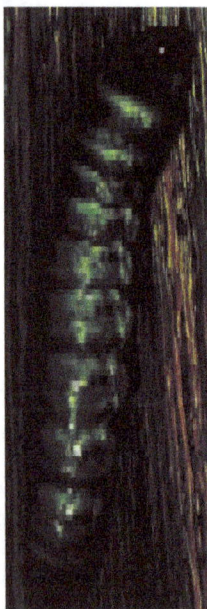

**Life Cycle**—Carpenterworms begin to appear in May in the Kansas and Nebras-

Figure 459. Carpenterworm adult. *Photo: James Solomon, USDA Forest Service, Bugwood.org.*

ka and in June or early July in colder regions. The female deposits eggs in groups in bark crevices near wounds or under vines, lichens, or moss. Young larvae bore directly into the inner bark or enter it through openings and then bore directly into the wood. Larval development may require 1-3 or 4 years, depending on latitude. The larva maintains an open tunnel to the exterior, through which it expels boring dust (fig. 461). Mature larvae pupate near the tunnel opening. The pupa wriggles to the mouth of the tunnel and continues until its head and thorax are protruding from the tree. Even after the adult moth has departed the tree, the pupal case usually remains in place, sticking out of the opening.

**Damage**—The earliest sign of attack by carpenterworms are sap spots and fine frass mixed with the sap. Carpenterworms tunnel in the sapwood and heartwood of trunks and branches of cottonwoods. Carpenterworms seldom kill trees outright, although

heavily riddled small trees may be broken off by the wind (fig. 462). Open-grown trees, or trees growing on poor sites with dry, shallow soils such as dry ridgetops or ridge slopes, are especially subject to attack and damage. Wounds, associate with the tunnel opening, usually heal in 1-2 years, leaving oval to irregular bark scars that remain as evidence of attack for 10-20 years.

Figure 460. Carpenterworm larva. *Photo: William H. Hoffard, USDA Forest Service, Bugwood.org.*

Figure 461. Carpenterworm tunnel and exit hole. *Photo: James Solomon, USDA Forest Service, Bugwood.org.*

**Management**—This insect mainly attracted attention due to its impact on shade, ornamental, and windbreak trees, but loss of oak timber value may also be significant. Natural enemies suppress the carpenterworm but do not keep damage at acceptable levels. Woodpeckers are able to excavate larvae from galleries, and many birds feed on the adults. Cultural practices that promote tree vigor, prevent bark injuries, and remove brood trees help to minimize damage. Trunk-applied insecticides timed with the use of sex attractants to correspond with egg hatch are effective in preventing infestation.

**Figure 462.** Carpenterworm damage to wood of host tree. *Photo: University of Idaho.*

**Reference: 161**

# Cottonwood Borer

## Striking black and white longhorned beetle

**Name and Description**—*Plectrodera scalator* (Fabricius) [Coleoptera: Cerambycidae]

The cottonwood borer is a common wood borer of cottonwood and poplars and infests the bases and roots of living trees. The insect is widely distributed throughout the eastern United States and the Midwest; however, it is most common in the Great Plains from Texas to South Dakota and throughout the Mississippi River Valley. Within Colorado,

**Figure 463.** Cottonwood borer adult. *Photo: Charles T. Bryson, USDA Agricultural Research Service.*

the cottonwood borer is restricted to the southeastern portion of the state. The adult is an elongate, robust, strikingly patterned beetle that is approximately 1-1 3/4 inches (2.5-4.4 cm) long and 1/2 inch (13 mm) wide (fig. 463). Mature larvae are approximately 2 inches (50 mm) long (fig. 464), and the pupa is approximately 1 1/2 inches (3.8 cm) long (fig. 465).

**Hosts**—Host trees include cottonwood, poplars, and, occasionally, willows. Eastern cottonwood is the preferred host tree throughout its range.

**Life Cycle**—Adults emerge in late June through mid-August in Kansas. Adults live about one month and feed on leaf petioles, new twig growth, and tender bark. Adult feeding may cause shoots to break or shrivel and die. Females dig shallow pits in the soil at the root collar and then shred the bark with their jaws to make an oviposition site. The female deposits a single egg at each site and then partially fills the soil pits. The young larvae mine downward in the inner bark and soon begin to etch the wood of the root. Galleries extend downward into the taproot of 1- and 2-year-old seedlings. In larger trees, larvae usually do not penetrate the wood deeper than 1 inch (2.5 cm) and often hollow out areas of 2-3 inches (5-7.5 cm) in diameter, particularly in large roots. The larva may take 1 or 2 years to complete development. The larvae pupate from April through early July.

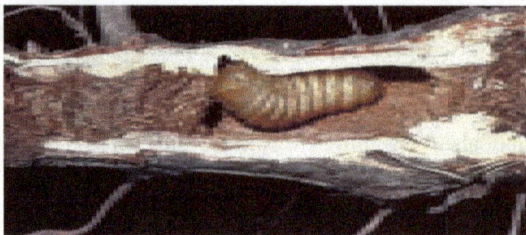

**Figure 464.** Cottonwood borer larva. *Photo: James Solomon, USDA Forest Service, Bugwood.org.*

**Damage**—The cottonwood borer attacks trees of all sizes. Cottonwood borers may infest nursery stock, causing young trees to die and shrivel, or, if they survive, the young trees often break at the root collar when handled. Injury to older trees is hard to detect unless the soil is removed to expose the root collar and the upper roots. The larvae bore into the heartwood of infested trees. In plantations, infested trees are often riddled with larvae, but they seldom die from the injury. Severely infested trees may break at the base during periods of high wind.

**Figure 465.** Cottonwood borer pupa. *Photo: James Solomon, USDA Forest Service, Bugwood.org.*

**Management**—Natural and planted cottonwood stands on poor sites such as sand flats and heavy clays are more heavily infested than stands on good sites. Also, shelterbelt plantings on the Great Plains have been severely damaged by the cottonwood borer. The cottonwood borer has few natural parasites and predators because most of its life cycle is spent in the tree below ground-level. Extended flooding in lowlands often kills many larvae.

**Reference: 161**

# Elm Borer

## Extensive mining loosens bark

**Name and Description**—*Saperda tridentata* Olivier [Coleoptera: Cerambycidae]

The elm borer is a common wood borer of North American elms. The adult beetle is $1/3$-$2/3$ inch (8-17 mm) long with three orange-red oblique crossbars on the wing covers and narrow stripes on the margins of the wing covers and pronotum (body segment behind the head). Three sets of two black spots occur on the pronotum, at the base of the wing covers, and near the apex of the wing covers (fig. 466). Mature larvae are white, legless, and approximately $1/2$-1 inch (13-25 mm) long.

**Figure 466.** Elm borer adult. *Photo: Pennsylvania Department of Conservation and Natural Resources, Forestry Archive, Bugwood.org.*

**Hosts**—American elm is the favored species, although slippery elm and cedar elm are hosts. The elm borer probably attacks other native elm species, but this beetle does not infest English elm, an introduced species.

**Life Cycle**—The elm borer has one generation per year. Adults appear in spring, with the males emerging slightly before females. The beetles mate and then feed extensively on foliage, petioles, and young twigs. At night, females carve out niches in the bark of host trees and deposit their eggs. After hatching, the larvae feed initially in the outer corky layer of bark and later move to the phloem layer, creating extensive galleries (fig. 467). In early August into October, larvae bore into the sapwood and create chambers in the woody tissue of the tree where they overwinter. In March and April, the larvae pupate, which requires 15-33 days. Adults emerge through a round exit hole (fig. 468) following pupation.

**Figure 467.** Elm borer larvae and galleries. *Photo: James Solomon, USDA Forest Service, Bugwood.org.*

**Damage**—Fresh sappy wood of trees weakened from drought, disease, or other causes are favored by ovipositing beetles. The first sign of attack is the appearance of thin, light-colored foliage followed by scattered dying of branches. Inspection of the trunk and branches reveals small egg niches in the bark. Within a few days, small pieces or ribbons of reddish frass are extruded from tiny openings in bark crevices. After attack has progressed, patches of the bark may be easily pulled from the tree. Removing the bark reveals a mass of mines or burrows in the inner bark. After a brood completes development, numerous round holes, about $1/8$ inch

**Figure 468.** Elm borer emergence hole. *Photo: James Solomon, USDA Forest Service, Bugwood.org.*

(3 mm) in diameter, are present on the surface of the bark. Larvae penetrate the wood only to pupate, doing so at depths of $1/8$-$1/4$ inch (3-6 mm).

**Management**—Because trees weakened by drought, mechanical damage, air pollution, disease, and insect defoliators are especially susceptible to attack, practices should be followed to keep trees vigorous. Severely infested trees should be removed and destroyed to reduce the borer population. Six species of wasp parasitoids help to reduce infestations but often do not prevent economic damage.

**References: 161**

# Flatheaded Wood Borers (Metallic Wood Borers)

## Oval exit holes on dead and dying trees

**Name and Description**—[Coleoptera: Buprestidae]

Flatheaded wood borer beetles attack stressed, dying, or dead trees. There are many species that belong to the beetle family Buprestidae. Adult flatheaded wood borers are small to relatively large beetles ($1/4$-2 $1/2$ inches [6-64 mm]) with small antennae and a characteristic oval body shape (figs. 469-470). Adult Buprestidae are called metallic wood borers because they are iridescent or metallic looking underneath and sometimes on top (fig. 470). Larvae are white, legless grubs similar to bark beetle larvae, but the body shape is elongate, and the head area is different than bark beetle larvae. The larval head is small, and the next body segment (thorax) is much broader than the following segments and usually has a hardened plate on the top and bottom, giving the appearance of a flat head (fig. 471). The larva can be distinguished from roundheaded wood borer larva by noting the flat head characteristic, which has been likened to a horse-shoe nail.

**Figure 469.** Adult large flatheaded pine heartwood borer, *Chalcophora virginiensis*. *Photo: Sheryl Costello, USDA Forest Service.*

**Figure 470.** Adult green flatheaded pine borer, *Phaenops gentilis*. *Photo: Sheryl Costello, USDA Forest Service.*

**Hosts**—Most western conifers, also found in hardwoods

**Life Cycle**—The life cycles of different species varies from 1 to many years. Flatheaded wood borer adults attack spring through fall, depending on the attacking species. Eggs are laid in the outer layers of the bark. Larvae develop under the bark in the phloem. Several species also tunnel into the sapwood and heartwood. Most overwinter as larvae under the bark.

**Damage**—Flatheaded wood borer beetles attack weakened, dying, and recently cut or killed trees. They can also attack freshly cut timber before it is dried. Larvae that tunnel into sapwood and heartwood can frequently damage logs and wood products. Adults aid in wood decomposition by introducing yeasts, bacteria, and wood-rotting fungi that lead to tree rot and checking in the wood. In some instances, these processes occur within a couple of years. The most obvious sign of a flatheaded wood borer attack is the wide, meandering galleries under the bark with tightly packed, fine boring dust. Holes that penetrate into the wood are most likely due to wood borer larvae. Emerging adults leave oval, cleanly cut exit holes (fig. 472).

**Management**—Because flatheaded beetles do not attack healthy conifers, management is focused on preventing attacks on recently dead or felled trees. Removing and processing wood quickly is the best way to prevent damage. Management can also be done through proper handling of wood products. Proper handling methods include milling or debarking susceptible logs prior to the attack period and storing logs in an area safe from attack.

*Melanophila* **spp.**—Several species of wood borers are attracted to fire. In fact, species of the genus *Melanophila* possess specific pit-sensing organs that detect infrared radiation produced by forest fires. As a result, these beetles are often seen by firefighters laying eggs on recently burned trees. Metallic wood borers can be responsible for biting firefighters. *Melanophila* spp. have been known to build up their numbers in fire-damaged hosts and emerge to attack adjacent, otherwise healthy, trees. Such "outbreaks" are generally short-lived.

**Emerald Ash Borer**—The emerald ash borer, *Agrilus planipennis* Fairmaire, is an exotic flatheaded borer that kills live ash trees but is not currently

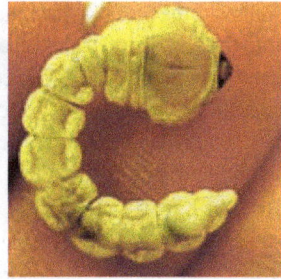

**Figure 471.** Flatheaded wood borer larva. *Photo: Sheryl Costello, USDA Forest Service.*

**Figure 472.** Exit holes of adult flatheaded wood borer. *Photo: Hannes Lemme, Germany, Bugwood.org.*

**Figure 473.** Adult emerald ash borer, *Agrilus planipennis*. *Photo: Marianne Prue, Ohio Department of Natural Resources, Division of Forestry, Bugwood.org.*

found in the Rocky Mountain Region (fig. 473). It was accidentally introduced into the Lake States and is responsible for killing millions of ash trees. The beetle can be found in live and recently dead ash trees. There is great potential for it to spread to other states by moving beetle-infested ash (e.g., moving infested firewood and nursery stock). Several native species resemble the emerald ash borer, and identification should be confirmed by a specialist.

**References: 29, 35, 50**

# Juniper Borers

## Juniper mortality or branch dieback

**Name and Description**—Black-horned juniper borer—*Callidium texanum* (Fisher)

Juniper borer—*Atimia huachucae* (Champlain and Knull)

Juniper twig pruner—*Styloxus bicolor* (Champain and Knull)

[Coleoptera: Cerambycidae]

Several roundheaded and flatheaded borers can kill drought-stressed junipers or cause noticeable dieback (fig. 474). Three common round-headed borers are discussed here. Larvae are white, cylindrical, legless grubs that vary in size. Often, only the wood mining and exit hole are apparent, and the specific species of borer that killed the tree cannot be determined. Species that have been verified to occur in the Rocky Mountain Region include the juniper borer and the black-horned juniper borer (fig. 475) in trunks and branches and the juniper twig pruner in twigs and small branches (fig. 476). Damage can be extensive before symptoms are apparent. Often, a

**Figure 474**. Juniper killed by wood borers. *Photo: Southwestern Region, USDA Forest Service.*

**Figure 475**. Black-horned juniper borer. *Photo: Southwestern Region, USDA Forest Service.*

**Figure 476**. Juniper twig pruner damage. *Photo: Robert Cain, USDA Forest Service.*

large portion of the tree or the entire tree dies before the insects' exit holes are noticed.

**Hosts**—Junipers

**Life Cycle**—Life cycles vary with species.

Black-horned juniper borer (*Callidium texanum*) larvae bore beneath the bark, making very wide, wavy galleries that distinctively score the outer sapwood. Older larvae excavate oval tunnels deep in the wood and spend the winter within affected branches or boles. Adults emerge throughout the warm months of the year. There is one generation per year. Adult beetles are rather short-horned for cerambycids, dark blue or black, and less than $1/2$ inch (1 cm) long. These roundheaded borers leave an oval or rectangular exit hole.

Juniper borers (*Atimia huachucae*) are small roundheaded borers about $1/4$ inch (6 mm) long and generally have a 1-year life cycle. Adults emerge from mid-May through June and feed on the foliage. Females lay eggs under bark scales, especially near wounds. The larvae feed just below the bark and pack fibrous frass behind them as they tunnel. Larvae overwinter as pupae under the bark.

The juniper twig pruner (*Styloxus bicolor)* is a slender roundheaded borer. Adults are $1/4$-$1/2$ inch (7-1 mm) long and have a reddish orange head and brownish to black body with narrow, tapered wing covers that do not completely cover the abdomen. Larvae mine small branches and cause twig dieback. Damage is most apparent during dry periods. Eggs are laid on branches, often near an intersection of twigs, 1-2 ft (30-60 cm) from the branch tip. Larvae kill twigs by boring through the center. The life cycle may take as long as 2 years to complete.

**Damage**—Larval feeding produces girdling wounds that kill branches or entire trees. Damage may occur gradually over several years before obvious symptoms are visible.

**Management**—No management strategies have been developed for these insects in natural areas.

**References: 22, 35**

# Lilac (Ash) Borer

## Pupal skins extrude from trunk

**Name and Description**—*Podosesia syringae* (Harris) [Lepidoptera: Sesiidae]

Adult lilac (ash) borers are moths that vary in color from brown to yellow to orange. They have clear wings with a span of about 1-1 $1/8$ inches (26-28 mm) and appear wasp-like (fig. 477). Larvae are about 1 inch (2.5 cm) long and are white with brown heads (fig. 478).

**Figure 477.** Adult ash borer.
*Photo: Daniel Herms, Ohio State University, Bugwood.org.*

**Hosts**—Ash and lilac

**Life Cycle**—There is one generation per year. Mature borer larvae overwinter in tunnels under the bark. Adult moths emerge from March through June to lay eggs on the bark of host trees. The larvae bore into trunks and branches of the sapwood of trees during the summer. Galleries can be up to 6 inches (15 cm) long.

**Figure 478.** Ash borer larva. *Photo: James Solomon, USDA Forest Service, Bugwood.org.*

**Damage**—The mining of the larvae causes branch dieback (fig. 479). It can also lead to broken branches. The leaves on affected branches turn brown as the branch dies. Extensive mining can also lead to tree death. Entrances to larval mines often appear as sunken or cankered areas on the bark of the trunk or branch. Dark, moist sawdust can be found around the gallery entrance (fig. 480). Pupal skin remaining in the bark is often also observed (fig. 481).

**Management**—Avoid damaging trees—maintaining trees in good health reduces their susceptibility to attack. There are chemical sprays that are highly effective at preventing attacks, but they must be used at the onset of moth flight. Trees and branches that have been attacked can be removed and destroyed to kill maturing larvae.

**References: 39, 174**

**Figure 479.** Ash borer damage. *Photo: James Solomon, USDA Forest Service, Bugwood.org.*

**Figure 480.** Ash borer gallery entrance. *Photo: James Solomon, USDA Forest Service, Bugwood.org.*

**Figure 481.** Pupal skin pulled out from trunk during adult emergence. *Photo: Whitney Cranshaw, Colorado State University, Bugwood. org.*

# Poplar Borer

## Oozing varnish on stressed aspen stems

**Name and Description**—*Saperda calcarata* Say [Coleoptera: Cerambycidae]

The poplar borer is a cerambycid, a member of a family commonly known as roundheaded wood borers or longhorned beetles. The adult beetle is elongate, 4/5-1 1/8 inches (2-3 cm) long, and grayish blue in color with fine, brown dots (fig. 482). Antennae are about as long as the beetle's body. Larvae are yellowish white, cylindrical, and about 1 1/8-1 1/2 inches (3-3.8 cm) long (fig. 482). The head is broader than the rest of the larva's body.

**Hosts**—Within the Rocky Mountain Region, the poplar borer typically attacks weakened aspen. Increases in poplar borer activity have been specifically noted in drought-stricken, defoliated, sun-scalded, and partially cut aspen stands. Borers favor large-diameter trees but can infest trees as small as saplings. Brood trees are common, where female borers lay eggs on the same trees from which they emerged.

**Figure 482.** Life stages (larva, pupa, and adult) and gallery pattern of the poplar borer. Excelsior-like frass is particularly indicative of advanced stages of poplar borer activity. *Photo: James Solomon, USDA Forest Service, Bugwood.org.*

**Life Cycle**—The life cycle typically takes 2-3 years. Adults emerge in July and August and feed on aspen leaves and new shoots. Female beetles chew slits in the bark of aspens (often the same trees the females emerged from) and deposit one or two eggs. Eggs hatch in about two weeks. Young larvae begin feeding on bark tissue and eventually mine the host's sapwood. Larvae expel frass through enlarged entrance holes along the host's trunk. Pupation occurs in pupal cells constructed near the lower end of larval mines. The poplar borer typically overwinters as a pupa, emerging as an adult beetle the following summer.

**Damage**—Wet spots with oozing sap mixed with frass along the trunk are signs of poplar borer (fig. 483). Frass becomes coarse and excelsior-like as larvae develop. As larval feeding advances, frass may be seen at tunnel entrances along the trunk, in bark crevices, and in piles at

**Figure 483.** Frass and oozing sap along the trunk of aspen indicates the presence of poplar borers. *Photo: Jim Worrall, USDA Forest Service.*

the base of trees. Galleries (tunnels) typically meander in the sapwood (fig. 482) and total length may approach 1 inch (2.5 cm). Adults emerge from oviposition sites. Small trees may be killed by larval girdling alone, but large trees are seldom killed by this alone. Poplar borer egg deposition, excavation, and emergence activities provide infection courts for numerous canker and decay fungi. Multiple borer attacks and resulting tunnels reduce the tree's stability, and heavily infested trees are prone to wind breakage.

**Management**—Maintenance of tree vigor reduces the likelihood of attack, as does prevention of mechanical injuries and diseases. Control includes a variety of egg and larval parasites. Fungi and bacteria readily colonize egg niches, often resulting in extensive mortality of larvae. Woodpeckers are particularly effective predators. Partial cutting of aspen to remove borer-infested individuals is not recommended.

**References: 50, 99, 159**

# Roundheaded Wood Borers (Longhorned Beetles)

## White, splinter-like boring frass under bark

**Name and Description**—[Coleoptera: Cerambydicae]

Roundheaded wood borer beetles attack stressed, dying, or dead trees. There are many species all belonging to the beetle family Cerambycidae. Adult roundheaded wood borers are small to relatively large beetles ($1/4$-$2$ $1/2$ inches [6-64 mm]) and often have long antennae. The long antennae give them the name longhorned beetles (fig. 484). Larvae are white, legless grubs similar to bark beetle larvae, but the body shape is elongate, and the head area is primarily mouthparts without the more defined head capsule seen on bark beetle larvae (fig. 485).

**Figure 484.** Adult pine sawyer, *Monochamus clamator. Photo: Sheryl Costello, USDA Forest Service.*

One common group of longhorned beetles is the pine sawyers, *Monochamus* spp. Eggs are laid in egg niches chewed by the females, and larvae bore into sapwood and heartwood. Adults are large (1 inch, 25 mm long) with large antennae (fig. 484) and can be mistaken for the similar-appearing exotic Asian longhorned hardwood beetle (not established in this Region). Distinctive spots and antennal patterns can be used to identify the species. Pine sawyers are also responsible for spreading the pinewood nematode that causes

**Figure 485.** Roundheaded wood borer larva. *Photo: Sheryl Costello, USDA Forest Service.*

**Figure 486.** White, splinter-like boring frass produced by roundheaded wood borers when feeding. *Photo: Sheryl Costello, USDA Forest Service.*

**Figure 487.** Exit hole of adult roundheaded wood borer. *Photo: Sheryl Costello, USDA Forest Service.*

pine wilt, a fatal disease of Scots, Austrian, and other non-native pines, that is currently affecting these species in the Midwest.

**Hosts**—Most western conifers; also found in hardwoods

**Life Cycle**—The life cycle frequently spans 1-3 years, but some species can take many more years to complete development. Roundheaded wood borer adults attack spring through fall, depending on species. Eggs are laid in the outer layers of the bark or by first chewing a slit (referred to as an egg niche) in the inner bark and depositing eggs into the moist inner layers of the bark. Larvae develop under the bark in the cambium. Several species also tunnel into the sapwood and heartwood. Most overwinter as larvae under the bark.

**Damage**—Roundheaded wood borer beetles attack weakened, dying, recently cut, and fire damaged or killed trees and can attack freshly cut timber before it is dried. Larvae that tunnel into the sapwood and heartwood can frequently damage logs and wood products. Adults aid in wood decomposition by introducing yeasts, bacteria, and wood-rotting fungi that lead to tree rot and checking in the wood. In some instances, these processes can occur within a couple of years. The most obvious sign of a roundheaded wood borer beetle attack is the egg niches that some beetles chew on the outer bark or the white, splinter-like boring frass that can be found under the bark (fig. 486). Larvae produce wide tunnels that meander under the bark. Holes that penetrate into the wood are most likely due to wood borer larvae. Emerging adults leave broadly oval to circular, cleanly cut exit holes (fig. 487).

**Management**—Because roundheaded beetles do not attack healthy trees, management should focus on preventing attacks on recently dead or felled trees. Removing and processing wood quickly is the best way to prevent damage. Management can also be done through proper handling of wood products. Proper handling methods can include milling or debarking susceptible logs prior to the attack period and storing logs in an area safe from attack.

**References: 29, 35, 50**

# Twolined Chestnut Borer

## Top-down attacks on oaks can take years to kill

**Name and Description**—*Agrilus bilineatus* (Weber) [Coleoptera: Buprestidae]

Adult twolined chestnut borers are slender, black beetles with a bluish to greenish hue that are $^1/_5$-$^1/_2$ inch (5-13 mm) long with two faint, yellowish stripes along their back (fig. 488). Larvae are legless, white, slender, and about 1 inch (2.5 cm) long when fully grown with two spines at the tip of the abdomen (fig. 489) and have an enlarged, flattened front end, characteristic of the Buprestidae. Twolined chestnut borer larvae live under the bark in cylindrical, winding tunnels tightly packed with fine grain sawdust and excrement (fig. 490). Pupae form in chambers at the end of these tunnels. Twolined chestnut borers occur in southern Canada and throughout the eastern United States, including in Kansas, Nebraska, and South Dakota.

**Figure 488.** Adult twolined chestnut borer. *Photo: Robert A. Haack, USDA Forest Service, Bugwood.org.*

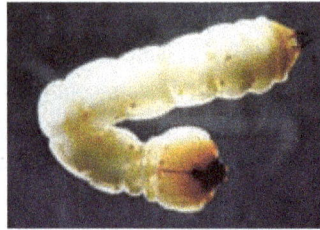

**Figure 489.** Twolined chestnut borer larva. *Photo: Robert A. Haack, USDA Forest Service, Bugwood.org.*

**Hosts**—Oaks, especially bur oak in South Dakota; the word chestnut in the common name refers to the beetle's past status as a principal pest of American chestnut, *Castanea dentata.*

**Life Cycle**—The twolined chestnut borer has one generation per year. Adults are active from April to August, depending on the location and temperature. After emerging, adults fly to the crowns of oak trees and feed on foliage before moving to the branches and trunks to mate. Females lay their eggs in small clusters in bark cracks and crevices. Larvae hatch within 1-2 weeks. The larvae burrow through the bark to the cambial region where they construct meandering galleries (fig. 490), lightly scoring the sapwood (fig. 491). Galleries may be straight rather than serpentine in host trees that are highly stressed. When fully mature, usually in August to October, larvae burrow into the outer bark and construct individual chambers in which they overwinter. If the bark is thin, the larvae construct

**Figure 490.** Twolined chestnut borer larvae with galleries on the inside of the bark. *Photo: James Solomon, USDA Forest Service, Bugwood.org.*

chambers in the outer sapwood. Larvae pass the winter in a doubled-over position. Pupation occurs the following spring. Adults emerge soon after through distinctive D-shaped exit holes about $1/5$ inch (5 mm) wide that they chew in the bark. In rare cases, it may take 2 years for larvae to complete their growth and development, especially when originating from eggs laid late in the growing season or at the northern extent of the insect's range.

The bronze birch borer, *Agrilus anxius* Gory, has a life cycle similar to that of the twolined chestnut borer. It can be a devastating pest of stressed or off-site mature birch trees, and it attacks young birch that have been recently transplanted.

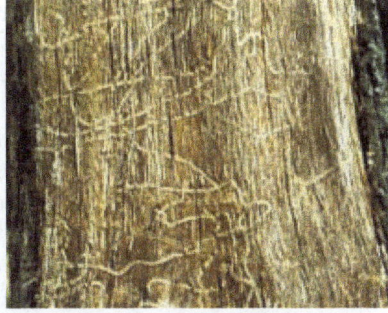

**Figure 491.** Outer sapwood on the tree trunk engraved by twolined chestnut borer larval galleries. *Photo: Minnesota Department of Natural Resources, Bugwood.org.*

**Damage**—Adult twolined chestnut borers primarily attack oaks that are weakened by drought or trees that are suppressed or declining. Urban oaks that suffer stress from trunk and root injury, soil compaction, and changes in soil depth are equally vulnerable to attack by this pest. Oaks that have been defoliated by insects or debilitated by root disease may also be attacked by the twolined chestnut borer. Although full-grown larvae are not much thicker than $1/13$ inch (1-2 mm), they are able to construct galleries of sufficient depth to girdle and kill branches and trees. The first symptom of borer attack is usually wilted foliage that appears on scattered branches during late summer. The foliage on infested branches wilts prematurely and turns brown but remains attached to the branches for several weeks or months before dropping (fig. 492). Such branches will die and produce no foliage the following year (fig. 492). Trees can be killed in the first year of attack; however, death usually occurs after 2-3 successive years of borer infestation. Typically, the crown is attacked during the first year, with the remaining live portions of the branches and trunk being infested during the second and third years. Practically nothing can be done to save infested portions of a tree once symptoms become visible because, at that time, the damage to the host tree is nearly complete.

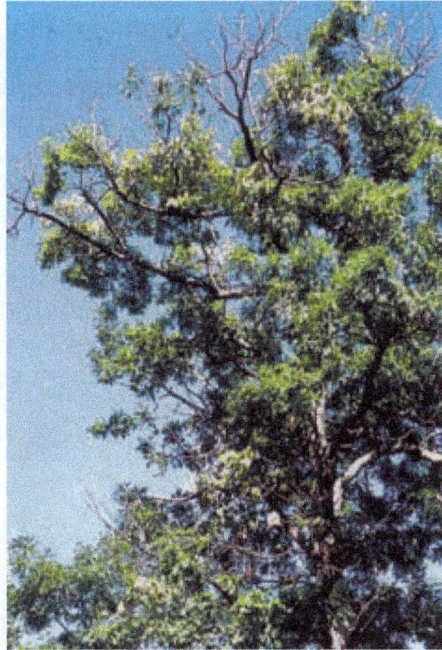

**Figure 492.** White oak infested by twolined chestnut borer; photo shows branch mortality and faded leaves. *Photo: Steven Katovich, USDA Forest Service, Bugwood.org.*

**Management**—Because this insect attacks stressed oaks, any action that reduces stress will lower the probability of infestation. This includes thinning overstocked stands to increase vigor, limiting oak defoliation with insecticide application, and watering during drought. Management programs should first attempt to prevent attack from occurring, but if it happens, several control options are available.

Natural controls include limited impact from a parasitic wasp attacking borer larvae and mortality from woodpecker predation that can cause significant population reduction within infested trees. Some direct cultural controls take advantage of the sensitivity of twolined chestnut borer larvae to rapid drying of the host tissues by appropriately timed cutting or pruning of infested material. Infested material can be chipped, burned, solarized, or treated in a way that kills resident borers, including treatment with pesticide. Girdled trees can be used as traps, attracting twolined chestnut borers that lay eggs on hosts that will dry out before larvae can mature. Preventive pesticide treatments can be used to protect high-value shade and ornamental trees from becoming infested but require multiple annual applications. Incorporating knowledge of twolined chestnut borer biology and the local timing of life cycle events into these control and prevention activities is essential for their success.

**References: 58, 174**

# Wood Wasps (Horntails)

## Stinger-like ovipositor drills into wood

**Name and Description**—[Hymenoptera: Siricidae]

There are several species of wood wasps, all belonging to the wasp family Siricidae. Adults are medium to large wasps, 1/2-1 1/2 inches (13-38 mm) long, and can be distinguished from common wasps by their thick waists (figs. 493-494). Identification is sometimes difficult due to the large range of the adult size. Adults are colored blue, black, or reddish brown and have red, ivory, or yellow markings. Flying adults make a noisy, buzzing sound. Both sexes possess a short, hornlike process at the end of their bodies. They do not possess a stinger; instead, females have a long stinger-like ovipositor for laying eggs under the bark of trees (fig. 493). Larvae are white, legless grubs similar to bark beetle larvae, except

**Figure 493.** Adult wood wasp, *Urocerus gigas*. Photo: *David Leatherman, Colorado State Forest Service, Bugwood.org.*

**Figure 494.** Adult wood wasp, *Sirex juvencus*. Photo: *Sheryl Costello, USDA Forest Service.*

the body shape is elongate (fig. 495). The larva has a small spine at the posterior end of the body.

**Hosts**—All except one western species attack conifers and most will attack trees in several different genera. The one exception, the pigeon tremex, attacks hardwoods.

**Life Cycle**—The life cycle spans 1 to several years, and many species take at least 2 years to complete development. Wood wasp adults attack summer through fall, depending on species. Females lay eggs deep in the wood by inserting their long, flexible ovipositor. Larvae develop under the bark in the cambium and tunnel into the sapwood and heartwood. Most overwinter as larvae under the bark.

**Figure 495.** Wood wasp larva. *Photo: Stanislaw Kinelski, Polish Forest Research Institute, Bugwood.org.*

**Damage**—Wood wasps most frequently attack weakened, dying, and recently cut or killed trees and can attack freshly cut timber before it is dried. They are particularly attracted to fire-damaged trees. For this reason, they are well-known to wildland firefighters. A female wood wasp will land on the substrate of potential host material and survey the site, using its antennae to determine the suitability of the material. Once an acceptable location is found, she will withdraw the ovipositor from its sheath and begin to "drill" down into the wood. At an acceptable depth, the egg is laid, and the female withdraws the ovipositor to begin the process anew. Larval tunnels in the sapwood and heartwood can frequently damage logs and wood products. Adults aid in wood decomposition by introducing yeasts, bacteria, and wood-rotting fungi that lead to decay and checking in the wood. In some instances, these processes can occur within a couple of years. The most obvious signs of a wood wasp attack are the meandering galleries under the bark packed with fine boring dust and finding larvae that have a spine on the end. Holes that penetrate into the wood are also signs of wood borer larvae. Emerging adults leave round, clean-cut emergence holes where they emerge.

**Management**—Because wood wasps do not attack healthy trees, management is focused on preventing attacks on recently dead or felled trees. Removing and processing wood quickly is the best way to prevent damage. Management can also be done through proper handling of wood products. Proper handling methods include milling or debarking susceptible logs prior to the attack period and storing logs in an area safe from attack.

**An Introduced Pest**—An exception to the "non-pest" status of siricids is the recent introduction of the European wood wasp, *Sirex noctilio* Fabricius into the eastern United States. This exotic species attacks living hosts, particularly pine trees. These insects have been introduced to a number of important pine-growing regions worldwide, including in Australia, New Zealand, and Brazil, where they have had an economic impact upon commercial softwood production. Although currently confined to the eastern United States, risk assessment studies have indicated that suitable conditions exist throughout the western United States, and several important tree species have been proven acceptable hosts (including lodgepole and ponderosa pine). Forest managers need to be aware of this situation. If siricids are found to be associated with living host trees, samples should be taken for expert identification.

**References: 29, 35, 50, 64**

# Introduction to Sap-Sucking Insects, Gall Formers, and Mites

## Honeydew, spittle masses, growth deformities, or dried out foliage

Sap-sucking insects feed on the sugary sap produced in foliage and transported in the soft phloem tissue beneath the bark. Some insects' feeding affects plant growth hormones, causing distinctive foliage or shoot deformities called galls. Gall formers feed protected within the gall. They may suck sap, as do gall-forming adelgids, or pierce and scrape the surface, as do some gall-forming midges. Mites (Acarina) are tiny arthropods that are more closely related to spiders and ticks than to insects. The class Acari contains 45,000 described species, but this is possibly only a portion of the total number of species yet to be described. Spider mites puncture individual cell walls on the outside of foliage and suck out the contents. This leads to a characteristic mottling of the foliage.

There are other types of mites that will be seen and noticed in a forested setting, but most are quite inconspicuous. One group of mites that may catch the observer's eye is the mites that are found associated with bark beetles. Opening up a bark beetle gallery or looking at an individual bark beetle may reveal the presence of mites. These bark beetle associates fulfill a wide array of functions, including being predators of bark beetle eggs and larvae, grazers of associated fungi, and even predators of other mites. Because mites are wingless, they must hitchhike to new bark beetle galleries on their bark beetle hosts. This pattern of movement is referred to as phoresy. These mites are an integral part of the bark beetle system.

**General Features—**

- Hosts and symptoms of sap feeding insects, gall formers, and mites are described in table 19 (see figs. 496-499).

- Sap-sucking insects like aphids and scales feed on foliage, twigs, branches, and, occasionally, on the trunks of trees. These insects often produce large amounts of sweet, sticky honeydew that can sometimes be seen as a fine mist coming from the trees. Honeydew can coat branches and objects

**Figure 496.** The boxelder bug is a seed-feeding insect that may overwinter in and around buildings. *Photo: William M. Ciesla, Forest Health Management International, Bugwood.org.*

**Table 19.** Common sap-sucking and gall forming insects and mites in the Rocky Mountain Region.

| Insect/mite | Host | Symptom |
|---|---|---|
| Black pine needle scale[a] (*Nuclaspis californica*) | Ponderosa and Austrian pine | Tiny, black discs on needles |
| Boxelder bug[a] (fig. 496) (*Leptocoris trivittatus*) | Seeds of boxelder and other maples | Notable home invader in fall |
| Cooley spruce gall adelgid (*Adelges cooleyi*) | Spruce and Douglas-fir | Galls on spruce; woolly material on Douglas-fir needles and twigs |
| Eriophyid mites (many species) | Aspen, oak, maples, and others | Areas of deformed foliage, often red or pink-colored |
| European elm scale (*Gossyparia spuria*) | Elms, primarily American elm | Honeydew, black sooty mold, and scales on branches |
| Giant conifer aphids(*Cinara* spp.) | Many conifers | Honeydew and aphid colonies |
| Hackberry budgall psyllid[a] (*Pachypsylla celtidisgemma*) | Hackberry | Enlarged, spherical-shaped killed buds |
| Hackberry nipplegall maker[a] (*Pachypsylla celtidismamma*) | Hackberry | Nipple-shaped swellings on leaves |
| Oystershell scale (*Lepidosaphes ulmi*) | Many hardwoods | Grey modeling on bark |
| Petiole gall aphid (*Pemphigus* spp.) | Aspen, cottonwood, and other poplars | Marble-sized galls on leaf petiole |
| Pine needle scale (*Chionaspis pinifoliae*) | Pines, spruce, and occasionally other conifers | Tiny, white, teardrop-shaped scales on needles |
| Pinyon needle scale (*Matsucoccus acalyptus*) | Pinyon pine | Tiny, black, bean-shaped bumps on year-old needles |
| Pinyon spindle gall midge[a] (*Pinyonia edulicola*) (fig. 497) | Pinyon pine | Football-shaped swelling joining needles with the fascicle |
| Pitch or Resin midge[a] (*Cecidomyia* spp.) | Ponderosa pine, primarily | Resin pits with many tiny, bright red midge larvae |
| Spider mites (*Oligonychus* spp., *Tetranychus* spp. and others) | Spruce, junipers, and others | Dried foliage; dusty appearance |
| Spittlebugs[a] (*Aphrophora* spp.) (fig. 498) | Primarily junipers, oaks, and herbaceous understory plants; also in southwestern dwarf mistletoe | Frothy masses on twigs and foliage |
| Western conifer seed bug[a] (*Leptoglossus occidentalis*) (fig. 499) | Seeds of pines and Douglas-fir | Damaged seed; can invade homes in the fall |

[a] Not discussed in this guide.

below the trees, which can subsequently be covered by black sooty mold.

- The majority of sap-sucking insects belong to the orders Hemiptera (true bugs) and Homoptera (aphids, leafhoppers, scales). Common gall formers belong to the insect orders Homoptera (adelgids), Hymenoptera (cynipid gall wasps), and Diptera (gall midges) and to the mite order Acari, which includes gall-forming mites (eriophyid mites) and spider mites.

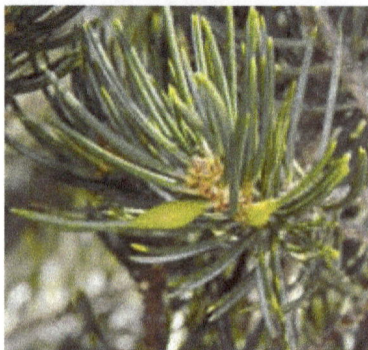

**Figure 497.** Pinyon spindle gall. *Photo: Brian Howell, USDA Forest Service.*

- Spider mite injury is usually related to drought conditions.

- Spider mites produce very fine webbing that causes a distinctive dirty appearance due to the accumulation of dust, cast mite skins, and eggs.

- There are many natural enemies that help control sap-sucking insects and mites. Insecticide use can sometimes prolong infestations by killing natural enemies of these insects.

- There are a number of non-native sap-sucking insects that have been introduced to other parts of the country that have become serious forest pests. The central Rocky Mountain Region does not have serious problems with non-native sap-sucking insects in its conifer forests, but several non-native species are found in deciduous landscape trees.

- Sap-sucking insects such as leafhoppers are known to vector certain plant diseases (e.g., X-disease on chokecherry).

**Figure 498.** Frothy mass covers spittlebug on juniper. *Photo: Whitney Cranshaw, Colorado State University, Bugwood.org.*

**Figure 499.** The western conifer seed bug may seek overwintering sites around buildings in the fall. *Photo: Whitney Cranshaw, Colorado State University, Bugwood.org.*

# Cooley Spruce Gall Adelgid

## Conspicuous galls on spruce

**Name and Description**—*Adelges cooleyi* (Gillette) [Homoptera: Adelgidae]

*Adelges cooleyi* are aphid-like insects whose activity causes conspicuous galls on spruce trees and premature needle cast of Douglas-firs.

**Hosts**—Blue and Engelmann spruces and Douglas-fir

**Life Cycle**—These adelgids have a complex, 2-year life cycle when both spruce and Douglas-fir are present.

Spruce—Immature females (with a "woolly" wax covering) overwinter beneath young branches. They mature in the spring and place large numbers of eggs near branch tips in proximity to developing buds. Eggs hatch near the time of bud break, and nymphs feed at the base of newly developing needles. Galls form as the tree's growth response to saliva that is introduced during this feeding activity (fig. 500). Adelgids grow and develop within the galls. During the summer, fully developed galls begin to dry out, and winged adelgids emerge. Most of these winged insects then leave the spruce tree in search of a suitable Douglas-fir host.

**Figure 500.** Galls formed on branch tips of spruce. *Photo: Whitney Cranshaw, Colorado State University, Bugwood. org.*

Douglas-fir—Female adelgids deposit eggs on Douglas-fir needles. Multiple generations of wingless adelgids are then produced. They appear as cottony tufts scattered among the foliage (fig. 501). In late summer, some woolly adelgids migrate to spruce trees and produce eggs. These give rise to the population that will overwinter among spruce branches. Wingless adelgids stay on Douglas-fir trees and produce offspring that will overwinter on those trees.

**Damage**—

Spruce—Where both spruce and Douglas-fir are present, galls are formed on branch tips

**Figure 501.** Wingless adelgids appear as cottony tufts on Douglas-fir. *Photo: Whitney Cranshaw, Colorado State University, Bugwood.org.*

of spruce. Impact to spruce in general forest settings is usually minimal. However, gall formation on spruce does cause the death of the branch tip, and, with extremely heavy infestations, some crown deformity may result. Galls range from $1/2$-3 inches (13-75 mm) long, and nymphs can be found within them when they are green. By late summer, galls dry and turn brown. They may persist for several years. Galls are often mistaken for cones. Many spruce trees express resistance to gall formation. For trees that are not resistant, gall formation is often concentrated on well-shaded sides (north and east) of the crown.

Douglas-fir—Nymphs pierce needles and feed on fluids. Needles then become discolored, often distorted, and drop from the tree prematurely. Also, as insects feed on needles, they secrete honeydew. Sooty mold often grows on the honeydew. Nymphs also feed on developing cones. Significant damage to cones can negatively impact seed production. Galls are not formed on Douglas-fir.

**Management**—Control of these insects and their resulting galls (spruce) or needle drop (Douglas-fir) is not necessary in forest settings. However, their impact may be significant to trees in nurseries, plantations, or ornamental settings. Old galls can be removed from ornamental spruce trees to improve aesthetics, but this does not influence existing or future adelgid populations. For ornamental trees, control efforts must occur before gall formation, typically as early as April. Several insecticides are registered for use to control adelgids and vary somewhat, depending upon host tree species (spruce or Douglas-fir).

**References: 33, 35, 50**

# European Elm Scale

## Elm branch dieback and shiny, sticky leaves

**Name and Description**—*Gossyparia* (= *Eriococcus*) *spuria* (Modeer) [Homotera: Eriococcidae]

European elm scale is often first noticed on urban elms when honeydew produced by feeding scales becomes a nuisance on parked cars or outdoor furniture. Branches on infested trees are typically blackened by sooty mold growing on the honeydew. Closer examination of the branches reveals female scales, which appear as gray or reddish brown discs surrounded by a white, waxy fringe less than $1/8$-$1/2$ inch (<10 mm) long (fig. 502). Tiny, orange crawlers can be seen on leaves along main veins in summer (fig. 503).

**Hosts**—Elms, primarily American elm

**Life Cycle**—Eggs hatch in June and July from beneath mature female scale coverings. Summer is spent as tiny, yellow nymphs on the undersides

Figure 502. European elm scale. *Photo: Whitney Cranshaw, Colorado State University, Bugwood.org.*

of leaves. The insects crawl back to branches in late summer and attach themselves to twigs and branches. Female scales cover their bodies with a gray, waxy covering, and they overwinter in the second instar. Winged and unwinged males, when present, overwinter in small white cocoons. In spring, females grow and become a dark red-brown and produce copious amounts of honeydew. Reproduction can be sexual or asexual.

**Figure 503.** Newly hatched crawlers emerging from female scales. *Photo: Whitney Cranshaw, Colorado State University, Bugwood.org.*

**Damage**—Honeydew from European elm scale feeding can be a nuisance. Yellowing leaves may drop early, and heavy feeding can cause branch dieback or death.

**Management**—Natural enemies include parasitic wasps on female branch scales and predatory plant bugs, mites, and spiders on leaves. On landscape elms, properly timed application of horticultural oils or insecticides is effective. Soil applications of the systemic insecticide imidacloprid have been very effective.

**References: 35, 98**

# Gall (Eriophyid) Mites

## Unique and sometimes colorful leaf distortions

**Name and Description**—Eriophyidae family

Eriophyid mites are soft-bodied, tiny to microscopic mites that feed on many species of plants. They are unique among the mites and other arachnids because they have only two pairs of legs instead of four, and they are shaped more carrot-like than spider-like. Eriophyid mites pierce the plant cells and suck up the plant's juices. Many are found on leaves, but some feed on flowers, fruits, or buds. Most species are restricted to feeding on one plant species and many are undescribed.

**Hosts**—Hardwoods and conifers

**Life Cycle**—There are several overlapping generations during the plant growing season. Growing-season forms include both

**Figure 504.** *Aceria macrorrhyncha* on maple. *Photo: Milan Zubrik, Forest Research Institute, Slovakia, Bugwood.org.*

**Figure 505.** *Eriophyes celtis* on hackberry. *Photo: Whitney Cranshaw, Colorado State University, Bugwood.org.*

**Figure 506.** *Eriophyes parapopuli* on aspen. *Photo: Whitney Cranshaw, Colorado State University, Bugwood.org.*

**Figure 507.** *Eriophyes calcercis* on Rocky Mountain Maple. *Photo: William M. Ciesla, Forest Health Management International, Bugwood.org.*

**Figure 508.** *Colomerus vitis* on grape. *Photo: Jody Fetzer, New York Botanical Garden, Bugwood.org.*

sexes. Eriophyid mites develop through four stages: egg, first nymph, second nymph, and adult. Dormant-season forms are fertile females that hide under bud scales or other protected sites. Species are often hard to identify because growing and dormant season forms can be different shapes and sizes.

**Damage**—Infestations of eriophyid mites produce a variety of symptoms on plants. The saliva in some mites produce bladder, bead, pouch, or finger galls of bright colors; others may cause leaf bronzing or brittleness. Some induce irregular leaf blisters and some stunt or distort buds or flowering parts (figs. 504-506). In fact, it is often possible to identify a causal organism by the type of gall induced on the host. Eriophyid mites can induce changes in plant hairs that produce velvety or cottony patches called erineum (fig. 507-508).

**Management**—Plants can tolerate large populations of mites, and management is often not needed. When management is desired, affected leaves or plant parts can be pruned or destroyed. A horticultural oil spray can be applied in the fall to smother overwintering females of some species. However, the horticultural oil may also remove beneficial predaceous mites.

**References: 35, 41, 50**

# Giant Conifer Aphids

## Feed on sap of twigs, branches, and roots

**Name and Description**—*Cinara* spp. [Homoptera: Aphidae]

Giant conifer aphids are soft-bodied, generally gregarious insects that are often found in large groups. They have piercing-sucking mouth parts that are used to feed on the sap in twigs, branches, trunks, and roots. Most species are restricted to feeding on one genus of tree; many attack only one tree species. Giant conifer aphids are large aphids, up to $1/4$ inch (6 mm) long, and are long legged and dark colored (figs. 509-510). Aphids have a pair of tube-like structures on their abdomen and can be winged or wingless.

**Figure 509.** Powdery juniper aphids, *Cinara pulverulens*. Photo: Whitney Cranshaw, Colorado State University, Bugwood.org.

**Hosts**—Many different conifer species. Each *Cinara* species is specific to a tree genus or tree species. About three dozen species occur in the Rocky Mountain Region.

**Life Cycle**—There are several generations a year. Eggs hatch in the spring into small aphids. They molt through several stages, becoming larger aphids each time. Females give birth to live young, except with the last summer generation when eggs are laid on needles or bark. Aphids secrete honeydew as they feed and other insects, especially ants, bees, and wasps, feed on the honeydew (fig. 511). Honeydew is also a good growth medium for sooty mold.

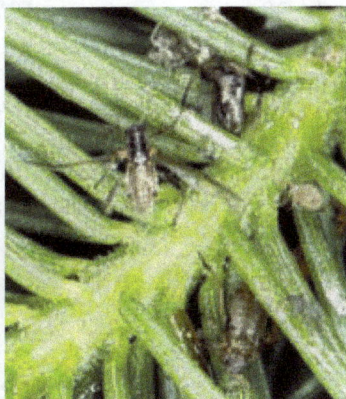

**Figure 510.** Spruce shoot aphid, *Cinara pilicornis*. Photo: Andrea Battisti, Università di Padova, Bugwood.org.

**Damage**—Heavy infestations of giant conifer aphids cause yellowing of the foliage and reduce tree growth, especially in young trees (fig. 512). Populations are usually highest in late spring and may crash by early summer. The best evidence of

**Figure 511.** Ants feeding on honeydew secretions. Photo: E. Bradford Walker; Vermont Department of Forests, Parks and Recreation; Bugwood.org.

an aphid population is the presence of honeydew (clear, sugary, sticky liquid), ants attracted to honeydew, or ladybeetles that feed on aphids. The juniper aphid, *C. sabinae* (Gillette and Palmer), is common on juniper and can be destructive.

Giant conifer aphids can also be found in Christmas tree plantations. Damage can be difficult to distinguish from damage caused by other sucking insects and needle diseases. Aphids can be confused with other sucking arthropods such as mites, mealybugs, or scale insects.

**Management**—Natural controls, including natural enemies such as ladybeetles, lacewings, syrphid flies, or parasitic wasps, usually bring aphid population under control. Often, damage is not noticed until the aphid population declines from natural controls. If direct control becomes necessary, aphids can

**Figure 512.** Aphid damage on juniper (photo taken from side of tree). *Photo: Sheryl Costello, USDA Forest Service.*

be controlled with insecticides. However, trees should be inspected for natural enemies before an insecticide is applied. Furthermore, because giant conifer aphid populations can build up by early spring, application timing is often critical.

**References: 35, 50**

# Oystershell Scale

## Tiny, armored scale on many hardwoods

**Name and Description**—*Lepidosaphes ulmi* (Linnaeus) [Homoptera: Diaspididae]

Oystershell scale is believed to have arrived in North America in the 1700s with European settlers. It now has a worldwide distribution and is one of the more common armored scales on hardwoods in the Rocky Mountain Region. Full-grown female scale coverings are comma-shaped and resemble miniature oyster shells

**Figure 513.** Oystershell scales and eggs revealed under a lifted scale covering. *Photo: Whitney Cranshaw, Colorado State University, Bugwood.org.*

attached to bark on twigs or branches (fig. 513). Mature scales are gray or brown, range from $1/16$-$1/8$ inch (1.5-3 mm) long, and have parallel, arcing ridges. Heavy infestations can impart a gray, mottled look to the bark (fig. 514). Older scales can stay attached to the tree for several years. Pale yellow crawlers are tiny and nondescript. Tiny adult males are winged and rarely seen.

**Hosts**—Oystershell scale has been recorded on over 125 species of plants, mostly hardwoods, and is most commonly found on aspen and other poplars, ash, maples, willows, and lilac.

**Figure 514.** Oystershell scale infestation on aspen. *Photo: Whitney Cranshaw, Colorado State University, Bugwood.org.*

**Life Cycle**—The insects overwinter in eggs beneath the female scale covering. Upon hatching in May or June, the young insect crawls about looking for a suitable site to begin feeding. Some may be carried by the wind to other trees at this time. This stage is called the crawler stage, and the young are only mobile for a few days. Once settled down to feed, they develop the waxy shell that covers their body. The scales have relatively long mouthparts for their size, and they feed on the contents of individual plant cells. There is one generation per year in the gray form of oystershell scale, with winged males appearing in summer and mating with sessile females. Two generations are reported from the brown race of oystershell scale, but this form of the insect is less common in the Rocky Mountain Region.

**Damage**—Heavy infestations of this insect can encrust entire branches and cause branch dieback and can kill plants.

**Management**—Natural enemies are generally adequate to reduce populations of this insect. Heavy infestations may warrant management on high-value trees.

Oystershell scales can be physically removed from branches and stems with a plastic dish scrubber, but be careful not to damage thin bark on the trees. Horticultural oil or insecticide application can also give good control if timed properly. Insecticides are only effective against the crawler stage, which normally lasts a few days but may be extended over weeks. Insecticide control is best targeted at the beginning of the emergence period and requires monitoring for the presence of crawlers.

**References: 34, 156**

# Petiolegall Aphids

## Swollen or disfigured leaves of poplars

**Name and Description**—*Pemphigus* spp. [Homoptera: Aphididae]

Adult petiolegall aphids are about 1/13 inch (2 mm) long, usually pale green with a dark thorax, and covered with a waxy substance. They have short, thread-like antennae and lack the terminal abdominal tubules called cornicles that are characteristic of other aphids. At certain times of the year, adults have four clear wings. Immatures, called nymphs, look like smaller versions of the adults. *Pemphigus* spp. aphids are cryptic; they are found within galls on trees (fig. 515) or on roots of their alternate hosts. Eggs are relatively large, given the size of laying females but are less than 1/25 inch (1 mm).

**Figure 515.** Characteristic marble-sized galls on the petiole of cottonwood leaves induced by the poplar petiolegall aphid, *Pemphigus populitransversus*. *Photo: Herbert A. "Joe" Pase III, Texas Forest Service, Bugwood.org.*

**Hosts**—Aspen, cottonwood, and other *Populus* spp. trees in winter, alternating with annual herbaceous host plants in summer

**Life Cycle**—Petiolegall aphids have complex, 1-year life cycles that alternate between two sets of hosts. Although probably similar in life cycle, considerable confusion exists regarding species' identities within the genus *Pemphigus*, so there may be more species involved than are named here. The poplar petiolegall aphid (*P. populitranversus* Riley) (fig. 515) life cycle is used as an example. Eggs are laid in fall in bark cracks on *Populus* spp. trees. In spring, more or less coincident with the flushing of the poplar foliage, eggs

**Figure 516.** When a gall is cut open, numerous petiolegall aphids can be found inside. *Photo: Herbert A. "Joe" Pase III, Texas Forest Service, Bugwood.org.*

hatch into nymphs that will all become asexually reproducing females. Nymphs feed on developing leaf petioles through tubular, sucking mouthparts. Feeding induces the host plant to produce a swollen growth, called a gall that envelops the developing aphid (fig. 516). As this overwintering form matures, it produces young that remain within the gall until full-grown. These new adult females have wings. The galls split open, and the winged adults fly during late June and July to their summer hosts. Using cracks in the soil, they locate and colonize the roots of plants in the cabbage family. Colonies of nymphs secrete a waxy substance that is believed to protect them from excess moisture. Several asexual female generations may be produced on these summer hosts. At the end of summer, winged adults are produced that fly back to their winter hosts where they give birth to small, mouthless males and females that mate, and then the males die. After mating, each female, which is less than 1/25 inch (1 mm) long, lays one egg that is almost as large as she is and then dies.

The poplar petiolegall aphid forms a spherical green gall with a transverse slit on the petiole of plains cottonwood (fig. 515). It is also a root-infesting pest on cruciferous crops in summer, where it is known as the cabbage root aphid.

The lettuce root aphid, *P. bursarius* (Linnaeus), is a European insect that was accidentally introduced into North America. It produces a flask-like gall on the petioles of Lombardy poplar and hybrid poplars (fig. 517). It is an occasional pest of lettuce crops and is known from lambsquarter and carrot.

**Figure 517.** Gall produced by the lettuce root aphid, *Pemphigus bursarius*. *Photo: Gyorgy Csoka, Hungary Forest Research Institute, Bugwood.org.*

The sugarbeet root aphid, *P. populivenae* Fitch, forms an elongate gall on the mid-vein on the upper side of *Populus* spp. leaves, such as narrowleaf cottonwood, and infests the roots of sugarbeets and other garden plants. Gall shape and location are variable.

The poplar vagabond aphid, *Mordwilkoja vagabunda* (Walsh) [Homoptera: Aphididae], may be confused with *Pemphigus* spp., although its characteristic leathery, distorted galls differ greatly in shape (figs. 518-519). Poplar vagabond aphids have a life cycle similar to other gall-making aphid species in that it over-winters as an egg and alternates between winter and summer host plants. Newly hatched aphids feed on developing leaves of aspen and certain cottonwoods, which produce an enlarged, twisted clump of growth at the leaf base (fig. 518). Several generations of aphids occur within the folds of such galls, which may contain as many as 1600 individuals each. The green galls fade in color and harden, tend to remain on trees, and may not be visible until after normal leaf fall (fig. 519). Galling is concentrated in the upper third of the tree. This Region's summer host plants are not known, although loosestrife is an important summer host in other areas.

**Figure 518.** A poplar vagabond aphid that was feeding on developing aspen leaf produced a gall of distorted growth that will dry and remain after leaf fall. *Photo: Dave Powell, USDA Forest Service, Bugwood.org.*

**Figure 519.** Dried gall from poplar vagabond aphid feeding. *Photo: Whitney Cranshaw, Colorado State University, Bugwood.org.*

**Damage**—None of the *Pemphigus* species are economic pests for foresters or arboriculturists. Galled leaves tend to fall from the tree prematurely. Severe infestations can render trees unsightly but cause little to no other injury. There are many other organisms, including aphid, fly, and mite species, that cause galls to form on *Populus* spp. leaves and branches, each with characteristic gall shape and location. Other than aesthetic considerations, these organisms are not known to damage their tree hosts. Serious economic damage can occur on the alternate annual herbaceous summer hosts.

**Management**—Excessive gall formation on host trees can be indicative of an unhealthy condition, relief from which might reduce galling. In some cases, gall-forming insects prefer hosts that are given abundant moisture and/or fertilization, so galls can be a problem in well-tended landscape settings. Given the lack of injury, active management of petiolegall aphid populations is not indicated. If control is desired, application of horticultural oil in spring as leaves begin to expand should kill the overwintered eggs.

**References: 35, 50, 88**

# Pine Needle Scale

## Tiny, white, tear drop-shaped scales on conifer needles

**Name and Description**—*Chionaspis pinifoliae* (Fitch) [Hemiptera: Diaspidae]

Adult female pine needle scale insects are about $1/8$ inch (3 mm) long, dark orange, and wingless under an armor scale covering they secrete that is about $1/3$ inch (8 mm) long. They are the most conspicuous form, with the scale being almost pure white, elongate oval, and yellow at the apex (fig. 520). Developing males are similar to developing females, though smaller at $1/25$ inch (1 mm) long and more slender. The minute adult males have wings when mature, but are rarely present in western populations. The immature crawler stage is generally oval, light purple to reddish brown, and tiny (fig. 520). Rusty brown-colored eggs are deposited under the scale covering. The characteristic white armor coverings can be seen on needles any time of the year, although the insects beneath the coverings may not all be alive. Pine needle scale is distributed throughout North America but is most common in the northern half of the United States and southern Canada.

**Hosts**—Pines, spruces, white fir, Douglas-fir, and cedar; may be

**Figure 520.** Close-up of small immature and larger mature female pine needle scales with newly-hatched brownish crawlers. *Photo: E. Bradford Walker, Vermont Department of Forests, Parks and Recreation, Bugwood.org.*

common on shelterbelt and ornamental plantings of native and introduced pines, especially mugo pine

**Life Cycle**—The pine needle scale has two generations per year in much of Colorado. Twenty to 30 eggs are laid in the fall by the female and overwinter under her scale covering. Some females may survive winter and lay eggs in spring. As they lay eggs, the females gradually shrink in size while the scale fills with eggs. Regardless of overwintering stage, egg hatch occurs over a relatively brief period sometime from late April to mid-June, depending upon local weather and location. The nymphs, or crawlers, then wander over the needles for a few days, select a needle, insert their sucking mouth parts, and begin feeding on sap. Wind is responsible for most of the dispersal of crawlers to uninfested trees. Females are sessile after settling and will remain on one needle for the rest of their lives. Feeding nymphs turn yellow and lose their appendages by molting. Combining their cast skins with waxy secretions, the pine needle scales create the characteristic protective white covering. When present, males also remain in place, developing as a scale and feeding until they emerge from the covering as minute, winged adults, flying off in search of females. Pine needle scales mature by early July. Clusters of eggs are then laid with or without mating in a white, waxy sack under the scale covering. Scales of this second generation mature by fall and lay overwintering eggs under the scale covering. In areas where there is one generation per year, scales mature by the end of summer and lay overwintering eggs.

**Damage**—Pine needle scale feeding causes a yellowing of the foliage in the area surrounding each sessile scale, and these areas coalesce when trees are heavily infested. Consequently, the foliage on heavily infested trees becomes discolored. The waxy, white, secreted coverings of the scale insects and black mold that grows on scale exudates often combine to give infested trees a grayish appearance (figs. 521-522). Heavy infestations can result in premature needle shed. On pines, infestations can result in a marked reduction in needle length and a diminished growth rate. Outbreaks are frequently confined to limited areas on a given tree. Prolonged infestations, rare in natural forests in this Region, can kill branches and even young trees and can weaken larger trees, predisposing them to attack by opportunistic insects and diseases. Planted pines can be severely damaged. Scale infestations are often associated with factors that negatively affect host tree vigor. In addition, conditions that adversely affect scale predators and parasites such as road dust or pesticide application can result in significant infestations.

**Figure 521.** Heavy infestations of pine needle scale on blue spruce. *Photo: Whitney Cranshaw, Colorado State University, Bugwood.org.*

**Management**—Being sessile, pine needle scales are the target of many predatory and parasitic insects and other natural enemies that frequently exert a high degree of control. Of particular note is a lady beetle species whose annual life cycle is synchronized with that of the scale. Severe cold temperatures are thought to limit outbreaks by reducing survival of the overwintering stages.

**Figure 522.** Heavy infestations of pine needle scale on pine. *Photo: Scott Tunnock, USDA Forest Service, Bugwood.org.*

In planted settings, proper tree care is a primary management strategy against scale insects, including pruning heavily infested branches. Refrain from planting pines or other hosts along dusty roads or in areas of heavy air pollution. Minimizing tree stress is key to keeping scale populations low, which will also lower susceptibility to other insect and disease pests.

Direct control options include the use of horticultural oils and insecticidal soaps, which are not toxic to and, therefore, will help conserve insect natural enemies active against the scales. Additional registered insecticides are also available, but they have wider non-target insect toxicity. Armored scales are more difficult to kill with insecticide due to their protective covering. Applications should be carefully timed against the crawler stage, which immediately follows egg hatch.

**References: 32, 35, 50, 88**

# Pinyon Needle Scale

## Black bumps on pinyon needles

**Name and Description**—*Matsucoccus acalyptus* (Herbert) [Homoptera: Margarodidae]

Pinyon needle scales appear as small, black, bean-shaped bumps on the surface of 1-year-old pinyon needles (fig. 523). These tiny, sap-sucking insects kill the needles and seriously weaken pinyon pines in woodlands and in urban landscapes.

**Host**—Pinyon pine

**Figure 523.** Scales infesting one-year-old needles. *Photo: Robert Cain, USDA Forest Service.*

**Life Cycle**—Adult, wingless females emerge from scale coverings in late winter or early spring and mate with winged males (fig. 524). Emergence time varies with temperature across the range of the insect. In Colorado, emergence occurs in late March or early April. Most males emerge from scale coverings the previous fall and spend the winter in silk webs in litter beneath the tree. A few males don't enter this stage until early spring. Mated females crawl along the bark to egg-laying sites around the root collar, on the undersides of large branches, in branch crotches, or in cracks of rough bark. Yellow eggs are laid in clusters held together by white, cottony webbing (figs. 525-526). Occasionally, egg masses are found several feet from the base of the tree on a rock or log. About four weeks after eggs are laid, tiny, red eye spots can be seen in the eggs with the aid of a hand lens. Nymphs, called crawlers, emerge about 7-10 days after eye spots appear. They climb to the ends of

**Figure 524.** Winged male mating with emerging female pinyon needle scale. *Photo: USDA Forest Service.*

**Figure 525.** Cottony webbing and eggs deposited under branches and in branch crotches. *Photo: Robert Cain, USDA Forest Service.*

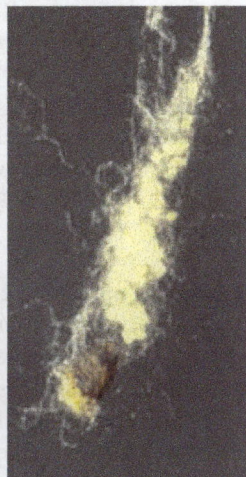

**Figure 526.** Eggs, webbing, and expired female scale. *Photo: Southwestern Region Archives, USDA Forest Service.*

branches and settle on the previous year's new growth. After inserting tube-like mouth parts into the needle, they become immobile, cover the body with wax, and turn black. By late August, they will molt in place to the more visible, bean-shaped second stage, which is about 1/16 inch (1.5 mm) long.

**Damage**—Reduced new growth and stunted needles are common on trees suffering repeated attacks (fig. 527). Heavy infestations frequently kill small trees and predispose weakened larger trees to attack by other insects, especially bark beetles, which can kill trees.

**Management**—No control strategies have been developed for woodlands. However, potential damage from these pests on landscape pinyons can be drastically

**Figure 527.** Only the new growth remains green on heavily infested pinyons. *Photo: Robert Cain, USDA Forest Service.*

reduced by destroying eggs before they hatch. Dislodge egg masses from the tree with a strong stream of water from a garden hose. After washing down the tree, rake up all the material around the base of the tree and destroy or remove it. Chemical insecticides are registered to control the pinyon needle scale, but timing of the spray application is critical for success. Apply insecticides to the bark and branch crotches when females are moving to egg-laying sites or as soon as crawlers begin to emerge from eggs. Examine eggs with a hand lens, and be ready to spray shortly after the crawler's red eve spots are visible. Once scales have established themselves on the needles, they become more difficult to control.

**Reference: 22**

# Spider Mites

## Dry or dusty looking foliage

**Name and Description**—Family Tetranychidae

The spider mites family includes about 1600 species, some of which are well-known due to their habits that are often injurious to plant life. These mites range in size from $1/25$-$1/20$ inch (0.9-1.25 mm) long and are often visible to the naked eye. They are spider-like in appearance with two body parts and four pairs of legs. The common name, spider mites, is derived from mites' habits of spinning webbing on the plants they have colonized. Most frequently, it is the discovery of the webbing that will lead an observer to determine the infestation of spider mites (fig. 528).

**Hosts**—Hardwoods and conifers

**Life Cycle**—Spider mites typically develop through five stages: egg, nymph, two immature stages, and adult. Many overwinter in the egg or adult female stage, and there are often several generations a year, with a generation being completed in as little as a week.

**Damage**—Unlike spiders, spider mites are herbivorous and feed upon host plants by piercing individual plant cells with their mouthpart and sucking out the fluids. The damage takes on a stippled yellow or brown appearance as large numbers of cells are affected on an infested plant part. Some species will feed upon a wide range of host plants, while others have a very limited host range. The presence of webbing,

**Figure 528.** Spruce spider mites, *Oligonychus ununguis* (Tetranychidae). Note the fine webbing. *Photo: USDA Forest Service, Intermountain Region, Bugwood.org.*

which may contain a large number of visible mites, and damage to plant tissue are indicative of spider mite activity.

Spider mites rarely achieve pest status in a forested setting. However, in landscape situations, greenhouses, and, at times, in forest plantations or other planted areas, spider mites can damage high-value plants. Even in these circumstances, an outbreak of spider mites is indicative of another condition that is allowing the spider mites to thrive. Outbreaks of spider mites are frequently associated with hot or dry conditions, and, even in forested settings, their numbers increase during periods of drought. Another situation in which spider mites can reach outbreak levels is when host plants are subjected to large amounts of dust like what might be found along a dirt road. Finally, spider mite outbreaks can be caused by the use of insecticides or fungicides that have been employed to deal with another pest. The use of these chemicals can kill off the organisms that normally keep spider mites under control, resulting in a secondary outbreak of the spider mites.

**Management**—Control of spider mites needs to start with investigation of the cause of an infestation. Often, removal of a causal factor results in the rapid and complete collapse of a spider mite infestation. Sometimes these causal factors can be somewhat obscure; in one case, spruce spider mites were infesting large numbers of host trees in an urban setting. Only later was it discovered that frequent spraying for mosquitoes contributed to the outbreak of spider mites. Adjustment of the spray schedule caused the spider mite population to decline.

**Reference: 35, 41, 50**

# Introduction to Bud, Shoot, Branch, and Terminal Insects

## Damaged tips and branches

The insects presented in this section are those that feed only within or on buds, shoots, or branches (table 20). These include insects primarily in the orders Lepidoptera (moths and butterflies) and Coleoptera (beetles). Bark beetles and twig beetles that are found in twigs and branches are presented in the Introduction to Bark Beetles entry. Aphids, scales, and midges that feed on shoots and branches are presented in the Introduction to Sap-Sucking Insects, Gall Formers, and Mites entry.

**General Features—**

- Twig-boring insects often mine out the pith or center of developing shoots.
- Resin may or may not be present.
- Pitch mass-forming insects feed protected within the soft pitch mass.
- The most serious damage, stunting, forked tops, or multiple leaders occurs to sapling conifer trees when the main stem or leader is affected.
- These insects can be important in young, intensively managed stands or replanted forested areas.

**Table 20.** Common insects that damage buds, shoots, branches, and terminals.

| Insect | Host | Symptom |
|---|---|---|
| Pine tip moth | Pines | Hollowed out shoots often with undeveloped or partially developed needles |
| Juniper twig pruner (discussed in the Wood Borers section) | Junipers | Branch dieback; hollowed out twigs |
| Pitch moths | Pinyon pine and ponderosa pine | Large, irregular pitch masses |
| Pitch nodule moth | Pinyon pine | Flagging and nodules of pitch at small branch/twig intersections |
| Western spruce budworm (discussed in the Defoliators section) | Spruce, white fir, and Douglas-fir | Early instar larvae mine in buds |
| Terminal weevils | Spruce and lodgepole pine | Dying leader; often with a shepherd's crook |
| Twig beetles (discussed in the Bark Beetles section) | Pines and Douglas-fir | Tiny holes; tip or branch flagging |

# Pine Tip Moth

## Hollowed out shoots

**Name and Description**—*Rhyacionia* spp. [Lepidoptera: Olethreutidae]
 Western pine tip moth—*R. bushnelli* Busck
 Nantucket pine tip moth—*R. frustrana* Comstock
 Southwestern pine tip moth—*R. neomexicana* Dyar

Larvae of pine tip moths are yellowish with black heads and are less than 1/2 inch (13 mm) long. Adults, though rarely seen, have forewings mottled with yellowish gray and reddish brown (fig. 529). Hind wings are gray. Wingspans extend to about 3/4 inch (19 mm).

**Hosts**—Ponderosa pine

**Life Cycle**—A single generation per year occurs in most parts of the Intermountain West. Winter is passed as pupae in cocoons in the litter or soil beneath infested trees. Adult moths emerge by

**Figure 529.** Adult pine tip moth. *Photo: Scott Tunnock, USDA Forest Service, Bugwood.org.*

late May or early June, mate, then females deposit eggs on needles, buds, and shoots of young trees. Newly hatched larvae either feed between or mine the needles. Later instars feed inside needle sheaths or buds and then enter new shoots and mine within developing shoots. Larvae complete growth by midsummer or fall, emerge from shoots, and drop to the ground to pupate, where they spend the winter (fig. 530).

### Western Pine Tip Moth Life Cycle

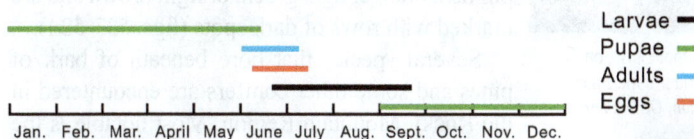

Larvae ▬
Pupae ▬
Adults ▬
Eggs ▬

Jan. Feb. Mar. April May June July Aug. Sept. Oct. Nov. Dec.

**Figure 530.** Western pine tip moth life cycle (from Johnson 1982).

**Figure 531.** Pine tip moth larva. *Photo: Whitney Cranshaw, Colorado State University, Bugwood.org.*

**Figure 532.** Pine tip moth damage. *Photo: John Guyon, USDA Forest Service, Bugwood.org.*

**Damage**—Larvae mine shoots and buds of young pines—this is especially damaging in plantations, even-aged natural stands, and ornamental plantings (fig. 531). Infested trees are often deformed and growth is reduced. Damage is unsightly but seldom fatal. Pine trees less than 25 ft (7.6m) tall are most commonly affected (fig. 532).

**Management**—Damage is seen in upper and mid-crowns of young ponderosa pines. Larval feeding activity distorts and kills both terminal and lateral shoots, which often stunts growth and deforms trees. No control is necessary in forest settings. Ornamental trees may be protected by applying insecticides to the foliage and soil in the spring.

**References: 50, 93, 167**

# Pitch Moths

## Oozing pitch masses on trunks and branches on pines

**Name and Description**—*Dioryctria* spp. [Lepidoptera: Pyralidae]

Pine pitch moth—*D. ponderosae* Dyar

Zimmerman pine moth—*D. zimmermani* (Grote)

Pitch moths attacks appear as large, oozing masses of pitch that form at the

**Figure 533.** Pitch moth larva on ponderosa pine under bark. *Photo: Southwestern Region, USDA Forest Service.*

**Figure 534.** Pitch moth larva in pitch mass on ponderosa pine. *Photo: Whitney Cranshaw, Colorado State University, Bugwood.org.*

wound site in response to larval feeding beneath the bark. Adult moths are rarely observed and are difficult to differentiate from other members of the genus. They are grey with white zig-zag markings and are up to 3/5 inch (15 mm) long. Full-grown larvae are 3/5-1 inch (15-25 mm) long, dirty white, pinkish, orange, light green, or light brown and are marked with rows of dark spots (figs. 533-534).

Several species that bore beneath of bark of pines and some other conifers are encountered in the Rocky Mountain Region. Most notable is the pinyon pitch mass borer, which causes large, oozing, soft pitch masses (fig. 535) on pinyon pine, and less noticeably, on ponderosa pine. This insect is also called the ponderosa twig moth and has caused notable damage on young pines in Nebraska. Other related species may produce less pitch or cause a clear sticky sap flow from branches or on the trunk of the tree. Zimmerman pine moth is an eastern species that has become established along Colorado's Front Range. It causes popcorn-like masses of pitch on the trunk and branches (fig. 536). Austrian pines are most

**Figure 535.** Pitch mass caused by *Dioryctria ponderosae* on pinyon pine. *Photo: Brian Howell, USDA Forest Service.*

**Figure 536.** Zimmerman pine moth damage. *Photo: Whitney Cranshaw, Colorado State University, Bugwood.org.*

frequently attacked, often at the junction of the trunk and a branch.

**Hosts**—Pinyon pine, ponderosa pine, lodgepole pine, Austrian pine, Scots pine, and, occasionally, Douglas-fir and true firs

**Life Cycle**—Pitch moth life cycles vary with species and can require 1 or 2 years for one generation. Pinyon pitch mass borer requires 2 years to complete its life cycle, and eggs are laid over an extended period from June through August. Eggs are laid in bark crevices or near bark wounds, including in previous attack sites and pruning cuts. Newly hatched larvae tunnel under the bark, forming irregular galleries or elongated gouges in the sapwood. Larvae pupate in silk-lined chambers in their tunnels or in the pitch mass in spring or early summer. The Zimmerman pine moth has one generation per year and does not bore under the bark until spring. They feed under bark, causing large amounts of pitch to be produced. Eggs are laid in August, and larvae hatch in late summer and seek overwintering sites under bark scales.

**Damage**—Larger branches, limbs, and trunks of young trees are attacked. Repeated attacks can seriously weaken trees and kill branches. The most severe damage is to trees less than 20 ft (6 m), especially in urban areas. The insects are rarely a problem on larger trees or in the forest environment.

**Management**—In forests and woodlands, no control strategies are warranted. Properly timed insecticide applications have successfully reduced damage on landscape trees. Individual larvae can be removed from the pitch mass or from under the bark with a knife or similar tool. Avoid pruning or mechanical injury to the bark during the summer months when adult moths are seeking egg-laying sites.

**References: 35, 50, 62, 63**

# Pitch Nodule Moths

## Tip dieback with pitch nodules on pines

**Name and Description**—Pinyon Pitch Nodule Moth—*Retinia* (= *Petrova*) *arizonensis* (Heinrich) [Lepidoptera: Tortricidae]
    Ponderosa pitch nodule moth—*R. metallica* (Busck)

Pitch nodule moths are seldom noticed, but the damage they cause is distinctive. Moths are rusty brown with mottled forewings and have a $3/4$-inch (19 mm) wingspan (fig. 537). Attacks are characterized by fading branch tips and nodules of pitch formed at the insects' feeding sites. The pitch nodules are hollow balls of pitch that are $1/2$-1 inch (10-25 mm) long, round, smooth, and that are often light purple or red. On pinyon pines, they are most often found at the crotch of two or more twigs. The fading twigs eventually lose their needles and fall off.

**Figure 537.** Wings of pinyon pitch nodule moth. *Photo: Gilligan, www.tortricidae.com.*

**Hosts**—Pinyon pine, ponderosa pine

**Life Cycle**—Pitch nodule moths have one generation per year. The small, rusty brown moths emerge through holes in the pitch nodule in July or early August. Eggs are laid on needle sheaths of the current year's growth. Newly hatched larvae feed on young needles before boring into the bark at nodes or whorls of twigs or branches. Pitch nodules form at feeding sites, and larvae feed and spend the winter protected within the distinctive nodule. Full-grown larvae are about $1/2$ inch (13 mm) long and reddish yellow with a black head and a dark area behind the head. Pupation occurs inside the pitch nodule in June. Pupae move just below the surface of the pitch before they emerge as adults.

**Damage**—Leaders are occasionally damaged, and trees with multiple terminals may result. Insect populations and damage can be abundant in localized areas. On ponderosa pine, attacks occur on the terminal growth (fig. 538).

**Management**—In the forest, top-killing of pinyon is rarely important, and damage to ponderosa pine is minor. On valuable ornamental trees, control the insect by pruning and destroying the infested tips as they fade in May or early June before the adult moths emerge. On branches and stems, destroy larvae by crushing them within the pitch nodule. Pinyons being dug from infested areas for landscape planting should be inspected, and pitch nodules should be crushed or removed.

**References: 22, 35, 50**

**Figure 538.** Pitch nodule moth in damaged ponderosa pine shoot. *Photo: Whitney Cranshaw, Colorado State University, Bugwood.org.*

# White Pine Weevil and Lodgepole Pine Terminal Weevil

## Drooping, dying leaders on spruce and pines

**Name and Description**—*Pissodes* spp. [Coleoptera: Curculionidae]

White pine weevil—*P. strobi* (Peck)

Lodgepole terminal weevil—*P. terminalis* Hopping

White pine weevil (*P. strobi*) and lodgepole terminal weevil (*P. terminalis*) infest terminals on spruce and lodgepole pine. Adults are rough-surfaced, brown to black beetles that usually have spots and patches of white, yellow, or brown scales and are about 1/4 inch (6 mm) long (fig. 539). The head is prolonged into a slender beak or snout used for feeding. Larvae are small, white, legless grubs and resemble bark beetle larvae (fig. 540).

**Figure 539.** White pine weevil adult, *Pissodes strobi. Photo: E. Bradford Walker, Vermont Department of Forests, Parks and Recreation, Bugwood.org.*

**Hosts**—White pine weevil: Engelmann and blue spruce and, occasionally, lodgepole pine

Lodgepole terminal weevil: lodgepole pine

**Life Cycle**—There is one generation per year. The adults emerge in April and May and make feeding punctures on terminal growth of the preceding year. Eggs are laid in May or June, and the developing larvae feed first in the cambium region and later mine and pupate in the pith. Larvae girdle and kill the terminals. White pine weevil larvae kill the terminals once the current year's growth has begun to elongate, resulting in the death of 2 years of growth, compared to lodgepole terminal weevils that only kill the current year's growth. Starting in August, adults emerge and feed on needles, buds, and twigs. Adults overwinter in forest litter; larvae and pupae overwinter in infested terminals.

**Damage**—Larvae attack and kill or seriously injure current and previous year's terminal shoots of young trees. This causes tops to die and leads to reduced height growth, forked tops, and crooked stems. Open-grown, even-aged stands of young spruce and lodgepole pine are highly susceptible. The first sign of attack occurs in the spring when copious resin flow from small punctures near

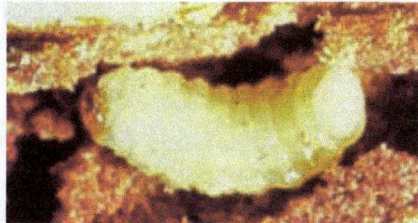

**Figure 540.** Lodgepole terminal weevil larva, *Pissodes terminalis. Photo: Intermountain Region, Ogden Archive, USDA Forest Service, Bugwood.org.*

the top of the leader can be seen. The leader begins to droop in midsummer and is brittle and brown by fall (fig. 541). Large holes made by emerging beetles may be visible on the dead leader in late summer and fall. One of the diagnostic characteristics of terminal weevils is the construction of chip cocoons. Chip cocoons are the oval, pupal cells that have been surrounded with shredded wood fiber and are constructed partly in the bark but mostly in the sapwood (fig. 542). These chip cocoons remain in the wood long after the beetles emerge.

**Management**—Severe problems with this insect are rare in forested areas in the Rocky Mountain Region, and direct control is not often warranted. However, damage is sometimes significant in landscape situations, and it is possible to apply insecticides during the spring when adults are active and laying eggs. Only the upper areas of the spruce or lodgepole pine tree need to be treated. Focus should be placed on trees growing in open-grown, even-aged stands.

**References: 35, 50**

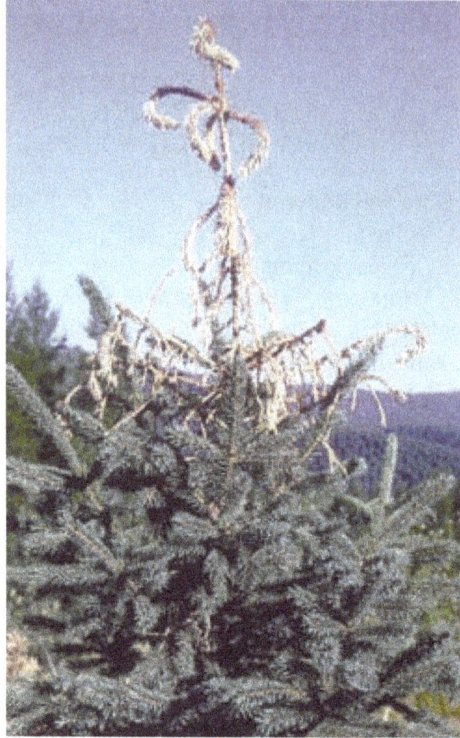

**Figure 541.** White pine weevil damage. *Photo: Scott Tunnock, USDA Forest Service, Bugwood.org.*

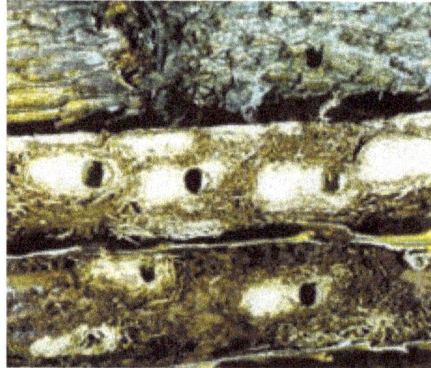

**Figure 542.** Pupal cells or "chip cocoons." *Photo: Ladd Livingston, Idaho Department of Lands, Bugwood.org.*

# Glossary

## A

**abdomen**—Posterior or hindmost portion of an insect's three main body divisions.

**abiotic**—Non-living factors such as temperature, moisture, and wind.

**adult**—Full-grown, sexually mature insect; usually has wings, in contrast to larvae, which lack wings.

**advanced decay**—Late stages of the decay process in which a fungus has produced its characteristic type of wood decay.

**aecium,** plural **aecia**—One of the many types of fruiting bodies formed during the life cycle of rust fungi; a cup-like structure producing aeciospores.

**alternate host**—One of two dissimilar plants infected by a different spore type of a rust or infested by a different life stage of an insect.

**anamorph**—Imperfect state of a fungus; produces asexual spores.

**annual**—An event that occurs once a year, or something that lasts 1 year or season; also, completing the life cycle in 1 year.

**antenna,** plural **antennae**—A pair of appendages used as sensory organs that are located on the head, usually above the eyes in insects.

**apothecium,** plural **apothecia**—Cup-shaped fruiting body of ascomycete fungi.

**arthropod**—An invertebrate creature characterized by an exoskeleton, segmented body, and paired, jointed limbs; includes insects, arachnids, and crustaceans.

**Ascomycetes**—A class of fungi that produces spores by sexual reproduction in a sac-like structure called an ascus. Many leaf and canker diseases are caused by Ascomycetes.

**ascospore**—Spore produced by sexual reproduction in Ascomycetes.

**asexual spores**—Spores produced vegetatively in fungi without going through genetic recombination.

## B

**Basidiomycetes**—A class of fungi that produces spores by sexual reproduction in or on the outside of a structure called a basidium. Rusts and most decay fungi are Basidiomycetes.

**biotic**—Relating to living organisms.

**blight**—A disease or injury that results in rapid discoloration, withering, cessation of growth, and death of parts without rotting.

**blue stain**—Coloration of wood infected by fungi with blue, brown, or black hyphae; a group of fungi that cause blue stain.

**boring dust**—Tiny particles of bark or wood produced by insects as they tunnel in woody plant structures.

**brood**—All of the bark beetle offspring produced by a single set of parents that hatch and mature at about the same time.

**broom**—An abnormal proliferation of branches or twigs on a single branch.

**brown rot**—A light to dark brown decay of wood that is friable and cracks into more or less cubical segments in the advanced stage; caused by fungi that attack mainly the cellulose and associated carbohydrates. Residue left is predominantly lignin.

**bug**—An insect in the order Hemiptera, the "true bugs."

**burl**—Abnormal proliferation of plant tissue, typically a globose growth on a tree stem or branch.

# C

**callow adult**—Young adult bark beetle that is pale brown in color and has not emerged from under the bark.

**callus**—Woody healing tissue that attempts to grow over wounds or otherwise damaged tissue.

**cambium**—Layer of actively dividing cells between the xylem (sapwood) and the phloem (inner bark) of trees that forms additional conducting tissue, therefore increasing the girth of a stem, branch, or trunk.

**canker**—An oval or circular killed area on a stem or branch; usually with a shrunken surface.

**chlorotic/chlorosis**—Yellow appearance of normally green foliage caused by loss or lack of chlorophyll.

**cocoon**—A covering spun or constructed by a larva as a protection to the pupa.

**conidia**—Asexual spores produced on microscopic structures called conidiophores.

**conk**—The large, often bracket-like fruiting bodies of wood-consuming fungi (Basidiomycetes).

**complement**—The set of foliage produced by evergreen trees during one growing season.

**cornicle**—Paired, secretory, tubular structures protruding from the top of an aphid abdomen.

**cortex**—Outer layer of bark.

**crawler**—The active first instar of a scale insect.

**cryptic**—Hidden or concealed; not readily apparent.

**cubical decay**—Decayed wood breaking up into distinct cubes.

# D

**DBH**—Diameter at breast height. Tree diameter measured at 4.5 ft from the ground.

**decay**—The decomposition of wood or other substrate by microorganisms; also called rot.

**decline**—Gradual reduction in health and vigor as a tree is in the process of dying slowly.

**declivity**—Slope of elytra at posterior end of the beetle.

**defoliation**—Premature removal of foliage.

**defoliator**—An insect or pathogen that feeds on foliage, physically removes needles or portions of needles, or causes tissue necrosis and premature needle drop.

**dessication**—Rapid drying of plant parts.

**diapause**—Dormant phase.

**dieback**—Dead apical parts, usually twigs or limbs; also the process of dying from the outside in.

**disease**—A prolonged disturbance of the normal form or function of a tree or its parts

**distal**—Near or toward the free end of an appendage; far from the point of attachment or origin.

**dorsal**—Refers to the upper surface of an organism, appendage, or part.

# E

**egg gallery**—Tunnel constructed under the bark of host trees by adult beetles for the purpose of laying eggs. Egg galleries maintain a fairly constant width with increasing length but are sometimes associated with wider constructions, such as entrance points and nuptial chambers.

**egg niche**—A small recess constructed in woody plant tissue by a female insect for the purpose of egg deposition.

**elytra**—Leathery front wings of beetles that cover the membranous hind wings when the insect is not in flight; not used for flying.

**elytral declivity**—The posterior, down-sloping portion of the elytra on bark beetles.

**endemic**—Characteristic of populations at normal background levels.

**epidemic**—Large-scale, temporary increase in population; characteristic of a population at an unusually high level.

**erineum**—An abnormal growth of hairs induced on the epidermis of a leaf by certain mites.

**etched**—Lightly imprinted on a surface.

**exotic**—An introduced, or non-native, insect or pathogen.

**exudate**—Material that has oozed to the outside surface through small openings.

# F

**fascicle**—An individual needle bundle or cluster on coniferous species.

**flagging**—Conspicuous recently dead or dying shoots or branches on a live tree that have discolored foliage still attached.

**forewings**—The first pair of wings from the head.

**frass**—Solid excrement of insects, particularly of larvae; particles are usually rounded pellets.

**fruiting body**—Any of a number of kinds of fungal reproductive structures that produce spores.

**fungus, plural fungi**—A member of the group of saprophytic and parasitic organisms that lack chlorophyll, have cell walls made of chitin, and reproduce by spores; includes molds, rusts, mildews, smuts, and mushrooms. Fungi absorb nutrients from the organic matter in which they live. Not classified as plants; instead, fungi are placed in the kingdom Fungi.

# G

**gall**—Abnormal proliferation of plant tissue stimulated by insects, pathogens, or abiotic influences.

**gallery**—A tunnel constructed by an insect in which it lives, feeds, or deposits eggs.

**generation**—Development of insects from egg to adult; a brood.

**genus**—A taxonomic grouping of species with some set of similar characteristics.

**girdle**—To destroy or remove the cambium around the complete circumference, typically resulting in death of all tissues beyond the point of girdling.

**gouting**—Abnormally thickened or swollen branch tips or branches, caused by insects, pathogens, or abiotic influences.

**gregarious**—Living or feeding in groups.

**grub**—Thick-bodied, usually sluggish, larva of a beetle.

# H

**heartrot**—A decay restricted to the heartwood.

**heartwood**—Central cylinder in tree trunks consisting of dead xylem tissue; functions as mechanical support.

**hibernaculum, plural hibernacula**—A tent-like structure or sheath constructed from a leaf or other material, in which a larva hides or hibernates.

**hindwings**—The second pair of wings from the head.

**hip canker**—A canker on one side of a pine stem with flared edges, caused by western gall rust.

**hindwings**—On insects having two pairs of wings, the pair of wings farthest from the head.

**honeydew**—Clear, sugary, liquid excretion of aphids or scales. Becomes sticky and shiny when dried.

**host**—Plant infected by a pathogen or infested by insects.

**hypha, plural hyphae**—Microscopic filaments of fungal cells.

**hysterothecium, plural hysterothecia**—Elongated fungal fruiting bodies that open with a slit, found in Ascomycete fungi. Common to many needle cast fungi.

# I

**incipient decay**—The earliest stages of wood decay in which the wood may show discoloration but is otherwise not visibly altered.

**infection**—The process or result of a pathogen establishing itself on or within a host.

**infection court**—The point of invasion of a disease organism into its host.

**instar**—Life stage of a larva between successive molts.

# L

**laminated decay**—Selective, more extensive decay in the spring wood than in summer wood, causing the wood to separate into sheets or laminae along annual rings.

**larva, plural larvae**—An immature form of an insect that undergoes complete metamorphosis, such as a caterpillar, grub, or maggot.

**larval gallery**—Tunnel constructed by developing beetle larvae as they feed under the bark. Larval galleries tend to grow wider with length and are often constructed at a right angle from the parent gallery.

**latent infection**—Infection by a disease organism that is not yet expressed by symptoms or signs.

**lateral**—to the side or on the sides, as in lateral shoot, lateral bud, lateral spread, lateral stripes.

**lesion**—A localized area of dead or dying tissue caused by a pathogen or insect.

**life cycle**—All stages in the development of an organism from the production of one stage through the production of the same stage again.

**longitudinal**—Placed or running lengthwise with the axis of the tree bole or limb.

# M

**maturation feeding**—A period of feeding by recently pupated, sexually immature adults of some insect species upon specific plants or plant parts that is essential for completion of reproductive development.

**metamorphosis**—Development of an insect as it goes through life stages from egg to adult.

**midge**—Adult of a group of small "flies" in the order Diptera.

**molt**—Process of shedding the exoskeleton between insect instars.

**monogamous**—Mating with only one individual.

**mushroom**—A fleshy, fruiting body of a fungus, often with a gilled pore surface.

**mycelium**—The collective mass of vegetative elements or hyphae, of a fungus.

**mycelial fan**—Dense mass of mycelium that takes the form of a thick mat that is fan-shaped.

# N

**necrotic**—Dead; usually refers to cells or tissue killed by injury or disease.

**necrosis**—Localized death of living tissue; usually results in darkening of the tissue.

**needle sheath**—The thin, somewhat papery, tubular covering that surrounds the base of a pine needle fascicle.

**node**—Location on a stem where branches or leaves normally originate.

**notch**—Abrupt, rounded protuberance of a bark beetle egg gallery wall; has no egg niches or larval galleries. This usually marks the point of entry to the egg gallery from the outer bark surface and sometimes is used as a nuptial chamber.

**nuptial chamber**—An open, cave-like area constructed in the inner bark beneath the entrance hole by the male of some bark beetle species where mating takes place and from which the egg galleries originate. When viewed with the bark removed, nuptial chambers appear as enlarged areas at junctions or at beginnings of egg galleries.

**nymph**—An immature form of an insect that undergoes incomplete metamorphosis; nymphs resemble the adults except for size and wing development.

# O

**obligate parasite**—An organism capable of living only as a parasite on a living host.

**overwinter**—To survive through winter; term commonly used when describing the life stage in which an insect passes the winter months; e.g., second instar western spruce budworm larvae overwinter.

**ovipositor**—A tubular or valved structure used by an insect to lay eggs (oviposition); usually concealed but sometimes extended far beyond the end of the body.

# P

**parasite**—An organism that lives at the expense of another living organism; usually by invading and subsequently causing disease.

**parenchyma cells**—The primary tissue of higher plants, composed of thin-walled cells and forms the greater part of leaves, roots, the pulp of fruit, and the pith of stems.

**parent gallery**—Tunnel constructed by invading parent beetles where reproduction takes place and eggs are laid.

**pathogen**—An organism that causes disease in another organism.

**pathogenic**—Being capable of causing disease.

**perennial**—An organism that continues to live from year-to year; persisting for 3 or more years.

**perithecium, plural perithecia**—Small fungal fruiting structure that is round or flask-shaped and contains asci.

**pheromone**—Substance secreted to the outside of an insect's body that serves as a chemical signal between individuals of the same species. Pheromones are usually airborne and act as attractants, repellants, or alarm signals.

**phloem**—Vascular tissue of higher plants that actively conducts the food produced in leaves and green stems through photosynthesis. Phloem is the inner bark of a tree.

**phoresy**—A form of symbiosis in which one organism is carried on the body of another larger-bodied organism; the former does not feed off of the latter.

**photosynthesis**—Process in green plants that uses sunlight and carbon dioxide to produce food.

**pitch**—A resinous exudate of various conifers.

**pitch tube**—A globular mass of resin, boring dust, and frass that forms on the bark of pine trees at bark beetle entrance holes.

**pitch streamers**—Long, thin flows of pitch on the tree bole.

**pith**—The soft, sponge-like center of stems and branches of woody plants.

**pocket rot**—A characteristic pattern of rot that occurs in distinct, scattered pockets rather than in columns.

**pore**—The open end of a tube in which certain spores of higher fungi are produced.

**pore layer or pore surface**—The surface of a fruiting body on which the pores are found.

**posterior**—The hind or rear part of the body.

**progeny**—Offspring or brood.

**proleg**—The fleshy, unjointed appendages found on the abdomens of caterpillars and some sawfly larvae; also called false legs.

**pronotum**—Dorsal (top) plate of an insect prothorax.

**prothorax**—The front-most portion of the three thoracic segments.

**pseudostroma, plural pseudostromata**—Small mass of vegetative fungal tissue mixed with host tissue, in or on which fruiting bodies are produced.

**pseudothecium, plural pseudothecia**—Tiny, round to flask-shaped fruiting bodies that contain asci and ascospores and are produced by certain Ascomycetes.

**punk knot**—Soft, decayed branch stubs that usually indicate the presence of decay in a tree.

**pupa, plural pupae**—The inactive transitional stage between the larval and adult life stages that occurs in insects with complete metamorphosis. Characterized by a covering formed by the larva as it changes into a pupa.

**pupal cell**—Cavity at the end of the larval gallery in which pupation occurs.

**pupate**—To change from a larva into a pupa; to pass through the pupal stage.

**pustule**—Blisters of an infecting fungus that mature into fruiting structures.

**pycnidium, plural pycnidia**—Very small fruiting structure, flask-shaped and hollow, that produces asexual spores.

# R

**radial**—In an outward direction from the center of the tree.

**resinosis**—Reaction of a tree to invasion by pathogens, insects, or abiotic injury that results in flow of resin on the outer bark or accumulation of resin within or under bark.

**resistant**—Characteristic of a plant that is able to prevent infection or limit disease.

**rhizomorph**—A specialized thread or cord-like structure made up of parallel hyphae with a protective covering.

**root collar or root crown**—The transition zone between the stem and root of a tree; located at the ground line.

# S

**saprobe or saprophyte**—An organism that uses dead organic matter as food; a saprotroph.

**saprot**—Decay of dead wood; organism that decays only dead material; also saprophyte.

**sapwood**—The outer conducting layers of wood that contain living cells and food reserves.

**sclerotium, plural sclerotia**—Small, dark, hardened mass of fungal tissue, capable of enduring adverse conditions and later growing again.

**score**—Mark deeply, forming a groove.

**secondary**—Organisms that are limited to invading hosts predisposed by stress or attack by some other more aggressive organism.

**seta, plural setae**—Hollow, often slender, hair-like projections.

**sexual fruiting bodies**—Fruiting structures that produce spores formed through the process of meiosis, providing for genetic recombination.

**sign**—The manifestation of a causal organism, such as beetle galleries or fungal conks. (Contrast with symptom.)

**slash**—Woody debris such as logs, bark, and branches left after logging activities.

**species, plural species**—A taxonomic grouping or organisms alike in appearance and structure that can mate and produce fertile offspring.

**spore**—A microscopic reproductive cell capable of dispersing.

**sporulate**—To release spores.

**stage**—A distinguishable period of development, as in a stage of decay, a spore type or stage of a rust, or an insect life stage.

**stoma, plural stomata**—Opening in the epidermis (outer, protective layer of cells) of a leaf or stem that allows gas exchange.

**stroma, plural stromata**—Small mass of vegetative fungal tissue, in or on which fruiting bodies are produced.

**stress cones**—A cone crop produced as the result of tree stress; often associated with root diseases. Cones may be produced relatively early in the tree's life and are often very numerous and small in size with high percentages of nonviable seeds.

**stringy rot**—Decay that results in the heartwood being reduced to fibrous material.

**strip attack**—An attack by bark beetles that does not completely girdle the tree but is concentrated along one side or within a relatively confined area on the bole; results in death of only a portion of the tree bole.

**sunscald**—Damage to the cambium of a thin-barked stem caused by overexposure to the heat from sunlight.

**suppressed**—Trees with crowns overtopped by larger trees that receive no direct sunlight from above or from the sides.

**symbiotic**—Characteristic of a beneficial relationship between two organisms of different species living in the same space; each participating organism is a symbiont.

**symptom**—The expression of disease or injury by the host plant, such as chlorosis or bark streaming.

# T

**target canker**—Canker in which the pattern of annual expansion by the pathogen and callus production by the host results in concentric ridges.

**telium, plural telia**—One of the many types of fruiting bodies formed during the life cycle of rust fungi; produce teliospores.

**tendril**—Mass of spores in a gelatinous matrix that oozes from a fruiting body in a long, curling string.

**terminal**—The main or primary growing tip of a conifer.

**thorax**—In insects, the middle body division located behind the head and in front of the abdomen. Bears the wings and legs of adult insects.

**tolerant**—Characteristic of a host that is able to survive infection, often with minimal or delayed development of symptoms.

**topkill**—Death of the upper crown of a tree; usually caused by insects, pathogens, or weather events.

**transverse**—Lying or running at right angles to the axis of the tree bole or limb.

**tubercle**—A small, rounded prominence.

# U

**uredinium, plural uredinia**—One of the many types of fruiting bodies formed during the life cycle of rust fungi; produce urediniospores.

# V

**vascular plant**—Higher plant with specialized tissues for conducting water and nutrients.

**vascular tissue**—The xylem, cambium, and phloem tissues that make up the vascular system of higher plants.

**vector**—An organism that spreads a pathogen.

# W

**webbing**—A mat or loose weave of silk strands produced by larvae; typically forming a protective nest in which larvae feed.

**white rot**—Decay caused by fungi that attack both chief constituents of wood, cellulose, and lignin and leave a whitish or light-colored residue.

**windthrown**—Uprooted and blown down by wind; as in windthrown trees.

**wingspan**—Width of wings when extended as if for flight.

**witches' broom**—An abnormal proliferation of branches or twigs on a single branch.

# X

**xylem**—The woody tissue of the stem, branches, and roots that transports water and minerals.

# Z

**zone line**—Narrow, black or brown layers of tough fungal tissue in decayed wood that presumably resist the advance of other fungi.

USDA Forest Service RMRS-GTR-241. 2010.

# References

1. Allen, E.A.; Morrison, D.J.; Wallis, G.W. 1996. Common tree diseases of British Columbia. Natural Resources Canada, Canadian Forest Service. 178 p.

2. Amman, A.G.; Amman, S.L.; Amman, G.D. 1974. Development of *Pityophthorus convertus*. Environmental Entomology 3(3):562-563.

3. Amman, G.D.; McGregor, M.D.; Dolph, Jr., R.E. 1985. Mountain pine beetle. Forest Insect and Disease Leaflet 2. Washington, DC: U.S. Department of Agriculture, Forest Service. 11 p.

4. Anderson, G.W.; Hinds, T.E.; Knutson, D.M. 1977. Decay and discoloration of aspen. Forest Insect and Disease Leaflet 149. Washington, DC: U.S. Department of Agriculture, Forest Service. 4 p.

5. Anderson, R.L.; Skilling, D.D. 1955. Oak wilt damage: a survey in central Wisconsin. Station Paper 33. St. Paul, MN: U.S. Department of Agriculture, Forest Service, North Central Forest Experiment Station. 11 p.

6. Arthur, J.C. 1934. Manual of the Rusts in the United States and Canada. Lafayette, IN: Purdue Research Foundation. 438 p.

7. Baker, C.J.; Harrington, T.C. 2001. *Ceratocystis fimbriata*. Crop Protection Compendium. CABI Publishing. Updated information at: http://www.public.iastate.edu/~tcharrin/CABIinfo.html.

8. Baranyay, J.A.; Ziller, W.G. 1972. Broom rusts of conifers in British Columbia. Forest Pest Leaflet 48. Victoria, BC: Natural Resources Canada, Canadian Forest Service.

9. Barnes, I.; Crouse, P.W.; Wingfield, B.D.; Wingfield, M.J. 2004. Multigene phylogenies reveal that red band needle blight is caused by two distinct species of *Dothistroma*, *D. septosporum* and *D. pini*. Studies in Mycology 50:551-565.

10. Barrett, D.K.; Uscuplic, M. 1971. The field distribution of interacting strains of *Polyporus schweinitzii* and their origin. New Phytologist 70:581-598.

11. Batzer, H.O.; Morris, R.C. 1978. Forest tent caterpillar. Forest Insect and Disease Leaflet 9. Washington, DC: U.S. Department of Agriculture, Forest Service. 8 p.

12. Beatty, J.S.; Mathiasen, R.L. 2003. Dwarf mistletoes of ponderosa pine. Forest Insect and Disease Leaflet 40. Washington, DC: U.S. Department of Agriculture, Forest Service. 8 p. Online: http://www.fs.fed.us/r6/nr/fid/fidls/fidl40.htm.

13. Beckwith, R.C. 1973. The large aspen tortrix. Forest Insect and Disease Leaflet 139. Washington, DC: U.S. Department of Agriculture, Forest Service. 5 p.

14. Bedker, P.J.; O'Brien, J.G.; Mielke, M.M. 1995. How to prune trees. NA-FR-01-95. Radnor, PA: U.S. Department of Agriculture, Forest Service, Northeastern Area, State and Private Forestry. 12 p. Online: http://www.na.fs.fed.us/Spfo/pubs/howtos/ht_prune/htprune.pdf.

15. Berry, F.H. 1985. Anthracnose diseases of eastern hardwoods. Forest Insect and Disease Leaflet 133. Washington, DC: U.S. Department of Agriculture, Forest Service. 7 p.

16. Blenis, P.V. 2007. Impact of simulated aspen shoot blight on trembling aspen. Canadian Journal of Forest Research 37:719-725.

17. Bolander, P., ed. 1999. Dust palliative selection and application guide. Project Report. 9977-1207-SDTDC. San Dimas, CA: U.S. Department of Agriculture, Forest Service, San Dimas Technology and Development Center. 20 p.

18. Boutz, G.A.; Stack, R.W. 1986. Herbicides (air pollution). In: Riffle, J.W.; Peterson, G.W., tech. coords. Diseases of trees in the Great Plains. Gen. Tech. Rep. GTR-RM-129. Fort Collins, CO: U.S. Department of Agriculture, Forest Service, Rocky Mountain Forest and Range Experiment Station: 33-35.

19. Boyce, J.S. 1961. Forest Pathology. New York, NY: McGraw Hill Book Co. 572 p.

20. Brookhouser, L.W.; Peterson, G.W. 1971. Infection of Austrian, Scots, and ponderosa pines by *Diplodia pinea*. Phytopathology 61:409-414.

21. Burns, K.S.; Schoettle, A.W.; Jacobi, W.R.; Mahalovich, M.F. 2008. Options for the management of white pine blister rust in the Rocky Mountain Region. Gen. Tech. Rep. RMRS-GTR-206. Fort Collins, CO: U.S. Department of Agriculture, Forest Service, Rocky Mountain Research Station. 26 p. Online: http://www.fs.fed.us/rm/pubs/rmrs_gtr206.pdf .

22. Cain, R.; Parker, D. 1998. Conifer pests in New Mexico, Rev. Albuquerque, NM: U.S. Department of Agriculture, Forest Service, Southwestern Region. 50 p.

23. Callan, B.E. 1998. Diseases of *Populus* in British Columbia: a diagnostic manual. Victoria, BC: Natural Resources Canada, Canadian Forestry Service, Pacific Forestry Centre. 157 p.

24. Carolin, V.M.; Knopf, J.A.E. 1968. The Pandora moth. Forest Insect and Disease Leaflet 114. Washington, DC: U.S. Department of Agriculture, Forest Service. 7 p.

25. Childs, T.W.; Kimmey, J.W. 1938. Studies on probable damage by blister rust in some representative stands of young western white pine. Journal of Agricultural Research 57:557-568.

26. Childs, T.W.; Shea, K.R.; Stewart, J.L. 1971. Elytroderma disease of ponderosa pine. Forest Pest Leaflet 42. U.S. Department of Agriculture, Forest Service. 6 p.

27. Ciesla, W.M.; Ragenovich, I.R. 2008. Western tent caterpillar. Forest Insect and Disease Leaflet 119. Washington, DC: U.S. Department of Agriculture, Forest Service. 8 p.

28. Cole, W.E. 1971. Pine butterfly. Forest Insect and Disease Leaflet 66. Washington, DC: U.S. Department of Agriculture, Forest Service. 4 p.

29. Costello, S.L. 2005. Wood borers in fire-damaged ponderosa pine forests of the Black Hills, South Dakota. Fort Collins, CO: Colorado State University. 71 p. Thesis.

30. Coutts, M.P.; Rishbeth, J. 1977. The formation of wetwood in grand fir. European Journal of Forest Pathology 7:13-22.

31. Crane, P.E.; Blenis, P.V.; Hiratsuka, Y. 1994. Black stem galls on aspen and their relationship to decay by *Phellinus tremulae*. Canadian Journal of Forest Research 24(11):2240-2243.

32. Cranshaw, W.S. 2004. Garden insects of North America: the ultimate guide to backyard bugs. Princeton University Press. 656 p.

33. Cranshaw, W.S. 2005. Cooley spruce galls. Pest Fact Sheet No. 5.534. Fort Collins, CO: Colorado State University, Cooperative Extension. 3 p.

34. Cranshaw, W.S. 2008. Oytershell scale. Extension Fact Sheet No. 5.513. Fort Collins, CO: Colorado State University, Cooperative Extension. 2 p.

35. Cranshaw, W.S.; Leatherman, D.A.; Jacobi, W.R.; Mannix L. 2000. Insects and diseases of woody plants of the central Rockies. Bulletin 506A. Fort Collins, CO: Colorado State University, Cooperative Extension. 284 p.

36. Davidson, R.W.; Hinds, T.E.; Hawksworth, F.G. 1959. Decay of aspen in Colorado. Station Paper 45. Fort Collins, CO: U.S. Department of Agriculture, Forest Service, Rocky Mountain Forest and Range Experiment Station. 14 p.

37. DeMars, C.J.; Roettgering, B.H. 1982. The western pine beetle. Forest Insect and Disease Leaflet Number 1. Washington, DC: U.S. Department of Agriculture, Forest Service. 6 p.

38. Derr, J.F.; Appleton, B.L. 1988. Herbicide injury to trees and shrubs. Virginia Beach, VA: Blue Crab Press. 72 p.

39. Dix, M.E.; Pasek, J.E.; Harrell, M.O.; Baxendale, F.P. 1986. Common insect pests of trees in the Great Plains. EC 86-1548. Nebraska Cooperative Extension Service. 44 p.

40. Dobson, M.C. 1991. De-icing salt damage to trees and shrubs. Forestry Commission Bulletin 101. London. 64 p.

41. Dreistadt, S. 1994. Pests of landscape trees and shrubs. Publication 3359. Berkeley, CA: University of California Press. 327 p.

42. Eager, T.J. 1999. Factors affecting the health of pinyon pine trees (*Pinus edulis*) in the pinyon-juniper woodlands of western Colorado. In: Monsen, S.B.; Stevens, R., eds, Proceedings: ecology and management of pinyon-juniper communities within the interior West; 1997 September 15-18; Provo, UT. Proc. RMRS-P-9. Ogden, UT: U.S. Department of Agriculture, Forest Service, Rocky Mountain Research Station: 397-399.

43. Etheridge, D.E.; Craig, H.M. 1976. Factors influencing infection and initiation of decay by the Indian paint fungus (*Echinodontium tinctorium*) in western hemlock. Canadian Journal of Forest Research 6:299-318.

44. Farr, D.F.; Rossman, A.Y. 2009. Fungal Databases. U.S. Department of Agriculture, Agricultural Research Service, Systematic Mycology and

Microbiology Laboratory. Online: http://nt.ars-grin.gov/fungaldatabases/. [December 18, 2009].

45. Fellin, D.G.; Dewey, J.E. 1986. Western spruce budworm. Forest Insect and Disease Leaflet 53. Washington, DC: U.S. Department of Agriculture, Forest Service. 10 p.

46. Fettig, C.J.; Allen, K.K.; Borys, R.R.; Christopherson, J.; Dabney, C.P.; Eager, T.J.; Gibson, K.E.; Hebertson, E.G.; Long, D.F.; Munson, A.S. 2006. Effectiveness of bifenthrin (Onyx) and carbaryl (Sevin SL) for protecting individual, high-value conifers from bark beetle attack (Coleoptera: Curculionidae: Scolytinae) in the Western United States. Journal of Economic Entomology 99(5):1691-8.

47. French, D.W.; Stienstra, W.C. 1980. Oak wilt. Folder 310. Minneapolis, MN: University of Minnesota, Agricultural Extension Service. 11 p.

48. Funk, A. 1981. Parasitic microfungi of western trees. Victoria, BC: Natural Resources Canada, Canadian Forestry Service, Pacific Forestry Centre. 190 p.

49. Funk, A. 1985. Foliar fungi of western trees. Victoria, BC: Natural Resources Canada, Canadian Forestry Service, Pacific Forestry Centre. 159 p.

50. Furniss, R.L.; Carolin, V.M. 1977. Western forest insects. Misc. Publ. 1339. Washington, DC: U.S. Department of Agriculture, Forest Service. 654 p.

51. Furniss, R.L.; Malcolm, M.; Kegley, S.J. 2008. Biology of *Dendroctonus murrayanae* (Coleoptera: Curculionidae: Scolytinae) in Idaho and Montana and comparative taxonomic notes. Annals of the Entomological Society of America 101(6):1010-1016.

52. Garbutt, R. 1992. Western balsam bark beetle. Forest Pest Leaflet. Victoria, BC: Natural Resources Canada, Canadian Forestry Service, Pacific Forestry Centre. 4 p.

53. Geils, B.W.; Cibrián Tovar, J.; Moody, B., tech. coords. 2002. Mistletoes of North American conifers. Gen. Tech. Rep. RMRS-GTR-98. Ogden, UT: U.S. Department of Agriculture, Forest Service, Rocky Mountain Research Station. 123 p.

54. Gibbs, J.N. 1978. Intercontinental epidemiology of Dutch elm disease. Annual Review of Phytopathology. 16:287-307.

55. Gibson, K.; Kegley, S.; Oakes, B. 1997. Western balsam bark beetle activity and flight periodicity in the Northern Region. Report 97-3. Missoula, MT: U.S. Department of Agriculture, Forest Service, Cooperative Forestry and Forest Health Protection. 5 p.

56. Gilbertson, R.L. 1974. Fungi that decay ponderosa pine. Tucson: University of Arizona Press. 197 p.

57. Gilbertson, R.L.; Ryvarden, L. 1986. North American polypores. Volumes 1 and 2. Oslo, Norway: Fungiflora. 885 p.

58. Haack, R.A.; Acciavatti, R.E. 1992. Twolined chestnut borer. Forest Insect and Disease Leaflet 168. Washington, DC: U.S. Department of Agriculture, Forest Service. 12 p.

59. Hadfield, J.S.; Mathiasen, R.L.; Hawksworth, F.G. 2000. Douglas-fir dwarf mistletoe. Forest Insect and Disease Leaflet 54. Washington, DC: U.S. Department of Agriculture, Forest Service. 9 p. Online: http://www. fs.fed.us/r6/nr/fid/fidls/douglasfirdm.pdf.

60. Hagle, S.K.; Gibson, K.E.; Tunnock, S. 2003. Field guide to disease and insect pests of northern and central Rocky Mountain conifers. Report R1-03-08. Missoula, MT: U.S. Department of Agriculture, Forest Service, Northern Region. 67 p.

61. Harestad, A.S.; Bunnell, F.L.; Sullivan, T.P.; Andrusiak, L. 1986. Key to injury of conifer trees by wildlife in British Columbia. WHR-23. Victoria, BC: Natural Resources Canada, Canadian Forestry Service, Pacific Forestry Centre. 38 p.

62. Harrell, M.L. 1993. Influence of pine host species on infestation and damage by Dioryctria borers in the central Great Plains. Environmental Entomology 22(4):781-783.

63. Harrell, M.L. 1996. Life histories and parasitoids of Dioryctria borers of pines in Nebraska. Journal of the Kansas Entomological Society 69(4):279-284.

64. Haugen, D.A.; Hoebeke, E.R. 2005. Sirex woodwasp—*Sirex noctilio* F. (Hymenoptera: Siricidae). Pest Alert. NA-PR-07-05. Newton Square, PA: U.S. Department of Agriculture, Forest Service, Northeastern Area. 2 p.

65. Hawksworth, F.G. 1961. Dwarf mistletoe of ponderosa pine in the Southwest. Tech. Bull. No. 1246. Washington, DC: U.S. Department of Agriculture, Forest Service. 112 p.

66. Hawksworth, F.G. 1977. The 6-class dwarf mistletoe rating system. Gen. Tech. Rep. RM-48. Fort Collins, CO: U.S. Department of Agriculture, Forest Service, Rocky Mountain Forest and Range Experiment Station. 7 p.

66a. Hawksworth, F.G.; Dooling, O.J. 1984. Lodgepole pine dwarf mistletoe. Forest Insect and Disease Leaflet 18. Washington, DC: U.S. Department of Agriculture, Forest Service. 11 p.

67. Hawksworth, F.G.; Geils, B.W. 1990. How long do mistletoe-infected ponderosa pines live? Western Journal of Applied Forestry 5(2):47-48.

68. Hawksworth, F.G.; Johnson, D.W. 1989. Biology and management of dwarf mistletoe in lodgepole pine in the Rocky Mountains. Gen. Tech. Rep. RM-169. Fort Collins, CO: U.S. Department of Agriculture, Forest Service, Rocky Mountain Forest and Range Experiment Station. 38 p.

69. Hawksworth, F.G.; Scharpf, R.F. 1981. *Phoradendron* on conifers. Forest Insect and Disease Leaflet 164. Washington, DC: U.S. Department of Agriculture, Forest Service. 7 p.

70. Hawksworth, F.G.; Wiens, D. 1996. Dwarf mistletoes: biology, pathology and systematics. Agricultural Handbook 709. Washington, DC: U.S. Department of Agriculture, Forest Service. 410 p.

71. Hedgcock, C.G. 1913. *Herpotrichia* and *Neopeckia* on conifers. Phytopathology 3:152-158.

72. Hepting, G.E. 1971. Diseases of forest and shade trees of the United States. Agricultural Handbook 386. Washington, DC: U.S. Department of Agriculture, Forest Service. 658 p.

73. Hessburg, P.F.; Goheen, D.J.; Bega, R.V. 1995. Black stain root disease of conifers. Forest Insect and Disease Leaflet 145 (revised). Washington, DC: U.S. Department of Agriculture, Forest Service. 9 p. Online: http://www.fs.fed.us/r6/nr/fid/fidls/fidl-145.pdf.

74. Himelick, E.B.; Neely, D. 1960. Juniper hosts of cedar-apple and cedar-hawthorn rust. Plant Disease Reporter 44:109-112.

75. Hinds, T.E. 1963. Extent of decay associated with *Fomes igniarius* sporophores in Colorado aspen. Res. Note RM-4. Fort Collins, CO: U.S. Department of Agriculture, Forest Service, Rocky Mountain Forest and Range Experiment Station. 4 p.

76. Hinds, T.E. 1964. Distribution of aspen cankers in Colorado. Plant Disease Reporter 48(8):610-614.

77. Hinds, T.E. 1976. Aspen mortality in Rocky Mountain campgrounds. Res. Pap. RM-164. Fort Collins, CO: U.S. Department of Agriculture, Forest Service, Rocky Mountain Forest and Range Experiment Station. 20 p.

78. Hinds, T.E. 1981. Cryptosphaeria canker and Libertella decay of aspen. Phytopathology 71(11):1137-1145.

79. Hinds, T.E. 1985. Diseases. In: DeByle, N.V.; Winokur, R.P., eds. Aspen: ecology and management in the western United States. Gen. Tech. Rep. RM-119. Fort Collins, CO: U.S. Department of Agriculture, Forest Service, Rocky Mountain Forest and Range Experiment Station. 283 p.

80. Hinds, T.E.; Hawksworth, F.G. 1966. Indicators and associated decay of Engelmann spruce in Colorado. Res. Pap. RM-25. Fort Collins, CO: U.S. Department of Agriculture, Forest Service, Rocky Mountain Forest and Range Experiment Station. 15 p.

81. Hinds, T.E.; Hawksworth, F.G.; Davidson, R.W. 1960. Decay of subalpine fir in Colorado. Station Paper 51. Fort Collins, CO: U.S. Department of Agriculture, Forest Service, Rocky Mountain Forest and Range Experiment Station. 13 p.

82. Hinds, T.E.; Wengert, E.M. 1977. Growth and decay losses in Colorado aspen. Res. Pap. RM-193. Fort Collins, CO: U.S. Department of Agriculture, Forest Service, Rocky Mountain Forest and Range Experiment Station. 10 p.

83. Hiratsuka, Y.; Zalasky, H. 1993. Frost and other climate-related damage of forest trees in the prairie provinces. Inf. Rep. NOR-X-286. Edmonton, Alberta: Natural Resources Canada, Canadian Forestry Service, Northern Forestry Centre. 26 p.

84. Holeski, L.M.; Vogelzang, A.; Stanosz, G.; Lindroth, R.L. 2009. Incidence of *Venturia* shoot blight in aspen (*Populus tremuloides* Michx.) varies

with tree chemistry and genotype. Biochemical Systematics and Ecology 37:139-145.

85. Holsten, E.H.; Their, R.W.; Munson, A.S.; Gibson, K.E. 1999. The spruce beetle. Forest Insect and Disease leaflet 127. Washington, DC: U.S. Department of Agriculture, Forest Service. 12 p.

86. Hornibook, E.M. 1950. Estimating defect in mature and over-mature stands of three Rocky Mountain conifers. Journal of Forestry 48:408-417.

87. Hunt, R.S. 1982. White pine blister rust control in British Columbia I. The possibilities of control by branch removal. The Forestry Chronicle 59:136-138.

88. Ives, W.G.H.; Wong, H.R. 1988. Tree and shrub insects of the Prairie Provinces. Inf. Report NOR-X-292. Edmonton, Alberta: Natural Resources Canada, Canadian Forestry Service, Northern Forestry Centre. 327 p.

89. Jacobi, W.R.; Shepperd, W.D. 1991. Fungi associated with sprout mortality in aspen clearcuts in Colorado and Arizona. Research Note RM-513. Fort Collins, CO: U.S. Department of Agriculture, Forest Service, Rocky Mountain Forest and Range Experiment Station. 5 p.

90. James, R.L.; Gillman, L.S. 1979. *Fomes annosus* on white fir in Colorado. Tech. Rep. R2-17. Golden, CO: U.S. Department of Agriculture, Forest Service, Rocky Mountain Region, Forest Insect and Disease Management. 9 p.

91. James, R.L.; Gillman, L.S. 1980. Incidence and distribution of conifer broom rusts in Colorado. Tech. Rep. R2-21. Golden, CO: U.S. Department of Agriculture, Forest Service, Rocky Mountain Region, Forest Insect and Disease Management. 30 p.

92. James, R.L.; Lister, C.K. 1978. Insect and disease conditions of pinyon pine and Utah juniper in Mesa Verde National Park, Colorado. Biological Evaluation R2-78-4. Lakewood, CO: U.S. Department of Agriculture, Forest Service, Rocky Mountain Region, Forest Insect and Disease Management.

93. Johnson, D.W. 1982. Forest pest management training manual. Lakewood, CO: U.S. Department of Agriculture, Forest Service, Rocky Mountain Region. 138 p.

94. Johnson, D.W. 1984. An assessment of root diseases in the Rocky Mountain Region. Tech. Rep. R2-29. Lakewood, CO: U.S. Department of Agriculture, Forest Service, Rocky Mountain Region. 20 p.

95. Johnson, D.W. 1986. Comandra blister rust. Forest Insect and Disease Leaflet 62. Washington, DC: U.S. Department of Agriculture, Forest Service. 8 p.

96. Johnson, D.W.; Beatty, J.S.; Hinds, T.E. 1995. Cankers on western quaking aspen. Forest Insect and Disease Leaflet 152. Washington, DC: U.S. Department of Agriculture, Forest Service. 8 p.

97. Johnson, J.A.; Harrington, T.C.; Engelbrecht, C.J.B. 2005. Phylogeny and taxonomy of the North American clade of the *Ceratocystis fimbriata* complex. Mycologia 97(5):1067-1092.

98. Johnson, W.T.; Lyons, H.H. 1988. Insects that feed on trees and shrubs, 2nd ed. Ithaca, NY: Cornell University Press. 556 p.

99. Jones, J.R.; DeByle, N.V.; Bowers, D.M. 1985. Insects and other invertebrates. In: DeByle, N.V.; Winokur, R.P., eds. Aspen: ecology and management in the western United States. Gen. Tech. Rep. RM-119. Fort Collins, CO: U.S. Department of Agriculture, Forest Service, Rocky Mountain Forest and Range Experiment Station: 107-114 p.

100. Juzwik, J.; Nishijima, W.T.; Hinds, T.E. 1978. Survey of aspen cankers in Colorado. Plant Disease Reporter 62(10):906-910.

101. Kearns, H.S.J.; Jacobi, W.R. 2005. Impacts of black stain root disease in recently formed mortality centers in the piñon-juniper woodlands of southwestern Colorado. Canadian Journal of Forest Research 35:461-471.

102. Kegley, S.J. 2006. Western balsam bark beetle, ecology and management. In: U.S. Department of Agriculture, Forest Service. Forest Insect and Disease Identification and Management Guide of Northern and Central Rocky Mountain Conifer and Hardwood. U.S. Department of Agriculture, Forest Service, Forest Health Protection, Region 1 and 4. Online: http://www.fs.fed.us/r1-r4/spf/fhp/mgt_guide/western_balsambark_beetle/index.html.

103. Kegley, S.J.; Livingston, R.L.; Gibson, K.E. 1997. Pine engraver, *Ips pini* (Say), in the western United States. Forest Insect and Disease Leaflet 122. Washington, DC: U.S. Department of Agriculture, Forest Service. 8 p.

104. Kern, F.D. 1973. A revised taxonomic account of *Gymnosporangium*. University Park, PA: Pennsylvania State University Press. 134 p.

105. Koch, P. 1996. Lodgepole pine in North America. Volume 1. Madison, WI: Forest Products Society. 343 p.

106. Kondo, E.; Foudin, A.; Linit, M.; Smith, M.; Bolla, R.; Winter, R.; Dropkin, V. 1982a. Pine wilt disease–nematological, entomological, and biochemical investigations. Publication SR 282. Columbia, MO: University of Missouri Agricultural Experiment Station. 56 p.

107. Kondo, E.S.; Hiratsuka, Y.; Denyer, W.B.G. 1982b. Proceedings of the Dutch elm disease workshop and symposium. Winnipeg: Manitoba Department Natural Resources. 517 p.

108. Krupinsky, J.M.; Johnson, D.W. 1986. Septoria leaf spots of cottonwood, caragana, and maple. In: Riffle, J.W.; Peterson, G.W., tech. coords. Diseases of trees in the Great Plains. Gen. Tech. Rep. RM-129. Fort Collins, CO: U.S. Department of Agriculture, Forest Service, Rocky Mountain Forest and Range Experiment Station. 149 p.

109. Kruse, J.; Ambourn, A.; Zogas, K. 2007. Aspen leaf miner. R10-PR-14. Juneau, AK: U.S. Department of Agriculture, Forest Service, Alaska Region, State and Private Forestry. 4 p. Online: http:\\www.fs.fed.us/r10/spf/fhp/leaflets/aspen_leaf_miner.pdf.

110. Landis, T.D.; Helburg, L.B. 1976. Black stain root disease of pinyon pine in Colorado. Plant Disease Reporter 60:713-717.

111. Lännenpää, A.; Aakala, T.; Kauhanen, H.; Kuuluvainen, T. 2008. Tree mortality agents in pristine Norway spruce forests in northern Fennoscandia. Silva Fennica 42(2):151-163.

112. Lawrence, W.H.; Kverno, N.B.; Hartwell, H.D. 1961. Guide to wildlife feeding injuries on conifers in the Pacific Northwest. Portland, OR: Western Forestry and Conservation Association 44 p.

113. Leatherman, D.; Lange, D. 1997. Western cedar bark beetles. Colorado State Forest Service. Unnumbered leaflet. 4 p.

114. Lewis, R., Jr. 1978. Influence of infection court, host vigor, and culture filtrates on canker production by *Botryodiplodia theobromae* conidia in sycamore. Plant Disease Reporter 62:934-937.

115. Lightle, P.C.; Weiss, M.J. 1974. Dwarf mistletoe of ponderosa pine in the Southwest. Forest Insect and Disease Leaflet 19. Washington, DC: U.S. Department of Agriculture, Forest Service. 8 p. Online: http://www.fs.fed.us/r6/nr/fid/fidls/fidl19.htm.

116. Mahalovich, M.F. 2000. Whitebark pine restoration strategy--some genetic considerations. Nutcracker Notes 11:6-9.

117. Malek, R.B.; Appleby, J.E. 1984. Epidemiology of pine wilt in Illinois. Plant Disease 68:180-186.

118. Mathiasen, R.L.; Beatty, J.S.; Pronos, J. 2002. Pinyon pine dwarf mistletoe. Forest Insect and Disease Leaflet 174. Washington, DC: U.S. Department of Agriculture, Forest Service. 7 p. Online: http://www.na.fs.fed.us/spfo/pubs/fidls/pinyon_mistletoe/ppd_mistletoe.pdf.

119. McDonald, G.I.; Richardson, B.A.; Zambino, P.J.; Klopfenstein, N.B.; Kim, M.S. 2006. *Pedicularis* and *Castilleja* are natural hosts of *Cronartium ribicola* in North America: a first report. Forest Pathology 36:73–82.

120. McMillin, J.D.; Allen, K.K.; Long, D.F.; Harris, J.L.; Negron, J.F. 2003. Effects of western balsam bark beetle on spruce-fir forests of north-central Wyoming. Western Journal of Applied Forestry 18(4):259-266.

121. Miller, P.R. 1993. Abiotic Diseases. In: Scharpf, R.F., tech coord. Diseases of Pacific Coast Conifers. Agricultural Handbook 521. Albany, NY: U.S. Department of Agriculture, Forest Service, Pacific Southwest Research Station. 199 p.

122. Moltzan, B.D.; Blenis, P.V.; Hiratsuka, Y. 2001. Effects of spore availability, spore germinability, and shoot susceptibility on gall rust infection of pine. Plant Disease 85:1193-1199.

123. Negron, J.; Wilson, J. 2003. Attributes associated with probability of infestation by the pinyon ips, *Ips confusus* (Coleoptera: Scolytidae) in pinyon pine, *Pinus edulis*. Western North American Naturalist 63(4):440-51.

124. Ostry, M.E.; Wilson, L.F.; McNabb H.S., Jr.; Moore, L.M. 1989. A guide to insect, disease and animal pests of poplar. Agricultural Handbook 677. Washington, DC: U.S. Department of Agriculture, Forest Service. 118 p.

125. Partridge, A.D.; Miller, D.L. 1974. Major wood decays in the Inland Northwest. Natural Resource Series No. 3. Idaho Research Foundation, Inc. 125 p.

126. Peterson, G.W. 1977. Infection, epidemiology, and control of Diplodia blight of Austrian, ponderosa, and Scots pines. Phytopathology 67:511-514.

127. Peterson, G.W. 1978. Effective and economical methods for controlling Diplodia tip blight. American Nurseryman 147(1):13, 66, 70, 72.

128. Peterson, G.W. 1981. Diplodia blight of pines. Forest Insect and Disease Leaflet 161. Washington, DC: U.S. Department of Agriculture, Forest Service. 7 p.

129. Peterson, G.W. 1982. Dothistroma needle blight of pines. Forest Insect and Disease Leaflet 143. Washington, DC: U.S. Department of Agriculture, Forest Service. 6 p.

130. Peterson, R.S. 1960. Western gall rust of hard pines. Forest Pest Leaflet 50. Washington, DC: U.S. Department of Agriculture, Forest Service. 8 p.

131. Peterson, R.S. 1961. Conifer tumors in the central Rocky Mountains. Plant Disease Reporter 45:472-474.

132. Peterson, R.S. 1963. Effects of broom rusts on spruce and fir. Res. Paper INT-7. Ogden, UT: U.S. Department of Agriculture, Forest Service, Intermountain Experiment Station.

133. Peterson, R.S. 1971. Wave years of infection by gall rust on pines. Plant Disease Reporter 55:163-167.

134. Petty, J.L. 1977. Bionomics of two aspen bark beetles. Great Basin Naturalist 37(1):105-127.

135. Price, R.A.; Liston, A.; Strauss, S.H. 1998. Phylogeny and systematics of *Pinus*. In: Richardson, D.M., ed. Ecology and Biogeography of *Pinus*. Cape Town, South Africa: Cambridge University Press. 524 p.

136. Rexrode, C.O.; Jones, T.W. 1970. Oak bark beetles--important vectors of oak wilt. Journal of Forestry 68:294-297.

137. Riffle, J.W. 1981. Cankers. In: Stipes, R.J.; Campana, R.J., eds. Compendium of elm diseases. St. Paul, MN: American Phytopathological Society. 96 p.

138. Riffle, J.W.; Conway, K.E. 1986. Phellinus stem decays of hardwoods. In: Riffle, J.W.; Peterson, G.W., tech. coords. Diseases of trees in the Great Plains. Gen. Tech. Rep. RM-129. Fort Collins, CO: U.S. Department of Agriculture, Forest Service, Rocky Mountain Forest and Range Experiment Station. 149 p.

139. Riffle, J.W.; Ostrofsky, W.D.; James, R.L. 1981. *Fomes fraxinophilus* on green ash in Nebraska windbreaks. Plant Disease 65:667-669.

140. Riffle, J.W.; Peterson, G.W., tech. coords. 1986. Diseases of trees in the Great Plains. Gen. Tech. Rep. RM-129. Fort Collins, CO: U.S. Department of Agriculture, Forest Service, Rocky Mountain Forest and Range Experiment Station. 149 p.

141. Riffle, J.W.; Sharon, E.M.; Harrell, M.O. 1984. Incidence of *Fomes fraxinophilus* on green ash in Nebraska woodlands. Plant Disease 68(4):322-324.

142. Riffle, J.W.; Walla, J.A. 1986. Perennial wood-rotting fungi that cause stem decays of hardwoods. In: Riffle, J.W.; Peterson, G.W., tech. coords. Diseases of trees in the Great Plains. Gen. Tech. Rep. RM-129. Fort Collins,

CO: U.S. Department of Agriculture, Forest Service, Rocky Mountain Forest and Range Experiment Station. 149 p.

143. Ross, W.D. 1976. Fungi associated with root diseases of aspen in Wyoming. Canadian Journal of Botany 54:734-744.

144. Samman, S.; Schwandt, J.W.; Wilson, J.L. 2003. Managing for healthy white pine ecosystems in the United States to reduce the impacts of white pine blister rust. Report R1-03-118. Missoula, MT: U.S. Department of Agriculture, Forest Service, Northern Region. 10 p.

145. Scharpf, R.F., tech. coord. 1993. Diseases of Pacific Coast conifers. Agricultural Handbook 521. Washington, DC: U.S. Department of Agriculture, Forest Service. 199 p.

146. Schipper, A.L.; Anderson, R.L. 1976. How to identify Hypoxylon canker of aspen. How To Leaflet. St. Paul, MN: U.S. Department of Agriculture, Forest Service, North Central Forest Experiment Station. 4 p.

147. Schmid, J.M. 1981. Spruce beetles in blowdown. Res. Note RM-411. Fort Collins, CO: U.S. Department of Agriculture, Forest Service, Rocky Mountain Forest and Range Experiment Station. 5 p.

148. Schmid, J.M.; Frye, R.H. 1976. Stand ratings for spruce beetles. Res. Note RM-309. Fort Collins, CO: U.S. Department of Agriculture, Forest Service, Rocky Mountain Forest and Range Experiment Station. 4 p.

149. Schmid, J.M.; Frye, R.H. 1977. Spruce beetle in the Rockies. Gen. Tech. Rep. RM-49. Fort Collins, CO: U.S. Department of Agriculture, Forest Service, Rocky Mountain Forest and Range Experiment Station. 38 p.

150. Schmitz, R.F.; Gibson, K.E. 1996. Douglas-fir beetle. Forest Insect and Disease Leaflet 5. Washington, DC: U.S. Department of Agriculture, Forest Service. 8 p.

151. Schoettle, A.W.; Sniezko, R.A. 2007. Proactive intervention to sustain high elevation pine ecosystems threatened by white pine blister rust. Journal of Forest Research 12(5):327-336. Online: http://www.springerlink.com/content/9v91t44278w74430/fulltext.pdf .

152. Schwandt, J.W. 2006. Whitebark pine in peril: a case for restoration. Report R1-06-28. Missoula, MT: U.S. Department of Agriculture, Forest Service, Northern Region, Forest Health Protection. 20 p. Online: http://www.fs.fed.us/r1-r4/spf/fhp/whitebark_pine/WBPCover_4.htm.

153. Sever, A.L. 2005. Shade tree insect pest management studies: I. Evaluation of non-target effects of pesticides on arthropods: II. Biology of two emergent insect pests of oak. Fort Collins, CO: Colorado State University. 57 p. Thesis.

154. Shaw, C.G., II; Kile, G.A., eds. 1991. Armillaria root disease. Agricultural Handbook No. 691. Washington, DC: U.S. Department of Agriculture, Forest Service. 233 p.

155. Shope, P.F. 1943. Some ascomycetous foliage diseases of Colorado conifers. Boulder, CO: University of Colorado Student Services 2:31-43.

156. Shour, M. 2007. Shucking oystershell scale. Iowa State University Extension News. Ames, IA: Iowa State University 2 p.

157. Simms, H.R. 1967. On the ecology of *Herpotrichia nigra*. Mycologia 59:902-909.

158. Sinclair, W.A.; Lyon, H.H. 2005. Diseases of trees and shrubs. 2nd ed. Ithaca, NY: Cornell University Press. 659 p.

159. Sinclair, W.A.; Lyon, H.H.; Johnson, W.T. 1987. Diseases of trees and shrubs. Ithaca, NY: Cornell University Press. 574 p.

160. Smith, R.H. 1971. Red turpentine beetle. Forest Insect and Disease Leaflet 55. Washington, DC: U.S. Department of Agriculture, Forest Service. 8 p.

161. Solomon, J.D. 1995. Guide to insect borers in North American broadleaf trees and shrubs. Agricultural Handbook 706. Washington, DC: U.S. Department of Agriculture, Forest Service. 747 p. Online: http://www.forestpests.org/borers.

162. Spielman, L.J. 1985. A monograph of *Valsa* on hardwoods of North America. Canadian Journal of Botany 63:1355-1378.

163. Stack, R.W.; Conway, K.E. 1986. Gloeosporium and Gnomonia leaf diseases of broadleaf trees. In: Riffle, J.W.; Peterson, G.W., tech. coords. Diseases of trees in the Great Plains. Gen. Tech. Rep. RM-129. Fort Collins, CO: U.S. Department of Agriculture, Forest Service, Rocky Mountain Forest and Range Experiment Station. 149 p.

164. Staley, J.M. 1964. A survey of coniferous foliage diseases (other than rusts) in Colorado. Plant Disease Reporter 48(7):562-563.

165. Staley, J.M. 1976. Notes and a key, by host species, to the common foliar fungi and pathogens of conifers in the northern, central and southern Rocky Mountains, Great Basin, and Arizona and New Mexico. Unpublished document on file at U.S. Department of Agriculture, Forest Service, Rocky Mountain Research Station, Fort Collins, CO.

166. Staley, J.M. 1979. Penetration and latency in needlecast fungi. In: Proceedings of the 26th Annual meeting of the Western International Forest Disease Work Conference: 136-140.

167. Stevens, R.E. 1971. Ponderosa pine tip moth. Forest Insect and Disease Leaflet 103. Washington, DC: U.S. Department of Agriculture, Forest Service. 5 p.

168. Stewart, J.L. 1965. *Fomes annosus* found in Nebraska. Plant Disease Reporter 49(5):456.

169. Stokland, J.; Kauserud, H. 2004. *Phellinus nigrolimitatus*—A wood-decomposing fungus highly influenced by forestry. Forest Ecology and Management 187:333-343.

170. Sullivan, K. 2003. Decay of living conifers in spruce-fir forests of the central Rocky Mountains. LSC-03-04. Golden, CO: U.S. Department of Agriculture, Forest Service, Rocky Mountain Region, Renewable Resources. 19 p. Online: http://www.fs.fed.us/r2/fhm/bugcrud/decays_lsc-03-04.pdf.

171. Taylor, J.E.; Mathiasen, R.L. 1999. Limber pine dwarf mistletoe. Forest Insect and Disease Leaflet 171. Washington, DC: U.S. Department of Agriculture, Forest Service. 7 p.

172. Tisserat, N.A. 2001. Juniper Diseases. Pub. No. C-711. Kansas State University Research and Extension. 12 p.

173. True, R.P.; Barnett, H.L.; Dorsey, C.K.; Leach, J.G. 1960. Oak wilt in West Virginia. Bull. 448T. Morgantown, WV: West Virginia University Agricultural Experiment Station. 119 p.

174. U.S. Department of Agriculture, Forest Service. 1985. Insects of eastern forests. Misc. Publ. 1426. Washington, DC: U.S. Department of Agriculture, Forest Service. 608 p.

175. U.S. Department of Agriculture, Forest Service. 2003. Forest insect and disease identification and management. U.S. Department of Agriculture, Forest Service, Northern Region; Idaho Department of Lands, Insect and Disease Control; Montana Department of State Lands, Division of Forestry. 223 p.

176. U.S. Department of Agriculture, Forest Service. January 2000. Ips bark beetle management guide. [Online]. U.S. Department of Agriculture, Forest Service, Rocky Mountain Region. Online: http://www.fs.fed.us/r2/fhm/bugcrud/silvips.htm.

177. van der Kamp, B. 1994. Lodgepole pine stem diseases and management of stand density in the British Columbia interior. The Forestry Chronicle 70(6):773-779.

178. Wagener, W.W. 1959. The effect of a western needle fungus (*Hypodermella medusa* Dearn.) on pines and its significance in forest management. Journal of Forestry 57:561-564.

179. Wall, R.E. 1971. Variation in decay in aspen stands as affected by their clonal growth pattern. Canadian Journal of Forest Research 1:141-146.

180. Walla, J.A. 1984. Incidence of *Phellinus punctatus* on living woody plants in North Dakota. Plant Disease 68:252-253.

181. Walters, J.W.; Hinds, T.E.; Johnson, D.W.; Beatty, J.S. 1982. Effects of partial cutting on diseases, mortality and regeneration of Rocky Mountain aspen stands. Res. Pap. RM-240. Fort Collins, CO: U.S. Department of Agriculture, Forest Service, Rocky Mountain Forest and Range Experiment Station. 12 p.

182. Whitney, R.; Fleming, R.; Zhou, K.; Mossa, D. 2002. Relationship of root rot to black spruce windfall and mortality following strip clear-cutting. Canadian Journal of Forest Research 32(2):283-294.

183. Whitney, R.D. 1995. Root-rotting fungi in white spruce, black spruce, and balsam fir in northern Ontario. Canadian Journal of Forest Research 25(8):1209-1230.

184. Whitney, R.D.; Denyer, W.B.G. 1968. Rates of decay by *Coniophora puteana* and *Polyporus tomentosus* in living and dying white spruce. Forest Science 14(2):122-126.

185. Wickman, B.E.; Mason, R.R; Trestle, G.C. 1981. Douglas-fir tussock moth. Forest Insect and Disease Leaflet 86. Washington, DC: U.S. Department of Agriculture, Forest Service. 10 p.

186. Wilson, L.F. 1972. Life history and outbreaks of an oak leafroller, *Archips semiferanus* (Lepidoptera: Tortricidae), in Michigan. Great Lakes Entomologist 55:71-77.
187. Wingfield, M.J.; Blanchette, R.A.; Nicholls, T.H. 1984. Is the pine wood nematode an important pathogen in the United States? Journal of Forestry 82:232-235.
188. Witcosky, J.J.; Schowalter, T.D.; Hansen, E.M. 1986. *Hylastes nigrinus* (Coleoptera: Scolytidae), *Pissodes fasciatus*, and *Steremnius carinatus* (Coleoptera: Curculionidae) as vectors of black stain root disease of Douglas-fir. Environmental Entomology (15):1090-1095.
189. Wollerman, E.H. 1971. Bagworm. Forest Insect and Disease Leaflet 97. Washington, DC: U.S. Department of Agriculture, Forest Service. 7 p.
190. Wood, S.L. 1982. The bark and ambrosia beetles of North and Central America (Coleoptera: Scolytidae), a taxonomic monograph. Great Basin Naturalist Memoir 6. 1359 p.
191. Worrall, J.J.; Egeland, L.; Eager, T.; Mask, R.A.; Johnson, E.W.; Kemp, P.A.; Shepperd, W.D. 2008. Rapid mortality of *Populus tremuloides* in southwestern Colorado, USA. Forest Ecology and Management 255(3-4):686-696.
192. Worrall, J.J.; Fairweather, M. 2009. Decay and discoloration of aspen. Forest Insect and Disease Leaflet 149. Washington, DC: U.S. Department of Agriculture, Forest Service. 7 p.
193. Worrall, J.J.; Harrington, T.C.; Blodgett, J.T.; Conklin, D.A.; Fairweather, M.L. 2010. *Heterobasidion annosum* and *H. parviporum* in the southern Rocky Mountains and adjoining states. Plant Disease 94:115-118.
194. Worrall, J.J.; Nakasone, K.K. 2009. Decays of Engelmann spruce and subalpine fir in the Rocky Mountains. Forest Insect and Disease Leaflet 150. Washington, DC: U.S. Department of Agriculture, Forest Service. 12 p.
195. Worrall, J.J.; Parmeter, J.R. 1982a. Formation and properties of wetwood in white fir. Phytopathology 72:1209-1212.
196. Worrall, J.J.; Parmeter, J.R. 1982b. Wetwood formation as a host response in white fir. European Journal of Forest Pathology 12(6):432-441.
197. Worrall, J.J.; Parmeter, J.R. 1983. Inhibition of wood decay fungi by wetwood of white fir. Phytopathology 73:1140-1145.
198. Worrall, J.J.; Sullivan, K.F. 2002. Discoloration of ponderosa pine on the San Juan National Forest, 1999-2001. Biological Evaluation R2-02-06. Lakewood, CO: U.S. Department of Agriculture, Forest Service, Rocky Mountain Region, Forest Health Management. 20 p. Online: http://www.fs.fed.us/r2/fhm/reports/be_r2-02-06.pdf.
199. Worrall, J.J.; Sullivan, K.F.; Harrington, T.C.; Steimel, J.P. 2004. Incidence, host relations and population structure of *Armillaria ostoyae* in Colorado campgrounds. Forest Ecology and Management 192:191-206. Online: http://www.fs.fed.us/r2/fhm/reports/ArmillariaCGCO.pdf .
200. Wu, Y.; Johnson, D.W.; Angwin, P.A. 1996. Identification of *Armillaria* species in the Rocky Mountain Region. Tech. Rep. R2-58. Lakewood, CO:

U.S. Department of Agriculture, Forest Service, Rocky Mountain Region. 26 p.

201. Ziller, W.G. 1974. The tree rusts of western Canada. Publication No. 1329. Victoria, BC: Canadian Forestry Service, Environment Canada. 272 p.

# Host-Pest Index

(not a complete pest list for each host)

**Alder**

Stem-

   Alder borer (238)

Foliage-

   Cottonwood leaf beetle (196, **201**)

   Eriophyid mites (263, 264, **267**)

   Fall webworm (196, **207**)

   Forest tent caterpillar (196, **209**)

   Large aspen tortrix (196, **212**)

   Western tent caterpillar (195, 196, **233**)

**Ash (green, white, black)**

Stem-

   Ash bark beetles (**156**)

   Lilac (Ash) borer (238, **253**)

   Carpenterworm (238, **245**)

   Emerald ash borer (251)

   Stem decays of hardwoods (**56**)

   Wetwood (**149**)

Branches and Terminals-

   Ash bark beetles (**156**)

   Oystershell scale (263, **270**)

Foliage-

   Anthracnose (**110**)

   Eriophyid mites (263, 264, **267**)

   Fall webworm (196, **207**)

   Forest tent caterpillar (196, **209**)

   Verticillium wilt (79, 80, **89**)

Seeds and Cones-

**Aspen**

Root-

   Armillaria root disease (23, **28**, 42, 192)

   White mottled rot (23, 24, **42**)

Stem-

   Abiotic damage (**145**)

   Animal damage (**135**)

   Aspen bark beetles (**157**)

   Aspen trunk rot (21, 24, **44**)

   Black canker (60-63, **64**)

   Bronze poplar borer (238, **242**)

   Brown crumbly rot (23, **46**)

   Burls, galls, and tumors (**134**)

Cryptosphaeria canker (24, 60-62, **69**)
Cytospora canker (60-63, 70, **71**, 129)
Hypoxylon canker (61-63, **75**)
Oystershell scale (263, **270**)
Poplar borer (75, 238, **255**)
Sooty bark canker (60-62, 69, 77)
Wetwood (**149**)
White trunk rot (24, **44**)
Branches and Terminals-
Animal damage (**135**)
Oystershell scale (263, **270**)
Petiolegall aphids (**272**)
Shepherd's crook (**132**)
Foliage-
Abiotic damage (**138**)
Aspen leafminer (196, **197**)
Eriophyid mites (263, 264, **267**)
Forest tent caterpillar (196, **209**)
Ink spot leaf blight (**120**)
Large aspen tortrix (196, **212**)
Marssonina blight (108, 110, **121**)
Melampsora rusts (92, **124**)
Septoria leaf spot and canker (**128**)
Shepherd's crook (**132**)
Western tent caterpillar (195, 196, **233**)
Seeds and Cones-
Eriophyid mites (263, 264, **267**)

**Birch (paper birch)**
Root-
Armillaria root disease (23, **28**, 42, 192)
Stem-
Abiotic damage (**145**)
Animal damage (**135**)
Stem decays of hardwoods (**56**)

**Chokecherry (and other *Prunus*)**
Stem-
Stem decays of hardwoods (**56**)
Branches and Terminals-
Black knot of cherry (**66**)
Foliage-
Western tent caterpillar (195, 196, **233**)
Eastern tent caterpillar (196)
Fall webworm (196, **207**)

**Cottonwood (Plains, narrowleaf)**
Root-
   Armillaria root disease (23, **28**, 42, 192)
   White mottled rot (23, 24, **42**)
Stem-
   Carpenterworm (238, **245**)
   Cottonwood borer (236-238, **247**)
   Cryptosphaeria canker (24, **69**)
   Poplar borer (75, 238, **255**)
   Stem decays of hardwoods (**56**)
   Wetwood (**149**)
Branches and Terminals
   Oystershell scale (263, **270**)
   Petiolegall aphids (**272**)
Foliage-
   Animal damage (**135**)
   Aspen leafminer (196, **197**)
   Cottonwood leaf beetle (196, **201**)
   Fall webworm (196, **207**)
   Forest tent caterpillar (196, **209**)
   Ink spot leaf blight (**120**)
   Marssonina leaf blight (108, 110, **121**)
   Melampsora rusts (92, **124**)
   Poplar blackmine beetle (198)
   Septoria leaf spot and canker (**128**)
   Western tent caterpillar (195, 196, **233**)

**Currant (*Ribes* spp.)**
Foliage-
   White pine blister rust (92, 93, **105**)

**Douglas-fir**
Root-
   Armillaria root disease (23, **28**, 35, 42, 192)
   Schweinitzii root and butt rot (23, 36, **40**)
Stem-
   Abiotic damage (**145**)
   Animal damage (**135**)
   Ambrosia beetles (236, 238, **240**)
   Brown crumbly rot (23, **46**)
   Douglas-fir beetle (42, 49, **162**, 164)
   Douglas-fir engraver beetles (**163**)
   Douglas-fir pole beetle (**163**)
   Flatheaded wood borers (236, 238, **250**)
   Gray-brown saprot (**48**)
   Red ring rot (24, 38, **51**)

        Roundheaded wood borers (238, **256**)
        Wood wasps (237, 238, **260**)
    Branches and Terminals-
        Cytospora canker (**73**)
        Douglas-fir dwarf mistletoe (6, **10**)
        Douglas-fir secondary beetles (**163**)
    Foliage-
        Abiotic damage (**138**)
        Cooley spruce gall adelgid (263, **265**)
        Douglas-fir tussock moth (168, 195, 196, **202**, 231)
        Giant conifer aphids (263, **269**)
        Melampsora leaf rust (92, **124**)
        Pine needle scale (263, **274**)
        Tiger moth (196, **225**)
        Western spruce budworm (168, 195, 196, 227, **231**)
    Seeds and Cones-
        Conifer seed bugs (263)
        Cooley spruce gall adelgid (263, **265**)

## Elm (American, Siberian, and others)
    Stem-
        Abiotic damage (145)
        Botryodiplodia canker (**67**)
        Carpenterworm (238, **245**)
        Elm bark beetles (81, 82, **165**)
        Elm borer (238, **248**)
        Pigeon tremex (237, 261)
        Spider mites (262-264, **278**)
        Wetwood (**149**)
    Foliage-
        Anthracnose (**110**)
        Bagworm (196, **199**)
        Dutch elm disease (79, 80, **81**, 166, 167)
        Elm leaf beetle (196, **204**)
        Elm leafminer (196)
        Eriophyid mites (263, 264, **267**)
        European elm scale (263, **266**)
        Fall webworm (196, **207**)
        Forest tent caterpillar (196, **209**)
        Verticillium wilt (79, 80, **89**)

## Fir (white, subalpine, corkbark)
    Root-
        Annosus root disease (23, **25**, 35, 38)
        Armillaria root disease (23, **28**, 42, 192)
        Coniophora root and butt rot (**35**)

Stem-
    Abiotic damage (**145**)
    Ambrosia beetles (236, 238, **240**)
    Burls, galls, tumors (**134**)
    Fir engraver beetle (26, **167**)
    Flatheaded wood borers (236, 238, **250**)
    Rust-red stringy rot (24, **53**)
    Roundheaded wood borers (238, **256**)
    Red heart rot (24, **53**)
    Red ring rot (24, 38, **51**)
    Western balsam bark beetle (**190**)
    Wetwood (**149**)
    Wood wasps (237, 238, **260**)
Branches and Terminals-
    Fir broom rust (92, **94**)
    Fir engraver beetle (26, **167**)
Foliage-
    Brown felt blight/snow mold (**113**)
    Douglas-fir tussock moth (168, 195, 196, **202**, 231)
    Giant conifer aphids (263, **269**)
    Melampsora rusts (92, **124**)
    Tiger moth (196, **225**)
    Western spruce budworm (168, 195, 196, 227, **231**)
    White fir needleminer (196, **214**)

## Hackberry
Branches and Terminals-
    Eriophyid mites (witches' broom) (263, 264, **267**)
    Spider mites (262-264, **278**)
Foliage-
    Hackberry bud gall psyllid (263)
    Hackberry nipplegall maker (263)

## Juniper (One-seed, Utah, Rocky Mountain, Eastern redcedar)
Root-
    Annosus root disease (23, **25**, 38)
    Armillaria root disease (23, **28**, 42, 192)
Stem-
    Cedar bark beetles (**161**)
    Juniper borers (238, **252**)
    Juniper twig pruner (**252**, 280)
Branches and Terminals-
    Cedar bark beetles (**161**)
    Giant conifer aphids (263, **269**)
    Gymnosporangium rusts (92, 94, **100**)

Juniper mistletoe (**18**)
Juniper twig pruner (**252**, 280)
Foliage-
Bagworm (196, **199**)
Cercospora blight (**114**)
Giant conifer aphids (263, **269**)
Tiger moth (196, **225**)
Spider mites (262-264, **278**)

## Maples (Boxelder, Rocky Mountain)
Stem-
Carpenterworm (238, **245**)
Pigeon tremex (237, 261)
Wetwood (**149**)
Branches and Terminals-
Oystershell scale (263, **270**)
Foliage-
Anthracnose (**110**)
Bagworm (196, **199**)
Eriophyid mites (263, 264, **267**)
Boxelder leafroller (196)
Boxelder leafminer (196)
Fall webworm (196, **207**)
Verticillium wilt (79, 80, **89**)
Seeds and Cones-
Boxelder bug (262, 263)

## Oak (Gamble, bur)
Root-
Armillaria root disease (23, **28**, 42, 192)
Stem-
*Agrilus quercicola* (238, **239**)
Botryodiplodia canker (**67**)
Oak wilt (79-81, **84**)
Twolined chestnut borer (**258**)
Branches and Terminals-
*Agrilus quercicola* (238, **239**)
Foliage-
Anthracnose (**110**)
Bagworm (196, **199**)
Forest tent caterpillar (196, **209**)
Oak leafroller (196, **216**)
Oak wilt (79-81, **84**)
Sonoran tent caterpillar (196)
Western tent caterpillar (195, 196, **233**)

**Pines (Pinyon, ponderosa, lodgepole, limber, bristlecone, whitebark, Scots, Austrian)**

Root-
 Armillaria root disease (23, **28**, 35, 42, 192)
 Annosus root disease (23, **25**, 38)
 Black stain root disease (22, **31**, 79, 81, 178)
 Coniophora root and butt rot (**35**)
 Schweinitzii root and butt rot (23, 36, **40**)

Stem-
 Abiotic damage (145)
 Animal damage (**135**)
 Ambrosia beetles (236, 238, **240**)
 Brown crumbly rot (23, **46**)
 Burls, galls, and tumors (**134**)
 Comandra blister rust (92-94, **97**)
 Dwarf mistletoes (**6, 12, 13, 15, 16**)
 Flatheaded wood borers (236, 238, **250**)
 Gray-brown saprot (**48**)
 *Ips* species (132, **169, 174, 177**, 180, 181, 194)
 Mountain pine beetle (31, 170, 171, **172**, 183, 189, 190, 191, 228)
 Limber pine engraver (**169**)
 Lodgepole pine beetle (**170**)
 Pine engraver beetles (132, **174**, 194)
 Pinyon ips (**177**, 180)
 Pinyon twig beetle (**179**, 190)
 Pitch moths (238, 280, **282**)
 Red ray rot (24, **49**)
 Red ring rot (24, 38, **51**)
 Red turpentine beetle (**181**, 194)
 Roundheaded pine beetle (**183**, 194)
 Roundheaded wood borers (238, **256**)
 Stalactiform rust (92, **97**)
 Western gall rust (91-93, **103**, 134)
 Western pine beetle (183, 184, **194**)
 White pine blister rust (12, 13, 91-94, **105**, 169, 170)
 Wood wasps (237, 238, **260**)
 Zimmerman pine moth (**282**)

Branches and Terminals-
 Comandra blister rust (92-94, **97**)
 Dwarf mistletoes (**6, 12, 13, 15, 16**)
 Eriophyid mites (263, 264, **267**)
 Giant conifer aphids (263, **269**)
 *Ips* species (132, **169, 174, 177**, 180, 181, 194)
 Limber pine dwarf mistletoe (**12**)
 Limber pine engraver (**169**)

Lodgepole pine dwarf mistletoe (8, **13**)
Lodgepole terminal weevil (**285**)
Pine engraver beetles (132, **174**, 194)
Pine shoot blight (**129**, 148)
Pine tip moths (280, **281**)
Pinyon dwarf mistletoe (**15**)
Pinyon ips (**177**)
Pinyon spindle gall midge (263, 264)
Pinyon twig beetle (**179**)
Pitch moths (238, 280, **282**)
Pitch nodule moths (280, **283**)
Southwestern dwarf mistletoe (6-9, 14, **16**, 263)
Twig beetles (105, **179**, **188**, 280)
Western gall rust (91-93, **103**, 134)
White pine blister rust (12, 13, 91-94, **105**, 169, 170)
Zimmerman pine moth (**282**)
Foliage-
Abiotic damage (138)
Bagworm (196, **199**)
Brown felt blight/snow mold (**113**)
Brown spot needle blight (**125**)
Davisomycella needle cast (108, 109, **115**)
Elytroderma needle cast (109, **118**)
Lophodermella needle cast (109, **115**)
Melampsora rusts (92, **124**)
Needle casts (107, 109, 110, **115**, 118, 126)
Needleminers (196, **214**)
Pandora moth (196, **218**)
Pine butterfly (196, **220**)
Pine needle scale (263, **274**)
Pine sawflies (196, **222**)
Pine wilt disease (79, 81, 87)
Pinyon needle scale (263, **276**)
Red band needle blight (108, **125**)
Tiger moth (196, **225**)
Western pine budworm (196, **227**)
Western pine tussock moth (196, **229**)

**Serviceberry**
Foliage-
Gymnosporangium rust (92, 94, **100**)

**Spruce (Blue, Engelmann, white)**
Root-
Armillaria root disease (23, **28**, 35, 42, 192)
Coniophora root and butt rot (**35**)

Root diseases with white pocket rots (21, 23, **38**)
Schweinitzii root and butt rot (23, 36, **40**)
Stem-
    Abiotic damage (**145**)
    Animal damage (**135**)
    Ambrosia beetles (236, 238, **240**)
    Blue spruce engraver (**159**, 188)
    Brown crumbly rot (23, **46**)
    Brown felt blight/snow mold (**113**)
    Burls, galls, and tumors (**134**)
    Cytospora canker (**73**)
    Flatheaded wood borers (236, 238, **250**)
    Gray-brown saprot (**48**)
    Redheart rot (24, **53**)
    Red ring rot (24, 38, **51**)
    Roundheaded wood borers (238, **256**)
    Spruce engraver beetles (**187**)
    Spruce beetle (159, 171, 172, **184**, 188, 191)
    Wood wasps (237, 238, **260**)
Branches and Terminals-
    Cooley spruce gall adelgid (263, **265**)
    Blue spruce engraver (**159**, 188)
    Broom rust (92, 93, **94**)
    Cytospora canker (**73**)
    Giant conifer aphids (263, **269**)
    Spruce engraver beetles (**187**)
    Twig beetles (**188**)
    White pine weevil (**285**)
Foliage-
    Abiotic damage (138)
    Bagworm (196, **199**)
    Douglas-fir tussock moth (168, 195, 196, **202**, 231)
    Pine needle scale (263, **274**)
    Spider mites (262-264, **278**)
    Broom rust (92, 93, **94**)
    Needleminers (196, **214**)
    Western spruce budworm (168, 195, 196, 227, **231**)

**Willows (multiple species)**
Stem-
    Cottonwood borer (236-238, **247**)
    Cytospora canker (60, 61, 63, 70, **71**, 129)
    Stem decays of hardwoods (**56**)

Branches and Terminals-
      Cytospora canker (60, 61, 63, 70, **71**, 129)
      Oystershell scale (263, **270**)
Foliage-
      Bagworm (196, **199**)
      Cottonwood leaf beetle (196, **201**)
      Eastern tent caterpillar (196)
      Fall webworm (196, **207**)
      Forest tent caterpillar (196, **209**)
      Melampsora rusts (92, **124**)
      Western tent caterpillar (195, 196, **233**)

# General Index

## A

Abiotic foliage damage (**138**)
Abiotic stem damage (**145**)
*Aceria macrorrhyncha* (267)
*Adelges cooleyi* (263, 265)
*Agrilus anxius* (259)
*Agrilus bilineatus* (258)
*Agrilus liragus* (238, 242)
*Agrilus quercicola* (238, **239**)
*Agrilus planipennis* (251)
Alder borer (238)
Ambrosia beetles (236, 238, **240**)
*Amylostereum chaillettii* (24)
Animal damage (**135**)
Annosus root disease (23, **25**, 38)
Anthracnose (**110**)
Antler rubbing/scraping (136)
*Antrodia serialis* (24)
*Aphrophora* spp. (263)
*Apiognomonia errabunda* (111, 112)
*Apiognomonia quercina* (110, 111, 112)
*Apiognomonia tiliae (*111)
*Apiognomonia veneta* (111)
*Apiosporina morbosa* (66)
Aphid (135, 262-265, 269, 270, 272-274, 280)
*Arceuthobium americanum* (6, 13)
*Arceuthobium cyanocarpum* (12)
*Arceuthobium divaricatum* (15)
*Arceuthobium douglasii* (10)
*Arceuthobium* spp. (6, 10, 12, 13, 15, 16)
*Arceuthobium vaginatum* subsp. *cryptopodum* (16)
*Archips semiferana* (196, 216)
*Armillaria ostoyae* (21, 28)
*Armillaria solidipes* (23, 28)
Armillaria root disease (23, **28**, 42, 192)
Artist's conk (23, **42**)
Ash bark beetles (**156**)
Ash borer (238, **253**)
Asian longhorned beetle (256)
Aspen bark beetles (**157**)
Aspen leafminer (196, **197**)
Aspen shoot blight (132)
Aspen trunk rot (21, 24, **44**)

*Atimia huachucae* (252, 253)

# B

Bagworm (196, **199**)
Bark beetle-associated mites (262)
Bastard toadflax (92, 97, 99)
Bear (136, 137)
Bearberry (92, 94)
Bear wipe (113)
*Bifusella linearis* (109)
*Bifusella pini* (109)
*Bifusella saccata* (109)
Black canker (60, 61, 63, **64**)
Black knot (**66**)
Black pine needle scale (263)
Black stain root disease (22, **31**, 79, 81, 178)
Black-horned juniper borer (**252, 253**)
Bleeding Stereum (24, 53, 54)
Blue spruce engraver (**159**, 188)
Botryodiplodia canker (**67**)
*Botryodiplodia hypodermia* (67)
*Botryodiplodia theobromae* (67)
*Botryosphaeria* spp. (67)
Boxelder bug (262, 263)
Boxelder leafminer (196)
Boxelder leafroller (196)
Bronze poplar borer (238, **242**)
Brown crumbly rot (23, **46**)
Brown felt blight (**113**)
Brown spot needle blight (**125**)
Browse/browsing (136, 138, 235)
Buprestidae (194, 237-239, 242, 250, 258)
Burls, galls and tumors (**134**)
*Bursaphelenchus xylophilus* (79, 87)

# C

*Callidium texanum* (252, 253)
*Camponotus* spp. (244)
Carpenter ants (**244**)
Carpenterworm (238, **245**)
*Castilleja* spp. (98, 105)
*Cecidomyia* spp. (263)
Cedar bark beetles (**161**)
Cenangium canker (77)
Cerambycidae (183, 194, 238, 247, 248, 252, 253, 255, 256)
*Cerastium* spp. (94)

Ceratocystis canker (64)
*Ceratocystis fagacearum* (79, 84)
*Ceratocystis fimbriata* (64)
*Ceratocystis populicola* (64)
*Ceratocystis ulmi* (79)
Cercospora blight (**114**)
*Chalcophora virginiensis* (250)
Chemical damage (126, 142, 143)
*Chionaspis pinifoliae* (263, 274)
*Choristoneura conflictana* (196, 212)
*Choristoneura lambertiana* (196, 227)
*Choristoneura occidentalis* (196, 227, 231)
*Chrysomela scripta* (196, 201)
*Chrysomyxa arctostaphyli* (92, 94, 96)
*Ciborinia whetzelii* (120)
*Cinara* spp. (263, 269)
*Cinara pilicornis* (269)
*Cinara pulverulens* (269)
Circinatus root rot (23, 38)
Clear-wing moth (237)
*Coleotechnites edulicola* (214)
*Coleotechnites piceaella* (214, 215)
*Coleotechnites ponderosae* (214)
*Coleotechnites* sp. (196)
*Colomerus vitis* (268)
*Coloradia doris* (218)
*Coloradia pandora* (196, 218)
Comandra blister rust (92-94, **97**)
*Comandra livida* (98)
*Comandra umbellata* (97)
Conifer-aspen rust (124)
Conifer-cottonwood rust (124)

Conifer seed bugs (263, 264)
*Coniophora puteana* (23, 35, 37)
Cooley spruce gall adelgid (263, **265**)
Cottonwood borer (236-238, **247**)
Cottonwood leaf beetle (196, **201)**
*Cronartium coleosporioides* (92, 97)
*Cronartium comandrae* (92, 97)
*Cronartium ribicola* (92, 105)
*Cryptoporus (Polyporus) volvatus* (48)
Cryptosphaeria canker (24, 60-62, **69)**
*Cryptosphaeria lignyota (populina)* (24, 69, 70)
*Cyclaneusma niveum* (109)
*Cytospora (Leucocytospora) kunzei* (73)

Cytospora canker (60, 61, 63, 70, **71**, **73**, 129)
*Cytospora chrysosperma* (70, 71, 73, 129)

# D

*Dasychira grisefacta* (196, **229**)
*Dasychira pinicola* (229)
*Dasychira plagiata* (229)
*Davisomycella medusa* (109, 115-117)
Davisomycella needle cast (108, 116)
*Davisomycella ponderosae* (109, 116, 117)
Deer (135, 136)
*Dendroctonus adjunctus* (183, 194)
*Dendroctonus brevicomis* (183, 193, 194)
*Dendroctonus murrayanae* (170)
*Dendroctonus ponderosae* (170-172, 183)
*Dendroctonus pseudotsugae* (162)
*Dendroctonus rufipennis* (159, 171, **184**, 188)
*Dendroctonus valens* (181, 194)
*Dichomitus squalens* (*Polyporus anceps*) (24, 49-51)
Diffuse canker (60, 62)
*Dioryctria ponderosae* (282, 283)
*Dioryctria zimmermani* (282)
Diplodia blight (**129**, 148)
*Diplodia pinea* (129, 131)
*Diplodia scrobiculata* (129)
*Discula* spp. (111)
*Discula quercina* (111)
Dothistroma needle blight (125)
*Dothistroma pini* (109, 125)
*Dothistroma septosporum* (109, 125)
Douglas-fir beetle (42, 49, **162**, 164)
Douglas-fir dwarf mistletoe (6, **10**)
Douglas-fir pole and engraver beetles (**163**)
Douglas-fir tussock moth (168, 195, 196, **202**, 231)
*Drepanopeziza populorum* (122)
*Drepanopeziza tremulae* (121)
*Dryocoetes confusus* (190)
Dutch elm disease (79, 80, **81**, 166, 167)

# E

*Echinodontium tinctorium* (24, 53)
Elk (135, 136, 242)
Elm bark beetle (81, **165**)
Elm borer (238, **248**)
Elm leaf beetle (196, **204**)
*Elytroderma deformans* (109, 118)

Elytroderma needle cast (109, **118**)
Emerald ash borer (251, 252)
*Encoelia pruinosa* (*Cenangium singulare*) (77, 78)
*Endothenia albolineana* (214-216)
Engraver beetles (132, 163, 174, 187, 194)
*Entoleuca* (*Hypoxylon*) *mammatum* (**75**)
*Epinotia meritana* (214)
*Eriophyes calcercis* (268)
*Eriophyes celtis* (268)
*Eriophyes parapopuli* (268)
Eriophyid mites (263, 264, **267**)
European elm scale (263, **266**)

# F

Fall webworm (196, **207**)
Fir broom rust (92, **94**)
Fire damage (145, 149, 163, 251, 257, 261)
Fir engraver (26, **167**)
Fir-willow rust (124)
*Flammulina populicola* (23)
Flatheaded borers (Buprestids) (236, 238, 239, 242, **250**, 252)
*Fomitiporia* (*Phellinus*) *punctata* (56-59)
*Fomitiporia hartigii* (24)
*Fomitopsis* (*Fomes*) *pinicola* (23, 36, 46-48)
Forest tent caterpillar (196, **209**)
Frost cracks (145, 147, 149)
Frost/freeze injury (141, 142, 145-147, 149)
*Fusicladium radiosum* var. *lethiferum* (*Pollacia radiosum*) (132)

# G

Gall mites (**267**)
*Ganoderma applanatum* (23, 42, 43)
Giant conifer aphid (263, **269**)
*Gloeosporium ulmicolum* (111)
*Gnathotrichus* (240)
*Gnomonia leptostyla* (111)
*Gossyparia* (*Eriococcus*) *spuria* (263, 266)
Gray-brown saprot (**48**)
*Gymnosporangium* spp. (92, 100, 102)
*Gymnosporangium juniperi-virginianae* (100)
Gymnosporangium rusts (92, 94, **100**)

# H

Hackberry nipplegall maker (263)
Hail (74, 130-132, 145, 148, 149)
*Halisidota ingens* (225)
Heat injury (145, 147, 149)

*Hemiphacidium planum* (109)
Herbicide damage (143, 144)
*Herpotrichia juniperi (Herpotrichia nigra)* (109, 113)
*Heterobasidion annosum* (23, 25-27)
*Heterobasidion parviporum* (23, 25-27)
Hip canker (103, 104)
Horntails (260)
*Hylesinus* spp. (156)
*Hylurgopinus rufipes* (81,165)
*Hyphantria cunea* (196, 207)
Hypoxylon canker (61-63, **75**)

# I

Indian paint fungus (24, 53-55)
Ink spot (107, **120**)
*Ips* spp. (132, **169**, **174**, **177**, 180, 181, 194)
*Ips borealis* (187)
*Ips confusus* (177, 180)
*Ips hunteri* (159, 160, 187, 188)
*Ips pilifrons* (187)
*Ips pini* (175, 176, 194)
*Ips tridens* (187)
*Ips woodi* (169)
*Isthmiella (Bifusella) abietis* (109)
*Isthmiella (Bifusella) crepidiformis* (109)

# J

Juniper bark beetles (**161**)
Juniper borers (238, **252**)
Juniper mistletoe (**18**)
Juniper twig pruner (**252**, 280)

# K

*Kabatiella apocrypta* (111)

# L

Large aspen tortrix (196, **212**)
*Laurilia sulcata* (24)
Leaf blight (110, 120-123)
Leaf spot (96, 98, 102, 105, 122-124, 128)
*Lecanosticta (Coryneum) cinerea* (109)
*Lentinellus montanus* (23)
Leopard spotting (78)
*Lepidosaphes ulmi* (263, 270)
*Leptographium wageneri* (31, 79)
Lettuce root aphid (273)
*Leucocytospora nivea* (71, 73)

*Leucostoma niveum* (71)
*Libertella* sp. (69)
Lightning (145-147, 194)
Lilac (ash) borer (238, **253**)
Limber pine dwarf mistletoe (**12**)
Limber pine engraver (**169**)
*Lirula abietis-concoloris* (109)
*Lirula macrospora* (*Lophodermium filiforme*) (109)
Lodgepole pine beetle (**170**)
Lodgepole pine dwarf mistletoe (8, **13**)
Lodgepole terminal weevil (285)
Longhorned borers (238, **256**)
*Lophocampa* spp. (196)
*Lophocampa ingens* (196, **225**)
*Lophodermella* (*Hypodermella*) *arcuata* (109, 115, 117)
*Lophodermella cerina* (109, 115, 117)
*Lophodermella concolor* (109, 115-117)
*Lophodermella montivaga* (109, 115-118)
Lophodermella needle cast (**115**)
*Lophodermium decorum* (109)
*Lophodermium juniperinum* (109)
*Lophodermium nitens* (109)
*Lophodermium piceae* (109)
*Lophomerum* (*Lophodermium*) *autumnale* (109)

# M

*Malacosoma californicum* (196, 233)
*Malacosoma disstria* (196, 209)
*Marssoniella juglandis* (111)
*Marssonina brunnea* (121-123)
Marssonina leaf blight (108, 110, **121**)
*Marssonina populi* (122)
*Matsucoccus acalyptus* (263, 276)
Mechanical damage (136, 145, 147-149, 157, 178, 182, 194, 250)
*Melampsora epitea* (124)
*Melampsora medusae* (*Melampsora albertensis*) (92, 124)
*Melampsora occidentalis* (124)
Melampsora rust (92, **124**)
*Melampsorella caryophyllacearum* (92, 94)
*Melanophila* spp. (251)
Mites (262-264, 267, 268, 270, 278, 279, 280)
*Monochamus* spp. (87, 256)
*Monochamus clamator* (256)
Mountain pine beetle (31, 170, 171, **172**, 183, 189, 190, 191, 228)
*Mycosphaerella acicola* (125)

*Mycosphaerella pini* (109, 125)
*Mycosphaerella populicola* (128, 129)
*Mycosphaerella populorum* (128, 129)

# N

Nantucket pine tip moth (281)
Needle cast (107-110, 115-118, 120, 126, 265)
Needleminers (196, **214**)
Nematode (79, 87-89, 256)
*Neodiprion autumnalis* (223, 224)
*Neodiprion gillettei* (223)
*Neodiprion* spp. (196, 222, 223)
*Neopeckia coulteri* (109, 113)
*Neophasia menapia* (196, 220)
*Nuclaspis californica* (263)

# O

Oak bark beetle (84, 86)
Oak leafroller (196, **216**)
Oak wilt (79-81, **84**)
*Oligonychus ununguis* (279)
*Onnia leporina* (23, 38)
*Onnia tomentosa* (23, 37-39)
*Ophiostoma* spp. (81, 82)
*Ophiostoma novo-ulmi* (81, 83, 167)
*Orgyia pseudotsugata* (196, 202)
Oystershell scale (263, **270**)

# P

*Pachypsylla celtidismamma* (263)
Pandora moth (196, **218**)
*Paraorgyia grisefacta* (229)
*Pedicularis* spp. (105)
*Pemphigus* spp. (263, 272-274)
*Pemphigus bursarius* (273)
*Pemphigus populitransversus* (272)
*Pemphigus populivenae* (273)
*Peniophora polygonia* (24, 45)
*Perenniporia fraxinophila* (56-59)
*Peridermium harknessii* (91, 92, 103)
Petiolegall aphids (**272**)
*Phaenops gentilis* (250)
*Phaeocryptopus gaeumannii* (109)
*Phaeocryptopus nudus* (*Adelopus balsamicola*) (109)
*Phaeolus schweinitzii* (23, 36, 40-42)
*Phellinidium ferrugineofuscum* (24)
*Phellinus igniarius* (56, 57, 59)

*Phellinus tremulae* (24, 44-46, 51)
*Phellopilus nigrolimitatus* (23, 38, 40, 53)
*Phlebiopsis gigantea* (28)
*Phloeosinus* spp. (161)
*Pholiota alnicola* (23)
*Pholiota squarrosa* (scaly Pholiota) (23)
Phoradendron *juniperinum*(18)
*Phyllocnistis populiella* (196, 197)
Pigeon tremex (237, 261)
Pine butterfly (196, **220**)
Pine needle scale (263, **274**)
Pine sawflies (195, 196, **222**)
Pine sawyer (87-89, 256)
Pine shoot blight (**129**, 148)
Pine tip moth (280, **281**)
Pine wilt disease (79, 81, **87**, 257)
Pine wood nematode (87)
Pinyon dwarf mistletoe (**15**)
Pinyon Ips (**177**)
Pinyon needleminer (214, 215)
Pinyon needle scale (263, **276**)
Pinyon pitch mass borer (280, 282, 283)
Pinyon pitch nodule moth (283, 284)
Pinyon spindle gall midge (263, 264)
Pinyon twig beetle (**179**, 190)
*Pinyonia edulicola* (263)
*Pissodes strobi* (285)
*Pissodes terminalis* (285)
Pitch midge (263)
Pitch moths (238, 280, **282**)
Pitch nodule moths (280, **283**)
*Pityogenes* spp. (188)
*Pityophthorus* spp. (179, 188)
*Platypus* spp. (240, 241)
*Platypus wilsoni* (241)
*Plectrodera scalator* (238, 247)
*Pleurotus populinus* (23)
*Podosesia syringae* (238, 253)
Ponderosa pine needleminer (214, 215)
Ponderosa pitch nodule moth (283)
Poplar blackmine beetle (198)
Poplar borer (75, 238, **255**)
Poplar petiolegall aphid (272, 273)
Porcupine (9, 100, 135-137)
*Porodaedalea* (*Phellinus*) *pini* (24, 38, 51-54, 96)

Pouch fungus (48, 49)
Powdery juniper aphid (269)
*Prionoxystus robiniae* (238, 245)
*Procryphalus mucronatus* (157-159)
*Pseudocercospora juniper* (*Cercospora sequoia* var. *juniperi*) (114)
*Pseudohylesinus nebulosus* (163, 164)
*Pseudopityophthorus* spp. (84, 86)
Punk knot (24, 52-54)
*Pyrrhalta luteola* (204)

## R

Red band needle blight (108, **125**)
Red belt (weather damage) (140, 141)
Red-belt fungus (36, 46)
Red-brown cubical root and butt rot (Schweinitzii) (40)
Red heart rot (24, **53**)
Red ray rot (24, **49**)
Red ring rot (24, 38, **51**)
Red root rot (23, 38-40)
Red turpentine beetle (**181**, 194)
Resin midge (263)
*Retinia* (*Petrova*) *arizonensis* (283)
*Retinia* (*Petrova*) *metallica* (283)
*Rhabdocline pseudotsugae* (109)
*Rhizosphaera kalkhoffii* (109)
*Rhyacionia* spp. (281)
*Rhyacionia bushnelli* (281)
*Rhyacionia frustrana* (281)
*Rhyacionia neomexicana* (281)
*Ribes*-willow rust (124)
Roundheaded borers (87, 238, 250, 252, 253, 255, **256**)
Roundheaded pine beetle (**183**, 194)
Rust-red stringy rot (24, **53**)
Rusty knot (54, 56)

## S

Salt (142-144)
*Saperda calcarata* (238, 242, 255)
*Saperda obliqua* (238)
*Saperda tridentate* (238, 248)
Saprot (46, 48, 49)
Sapsucker (46, 136, 137)
*Sarcophaga aldrichi* (211)
Schweinitzii root and butt rot (23, 36, **40**)
*Scirrhia* (*Systremma*) *acicola* (125, 126)
*Scirrhia pini* (125)

*Scolytus monticolae* (164)
*Scolytus multistriatus* (81, 165)
*Scolytus schevyrewi* (81, 165)
*Scolytus ventralis* (26, 167)
Septoria leaf spot and canker (**128**)
*Septoria musiva* (128)
*Septoria populicola* (128)
Shepherd's crook (**132**, 280)
Silver spotted tiger moth (225)
*Sirex juvencus* (260)
*Sirex noctilio* (261)
Siricidae (237, 238, 260)
*Sistotrema raduloides* (23)
Snake canker (69)
Snow mold (113)
Sooty-bark canker (60-62, 69, **77**)
Southwestern dwarf mistletoe (6-9, 14, **16**, 263)
Southwestern pine tip moth (281)
Sphaeropsis shoot blight (129)
*Sphaeropsis sapinea* (129)
Spider mites (262-264, **278**)
Spittlebugs (263, 264)
Spruce beetle (159, 171, 172, **184**, 188, 191)
Spruce broom rust (92, 93, **94**)
Spruce engraver beetles (**187**)
Spruce needleminers (214, 215)
Spruce shoot aphid (269)
Spruce spider mite (279)
Stalactiform rust (92, **97**)
*Stegophora ulmea* (111)
*Stellaria* spp. (94)
Stem decays of hardwoods in the plains (**56**)
*Stereum (Haematostereum) sanguinolentum* (24, **53**)
Striped ambrosia beetle (236, 240)
Stroma/stromata (61, 62, 74-76, 122)
*Styloxus bicolor* (252, 253)
Sugar pine tortrix (228)
Sugarbeet root aphid (273)
Sunscald (46, 145-147, 149)

# T

Target canker (60, 62, 64, 72, 103)
Termite (245)
*Thyridopteryx ephemeraeformis* (196, 199)
Tiger Moth (196, **225**)

*Trypodendron* spp. (240)
*Trypodendron lineatum* (240)
*Trypophloeus populi* (157-159)
Tumors (**134**)
Twig beetles (105, 179-181, **188**, 280)
Twolined chestnut borer (**258**)

# U

*Urocerus gigas* (260)

# V

*Valsa* (*Leucostoma*) *kunzei* (73)
*Valsa sordida* (71, 72)
*Veluticeps abietina/fimbriata* (24)
*Venturia* spp. (132)
*Venturia moreletii* (132)
*Venturia populina* (132)
*Venturia tremulae* var. *grandidentatae* (132)
*Verticillium albo-atrum* (79, 89)
*Verticillium dalhiae* (79, 89)
Verticillium wilt (79, 80, **89**)
*Vesiculomyces citrinus* (23)
*Virgella* (*Hypoderma*) *robusta* (109)
Vole (137)

# W

Western balsam bark beetle (**190**)
Western gall rust (91-93, **103**, 134)
Western pine beetle (183, 184, **193**)
Western pine budworm (196, **227**)
Western pine tip moth (281)
Western pine tussock moth (196, **229**)
Western spruce budworm (168, 195, 196, 227, **231**)
Western tent caterpillar (195, 196, **233**)
Wetwood (**149**)
White fir needleminer (214-216)
White mottled rot (23, 24, **42**)
White pine blister rust (12, 13, 91-94, **105**, 169, 170)
White pine weevil (**285**)
White trunk rot (24, 44, 45)
White-headed ambrosia beetle (241)
Willow rust (124)
Wind damage (148, 237)
Winter injury/winter desiccation (72, 126, 139, 140)
Woodpecker (45, 46, 173, 186, 244, 247, 256, 260)
Wood wasps (237, 238, **260**)

# X

*Xanthogaleruca luteola* (196, 204)
*Xyleborus* spp. (240)

# Z

*Zadiprion rohweri* (223, 224)
*Zadiprion* spp. (196, 222, 223)
*Zadiprion townsendi* (224)
*Zeugophora scutellaris* (198)
Zimmerman pine moth (282, 283)

# Instructions for Submitting Insect and Disease Specimens for Identification

Proper collection and shipping of specimens often make the difference between identifiable specimens and those no longer identifiable when they reach the specialists. It is important to follow the procedures listed below to ensure accurate and timely identification.

## Collection

- Collect adequate material. Failure to identify a specimen is often a result of insufficient material to indicate the cause of damage.
- Provide as much information as possible.
  - Describe appearance of affected tree—include species, age, and setting (most important item should be listed first).
  - Who submitted the specimen.
  - Date: when the species was collected.
  - Location: legal description or the county, city, road, and address if available.

Any information you think might be related to the problem, such as the number of trees affected, any human activity that may be related, weather, or other environmental conditions. Give your opinion concerning the identity of the problem.

## Shipment

- Plant materials such as wood, bark, foliage, roots, or conks should be wrapped in paper bags or newspapers and mailed in a box without any plastic. Plastic wrappings induce mold.
- Insects, except adult moths and butterflies, should be sent in leak-proof vials or bottles with 70 percent isopropyl (rubbing) alcohol. Place the vial or bottle in a box for mailing.
- Moths and butterflies can be killed by putting them in a jar in an oven at 140 °F for 10 minutes or by putting them in a freezer for 15 minutes. Place the specimen between folds of a paper to keep the wings flat and pack in a box for mailing.
- Mail specimens as soon as possible to prevent drying of foliage or insects or deterioration of conks.

# Technical Assistance Sources

## Federal Lands:

### Regional Office Staff
USDA Forest Service
Forest Health Management
740 Simms St.
Golden, CO 80401
(303) 275-5061

### Rapid City Service Center
USDA Forest Service
Rapid City Service Center
8221 S. Highway 16
Rapid City, SD 57702
(605) 716-2781

### Lakewood Service Center
USDA Forest Service
Lakewood Service Center
740 Simms St.
Golden, CO 80401
Phone: (303) 236-9541
Fax: (303) 236-9542

### Gunnison Service Center
USDA Forest Service
Gunnison Service Center
216 N. Colorado St.
Gunnison CO 81230
(970) 642-1133

## State and Private Lands:

In Colorado:
Colorado State Forest Service
Colorado State University
5060 Campus Delivery
Fort Collins, CO 80523-5060
Phone: (970) 491-7282
Fax: (970) 491-7737

In Kansas:
Kansas Forest Service
2610 Claflin Road
Manhatten, KS 66502
Phone: (785) 532-3309
Fax: (785) 532-3305

In Nebraska:
Nebraska Forest Service
University of Nebraska–Lincoln
PO Box 830815
Lincoln, NE 68501
Phone: (402) 472-6635
Fax: (402) 472-2964

In South Dakota:
South Dakota Dept of Agriculture
Resource Conservation and Forestry
Division
3305 ½ West South Street
Rapid City, SD 57702
Phone: (605) 394-2663
Fax: (605) 394-2549

In Wyoming:
Wyoming State Forestry Division
1100 West 22nd St.
Cheyenne, WY 82002
Phone: (307) 777-5495
Fax: (307) 777-5986

www.ingramcontent.com/pod-product-compliance
Lightning Source LLC
Chambersburg PA
CBHW080227270326
41926CB00020B/4171